Electrooptical Arrays

D.I. Voskresenskii A.I. Grinev E.N. Voronin

Electrooptical Arrays

Translated by Peter F.H. Priest

With 123 Illustrations

Springer-Verlag New York Berlin Heidelberg
London Paris Tokyo

Dmitrii Ivanovich Voskresenskii
Aleksandr Iur'evich Grinev
Evgenii Nikolaevich Voronin
Moscow, USSR

Translator
Peter F.H. Priest
Rose-Hulman Institute of Technology
Terre Haute, IN 47803, USA

Original Russian edition *Radiooptícheskie antennye reshetiki* published © 1986 by Radio i sviaz, Moscow.

Library of Congress Cataloging-in-Publication Data
Voskresenskiĭ, D.I. (Dmitriĭ Ivanovich)
[Radioopticheskie antennye reshetki. English]
Electrooptical arrays / Dmitrii I. Voskresenskii, Alexander Y. Grinev, Eugene N. Voronin ; translated by Peter F.H. Priest.
p. cm.
Translation of: Radioopticheskie antennye reshetki.
Bibliography: p.
Includes index.
ISBN-13: 978-1-4612-8120-7
1. Optical data processing. 2. Antenna arrays. I. Grinev, A. IU. (Aleksandr IUr'evich) II. Voronin, E.N. (Evgeniĭ Nikolaevich) III. Title.
TA1632.V6713 1989
621.38′0414 — dc19 88-39760

Printed on acid-free paper

Softcover reprint of the hardcover 1st edition 1989

Camera-ready copy provided in TEX by the translator.

9 8 7 6 5 4 3 2 1

ISBN-13: 978-1-4612-8120-7 e-ISBN-13: 978-1-4612-3484-5
DOI: 10.1007/978-1-4612-3484-5

Preface

Modern radar, telecommunication, sonar, and radio-astronomy systems use integrated systems, complex signals, and signal processing methods extensively. These systems require the development of new, more efficient systems.

One of the major ways to improve radar and sonar systems is to use multichannel systems, the most promising of which employ space–time signal processing.

In the last few years the intensive growth in electronics, information processing techniques, microwave and laser technology, electrooptics and holography, acoustics, and optoelectronics have resulted in new scientific and technological research fronts. One of these is the theory and technology of electrooptical array antennas—*a new class of receiving antennas whose pattern is controlled by means of coherent optics and holography.* Electrooptical array theory and technology have been involved in the study of the power, resolution, and range characteristics of arrays with various coherent optical processors. The practical applications of these systems and methods for building them using contemporary optoelectronics have also been studied.

Optical information processing allows parallel processing in large two-dimensional data bases in real time. Therein lies their main advantage over digital information processing. However, in terms of mathematical operations, algorithms, and in the accuracy of their computations, optical information processing is inferior not only to digital computers but also to analog computing techniques. This lack of versatility of optical processors can be overcome in two ways. One is by developing a hybrid optoelectronic computing system that uses a general purpose computer and a special optical computer that combines the rapid response time and the productivity of optical information processing methods with the flexibility and the precision of computers. The second method is to use problem-oriented analog optical processors for solving highly specialized tasks. Let us note that currently there is much promise in a third method, using an optical computer that processes digital information using an optoelectronic device.

When considering the suitability of using optical devices for processing radar and sonar signals, it is useful to distinguish two groups of tasks:

1. Those problems that are difficult to solve using traditional electronic methods (the clearest example is panoramic scanning over a broad range of frequencies and look-angles);

2. Those problems that can be solved by traditional electronic means as well as by optical ones. In this case, the major considerations are simplicity, compactness, cost, and so forth.

The general issues in optical information processing and holography are treated in monographs by Born and Wolf [1], Papoulis [2], Goodman [3], Kondratenkov [4], Soroko [5], Collier, Burckhardt, Lin [6], and also in articles in various references [7–38]. Articles by Russian and foreign authors devoted to optical signal processing techniques, for example in [39–46], treat only certain problems in processing space–time signals received by array antennas (the shaping of the array pattern in real time). The exception is synthetic aperture antennas that are examined in detail in, for example, works by Reutov, Kondratenkov and their colleagues [47, 48].

The potential advantages of an optical space–time signal processing device in radar and sonar applications are now drawing increased attention. However, a situation has developed where the existing proposals for equipping arrays with coherent optical signal processors are ahead of the current state of the art. At the same time we do not have any systematic and complete exposition of the coherent optical methods for producing the directional characteristics because these descriptions have been scattered in separate monographs, articles, conference proceedings, and patents. A number of important issues have reached the critical stage: in particular, how to maximize the scanning sector, how to combat interfering spatial signals, how to improve resolution, etc. These problems hamper the design of such systems. Therefore, one inevitably either overestimates the potential of optical methods in array signal processing or limits their use when the applications of digital and nonoptical analog methods are being rapidly developed, a situation which is not always justified by technological or economic considerations.

This book, based on the original work of the authors and on a review of Soviet and foreign publications, treats three aspects of electrooptical array antennas. First, it presents from a single point of view, a theory of electrooptical array antennas as a system which performs coherent optical processing of spatial signals and studies their properties and performance. Second, it proposes and describes some new coherent optical methods which significantly expand the practical applications of electrooptical arrays. Third, it treats various designs and discusses basic electrooptical array components and presents experimental data which illustrate the applications of arrays which produce their patterns by means of coherent optic and holographic techniques.

Finally, using a single electrooptical approach we will examine electrooptical arrays that produce either two-dimensional spatial directional characteristics or panoramic scanning along angular coordinates parallel with a spectral and correlation analysis. We will study nonplanar electrooptical arrays which make it possible to perform two-dimensional broad-band scanning. We will examine interference-resistant electrooptical arrays that use spectral techniques for rejecting interfering signals. In order to evaluate the potential for electrooptical arrays, we must study the influence of optical processor errors on the accuracy of the antenna pattern, on the dynamic range, and on the maximum gain. We will determine the electrooptical array sensitivity to thermal and quantum noise. Inasmuch as at certain stages "even a little knowledge is worth a great deal," we will, therefore, conclude with a description of a number of different coherent optical processors for electrooptical arrays that use existing and emerging technologies and which have different practical applications. We will also describe the results of experimental research and discuss some of the trends in electrooptical array development.

The range of applications for the material in this book is broad enough if we take into account the multiple applications of electrooptical arrays, for instance, in sonar technology, though specific features of such systems are beyond the scope of this work.

Contents

Basic Mathematical Terms and Abbreviations

$[C_{nm}]$	Matrix of channel coupling coefficients of a space–time light modulator.
D	Directivity.
d_{xy}	Distance between space–time light modulator channels.
d_{XY}	Distance between antenna elements.
$\dot{E}^{(1)}(x, y, t)$	Information part of optical signal entering the processor.
$\dot{E}_{\omega}^{(1)}(\mathbf{r}_{\perp})$	Frequency spectrum of the optical signal $\dot{E}^{(1)}(\mathbf{r}_{\perp}, t)$.
$\mathcal{E}(R, t)$	Space–time signal at point $\mathbf{R}$ in the antenna aperture.
$\dot{\mathcal{E}}_{\Omega}(R)$	Frequency spectrum of the space–time signal $\mathcal{E}(R, t)$.
$\mathcal{E}_{\lambda/2}$	Half-wave voltage of the space–time light modulator channel.
$\dot{e}^{(1)}\omega(\mathbf{k}_{\perp})$	Distribution of the frequency spectrum of the optical signal $E^{(1)}\omega(\mathbf{r}_{\perp})$ in the lens focal plane.
$\mathbf{F}(\mathbf{K}, \mathbf{K}')$	Vector radiation pattern of a nonplanar array antenna with narrow focus in the direction $\mathbf{K}'$.
$F(\mathbf{K}_{\perp})$	Radiation pattern of a planar array antenna.
$F_t(\omega_y)$	Imaging kernel of an optical spectrum analyzer.
ΔF	Bandwidth of the signal spectrum (frequency band).
$\mathcal{F}_e$	Radiation pattern of the receiving element of the array antenna.
δF	Resolution of the optical spectrum analyzer.
$\hat{\mathcal{F}}\{...\}, \hat{\mathcal{F}}^{-1}\{...\}$	Fourier and inverse Fourier transforms.
f	Focal length of a lens.
f_e	"Equivalent" focal length of the lens.
$\hat{I}\{...\}$	Limitedly reproducing operator.

I_q	Average intensity of the light flux in the exit plane of the processor induced by a point source q.
$I_{out}(\mathbf{K}')$	Distribution of the light intensity at the output plane of the processor.
$\mathcal{J}_t(...)$	Channel weighting function of the modulator in the direction of the scanning signal.
$\dot{\mathcal{J}}(\mathbf{R}_\perp)$	Complex planar array aperture exitation function of the array antenna (amplitude-phase distribution).
$\dot{\mathcal{J}}(\mathbf{R}, \mathbf{K}')$	Amplitude phase distribution of an arbitrary surface of an antenna used as a highly directional receiving antenna pointed in the direction, $\mathbf{K}'$.
$\mathbf{K}_\perp$	The vector space–time signal frequency.
$\mathbf{K}$	Propagation constant (number) of the space–time signal.
$K = \frac{2\pi}{\Lambda}$	Wave number of the signal.
$\mathbf{k}_\perp$	Vector space–time frequency of the optical signal.
$k = \frac{2\pi}{\lambda}$	Propagation constant of the optical signal.
$\hat{L}\{...\}$	ST signal processing operator in approximation of a continuous aperture.
$\hat{L}_{array}\{...\}$	Operator of the space–time signal of separate receiving apertures.
$l_k = C\tau_k$	Duration of the coherence of the laser beam.
m_x, m_y	Scaling coefficients of the model array antenna.
$\hat{m}$	Tensor of the affine conversion coordinates.
$M \times N$	Number of elements in the array.
O_{stlm}	Response function of the space–time light modulator.
R_y	Resolution of the space–time light modulator.
$\mathbf{R}$	Vector describing the position of points on a nonplanar array antenna.
$\mathbf{R}_\perp$	Vector describing the position of points at the aperture of a planar array.
$\mathbf{r}_\perp$	Vector determining the position of points in the space–time light modulator.

$\dot{\mathbf{S}}(K_\perp, \Omega), \dot{s}(\mathbf{K}_\perp, \Omega)$	
$\mathbf{S}(\mathbf{K}), \dot{S}(\mathbf{K})$	Vector (scalar) space–time (frequency-angular) spectrum of a signal.
T	Transmission function of the space–time light modulator.
ΔT	Signal processing time.
t	Time.
$T(\Omega_\phi, \Omega_z), \hat{T}(\Omega_\phi, \Omega_\phi, \Omega_{\tau\theta})$	
$\equiv \mathcal{T}(\Omega_\phi, \Omega_{\tau\theta})$	Complex transmission functions of a holographic mask.
$U(\mathbf{R}, t) = \frac{\pi}{2}\mathcal{E}(\mathbf{R}, t)/\mathcal{E}_{\frac{\lambda}{2}}$	Slowly changing normalized amplitude of the space–time signal.
$v(c)$	Speed of light in a medium.
v_p	Sweep velocity of a signal.
$W_\omega^{(1)}(\mathbf{k}_\perp)$	Average power spectrum of light beam output, produced by $e_\omega^{(1)}(\mathbf{k}_\perp)$.
X,Y,Z	Spatial coordinates of the array antenna scale.
x,y,z	Spatial coordinates of CO processor scale.
$\Delta x, \Delta y$	Dimensions of the space–time light modulator.
N_T	Thermal noise coefficient of electrooptical array.
N_V	Quantum noise coefficient of electrooptical array.
$N_{T,\nu} = N_T \times N_\nu$	Noise coefficient of electrooptical array.
a	Dispersion of the amplitude-phase distibution (phase error).
$\gamma(\mathbf{R}_1, \mathbf{R}_2; t_1, t_2)$	Normalized mutual coherence function.
$_2(\mathbf{R}_1, \mathbf{R}_2); \gamma_{11}(t_1, t_2)$	Extent of space–time coherence of a read light beam.
$\delta_{x,y}$	Size (pupil) of the space–time light modulator channel.
δ_y	Size of the 'write' light beam in the direction of the scan.
η_n	Coefficient of efficiency of the nth electrooptical array element.
R, θ, ϕ	Spherical coordinates.
$\Delta\theta_q, \Delta\phi_q$	Error of determination of the phase coordinates.

Λ	Signal (electro-, and acoustooptical) wavelength.
λ	Light wavelength.
$\lambda_i(\lambda_{X,Y})$	Characteristic number of $\Psi_i(\psi_0^{X,Y})$.
ν	Frequency of light wave.
$\rho = \mathbf{K}_\perp/K$	Normalized vector defining the position of the radiating source.
Σ	Surface area of the array antenna.
σ_{ac}	Area of coherence.
τ_q	Pulse duration.
$\tau_x = \sin\theta_q \Delta X/c$	Filling time of the array aperture.
$\Delta\tau$	Filling time of the acoustic medium.
$\tau_\theta = \sin\theta_0 R_0/c$	Filling time of a circular array aperture of radius, R_0.
$\tau(\mathbf{K}_s)$	Transmission function of a noise suppression mask.
$\Phi(\mathbf{R},t)$	Slowly changing phase of normalized signal $U(\mathbf{R},t) = \frac{\pi}{2}\mathcal{E}(\mathbf{R},t)/\mathcal{E}_{\frac{\lambda}{2}}$.
ω_x, ω_y	Spatial frequency of the optical signal.
$\omega = 2\pi\nu$	Radial frequency of the light wave.
Ω_D	Doppler shift of the light wave.
Ω_{scan}	Spatial sector of the array antenna scan.
$\Omega_X, \Omega_Y, \Omega_Z$	Spatial frequency of the signal.
$\delta\Omega_{X,Y}$	Beamwidth of the array antenna.

AA	Array antenna.
AOM	Acoustooptical light modulator.
APAA	Active phased array antenna.
APD	Amplitude-phase distribution.
CC	Contrast coefficient.
CCD	Charge-coupled device.
CM	Controlled mask.
CO	Coherent optics.
COP	Coherent optical processor.
D	Directivity.
DR	Dynamic range.
DLIS	Diffraction–limited imaging system.
DF	Defocusing.
DE	Diffraction efficiency.
e	efficiency.
EOAA	Electrooptical antenna array.
FFT	Fast Fourier transform.
FCC	Frequency contrast characteristics.

G	Gain.
LC	Liquid crystal.
MLPD	Multielement linear photodetector.
M	Mask.
OSA	Optical spectrum analyzer.
OCM	Optically controlled mask.
PAA	Phased antenna array.
PM	Passive mask.
RP	Radiation pattern.
SBW	Spatial bandwidth.
SLL	Side-lobe level.
ST	Space–time.
STBW	Space–time bandwidth.
STLM	Space–time light modulator.
SU	Coefficient of surface usage.

Dirac delta or impulse train: $\text{comb}(x) = \sum_{n=-\infty}^{n=\infty} \delta(x-n)$

Circle function: $\text{circ}(\sqrt{x^2+y^2}) = \begin{cases} 1 & \text{if } \sqrt{x^2+y^2} \leq 1 \\ 0 & \text{otherwise} \end{cases}$

Rectangle function: $\text{rect}(x) = \begin{cases} 1 & \text{if } |x| \leq 1 \\ 0 & \text{otherwise} \end{cases}$

Sinc function: $\text{sinc}(x) = \sin(\pi x)/(\pi x)$.

Signum function $\text{sgn}(x) = \begin{cases} 1 & \text{if } x > 0 \\ 0 & \text{if } x = 0 \\ -1 & \text{if } x < 0 \end{cases}$

Chapter 1

Array Antennas with Coherent Optical Signal Processing

This chapter discusses the suitability of using coherent optical methods for processing signals with a large information content and examines a generalized block diagram of an array antenna which forms its radiation pattern by coherent optical methods. We will discuss different approaches for analyzing these systems and examine several ways of constructing electrooptical array antennas using various input devices to a coherent optical space–time signal processor. We will discuss the components of an electrooptical array antenna.

1.1 The Potential for Parallel Processing of Space–Time Signals with Coherent Optical Techniques

The need for increased processing speed and capacity, improved interference suppression, and multi-purpose adaptability has led to the appearance of phased array antennas in radar systems and communication links, while the desire to improve reliability, amplification, radiated power, and the signal-to-noise ratio has led to the development of active phased array antennas [49–55]. Phased array antenna technology is now mature and its development is now directed toward improving its performance and finding new uses for it. At the same time, in the case of coherent summing of fields received by the separate array receiving elements by a common feed, not all the information contained in the field arriving at the receiving array is used. Further development of the potential for radar array systems necessitates the development of signal processing techniques (primarily space–time ones) for transmitting and receiving antennas [24, 26, 56, 57].

It is generally recognized that solving the problems of processing radar and sonar signals with a wide frequency band ΔF and duration ΔT, for a wide dynamic range necessitates the use of various kinds of digital computers with large memories and very high processing speeds [57–59]. Thus, in particular, a synthetic-aperture system needs to have a processing speed of 10^9 to 10^{11} operations per second and an on-line memory of 10^8 bits [47]. A sonar detection system using a linear array with $\Delta F = 2000$ Hz, $\Delta T = 2$s, 200 elements in the array, and an eight bit analog-digital converter, the number of readings N is in the order of 10^6 and the average size of the on-line memory is approximately $6N$ bits. To do a fast Fourier transform on the computer (with a speed of approximately 10^6 operations per second and 128 Kilobytes of on-line memory) requires 25 sec. [19]. The information stream coming from a modern radar and communication system may be as high as 10^{11} bits/sec. at a signal bandwidth of approximately 1 GHz. These examples show that it is not always possible to solve signal processing problems directly on the computer.

The shortcomings of the computer is its high cost, the size of its components, high power consumption, insufficient reliability, as well as its limited bandwidth which is determined mainly by the capabilities of the analog-digital converter (the predicted potential frequency band is 250 MHz or higher). Therefore, despite the obvious merits of digital signal processing one must always bear in mind that it is not, by any means, the only method.

Studies done by Soviet and other specialists (see, for example, [22, 24, 32, 42]) show that CO data processors are very good for processing large quantities of space–time signals. They are noted for their high speed information processing and are limited only by the data input-output rate; they are also noted for their large parallel processing capacity using a two-dimensional optical system and for the possibility of processing signals from a wide range of frequencies. The ease with which CO processors perform a number of integral mathematical operations, such as Fourier, Fresnel, Laplace, and Hilbert transforms, convolutions, and correlations, as well as a number of others, make them especially attractive for radar and sonar signal processing, microwave technology, radio astronomy, and image processing. Here it is worthwhile to note that the idea of employing optical systems for processing radar signals was evidently first proposed in its fullest form by Cutrona et al [61].

We can see the value of optical data processing by looking at the example of Fourier transforms. The operation of a two-dimensional Fourier transform of a coherent optical signal is performed by a spherical lens in a time equal to the time it takes for the light to propagate in the optical system, i.e. in $2 \cdot (10^{-9})$ sec. for an analyzer's optical path length of 60 cm. If we assume that a 1 cm^2 information carrier (for example, a photographic film) may contain more than 10^8 bits of information (when the carrier has a resolving power of 1,000 mm^{-1}), then the potential output of such a Fourier processor is $2 \cdot 10^{17}$ bits/s $\cdot$ cm^2. Although in practice the processing speed is limited by the input-output speed, and not by the optical signal processing time, there are currently specialized analog optical computers with a speed of 10^{11} to 10^{12} bits/sec. which is 4 to 5 times faster than that of modern computers [42, 60].

The place of optical data processing devices in a number of systems is determined by the two basic parameters ΔF and ΔT (or by the product $\Delta F \Delta T$), although this comparison must be made with regard to still other parameters: the dynamic range, the side lobe levels, the accuracy, the resolving power, etc. A comparison of an optical system with a digital type shows that an optical system can be used to process signals with an information capacity $\Delta F \Delta T > 10^4$ resulting in a gain in the overall dimensions, weight, and cost of the processing equipment [24, 42].

As for spatial processing of signals received from array antennas by coherent optical techniques, it is necessary to note that they are intimately and organically connected to the problems of shaping and transforming wave fields, problems which must be solved in antenna technology. In this case, the criteria noted above for using coherent optical processors may be amended by making use of the following obvious condition. In the case of a coherent optical processor which performs a spectral analysis of incoming signals, the quantity $\Delta F \Delta T = \Delta F / \delta F$ characterizes the number of resolution points (δF is the frequency resolution). Thus even a one channel acoustooptic spectral analyzer with a bandwidth of 100 MHz is equivalent to an electronic analyzer with 10^4 parallel channels. Using the analogy of an antenna with a spatial frequency filter, the value $\Omega_{scan}/\Delta\Omega_{0.5} \approx N$ (Ω_{scan} is the array scanning angle; $\Delta\Omega_{0.5}$ is the solid angle of the main maximum of an axially symmetric radiation pattern; N is the number of elements in the array antenna) characterizes the number of parallel spatial channels. Therefore, with respect to space–time signal processing from linear arrays, the value of using a CO processor is determined by the condition $\Delta F \Delta T > 10^4$ (here it is necessary to consider the potential frequency resolution, signal duration, dynamic range, etc.). In the case of forming the radiation pattern of an array, the criterion $N > 10^4$ is not sufficiently grounded and it is necessary to undertake a complex comparative evaluation with a phased array antenna (digital array antenna) with respect to the scanning time, the surface forms of the array antenna, the potential for filtering spatial interference, etc.

The basic problem in radar and sonar is to evaluate and analyze a situation that constitutes a multi-dimensional distribution (angular, range, frequency) of active and passive sources undergoing restoration, a distribution that is referred to below as the *radio scene*. This single problem involves the scanning of a given region in the radio scene, searching for and detecting objects, determining their number while simultaneously evaluating their coordinates and their spectral parameters, and classifying and determining precisely their trajectory.

It is customary to distinguish passive and active modes of radar and sonar operation. In the first case the information being extracted from the radio scene is data on the spectral composition and on the angular coordinates of the radiation sources (noise), which are used to analyze and, possibly, to identify them. What has been said determines the need to provide a high degree of reliability in detecting space–time signals over a broad range of frequencies in a wide scanning area under conditions of complex interference when continuous and pulse signals interfere.

Active detection and ranging is accomplished by irradiating (illuminating) a radio scene with a complex coded signal and by extracting information (about the angular coordinates, range, speed, acceleration, etc.) from the reflected signal. In this case, one of the basic operations performed by practically all radar systems is the correlation processing of the transmitted and the received signal in order to generate an ambiguity function to increase the sensitivity and to achieve a high resolution for range and speed. Depending on the type of radar system it is also possible to determine the carrier frequency and the pulse repetition frequency, the azimuth and the angle of elevation, or the carrier frequency and one of the angular coordinates.

CO processors, thanks to their two dimensionality, make it possible to simultaneously (panoramically) process these types of signals.

Antenna arrays which produce their radiation pattern by coherent optical processing allow parallel scanning by placing receiving elements on planar and non-planar surfaces (this is especially effective for rotating cylindrical and spherical surfaces—with the use of holographic masks and random axially-symmetric ones—with the use of three-dimensional holographic filters and a number of others), space–time signal processing (spectral or correlation analysis, etc.) and simultaneously, panoramic scanning along one spatial coordinate, effectively suppressing interference experienced by the incoming signal. In comparison with phased array antennas and multiple beam antennas it is not necessary to have phase shifters (or other phasing devices), a computer controller, or a large feed system; a comparison with numerical methods shows the advantages of specialized optical processors and the techniques for using them.

1.2 Electrooptical Array Antennas. Background and State of the Art

The term *electrooptical array* antenna[1] [40, 62] as compared to others we have seen (*quasi-holographic array* antennas [27, 32] and *dynamic microwave real-time* holograms [13]) emphasizes three factors. First, the only method of electrooptic array analysis is an electrooptical one "which employs a particular approach that combines a well-developed device for transforming signals and the spectral analysis with optical applications" [20].[2] Second, receiving radar and sonar arrays operate in the microwave range, but the signals are processed in the optical range. Third, research has been primarily directed at controlling the radiation pattern (beam-forming properties); problems related to the real-time properties of signals are dealt with only insofar as they influence the formation of two-dimensional spatial characteristics and the need to preserve them at the processor output or when using one spatial measurement of the optical processor for real-time processing. Basic attention is further devoted to the electrooptic arrays although many results can be used for solving complex problems, for example, in sonar (some sections below will specifically deal with the processors used for this).

1.2.1 Structure of the electrooptical array

An idealized block diagram of an electrooptical array is shown in Fig. 1.1. An electrooptical array generally consists of a receiving array antenna, 1 (the receiving elements and the amplifier); the signals, which are amplified and transformed into an intermediate frequency, control the corresponding channels of a space–time light modulator, 2. This modulator performs space–time modulation of the phase (amplitude) of the coherent light wave from laser, 3, in accordance with changes in the parameters (amplitudes, frequency, phases, beginning of the time reading, two angular coordinates, two polarization values) of the controlling signal in the form of a wave field incident on the array, arriving from the objects thereby forming at its output an optical model of the radiation being received. The latter undergoes transformation in a CO processor, 4[3], which generally includes lenses, holographic filters, diffraction gratings, deflectors, optically controlled masks, acousto- and electrooptic light modulators, etc. Therefore, depending on the type of multichannel space–time light modulator at the output of the processor,

[1]In Russian *Radiooptícheskie antennye reshetki* or, literally, *radio optical arrays.*

[2]And vice versa, making it possible to transmit optical circuits and principles into other frequency ranges and also for the case of different wave fields.

[3]In a broad sense the CO processor encompasses elements 2, 4, and 5.

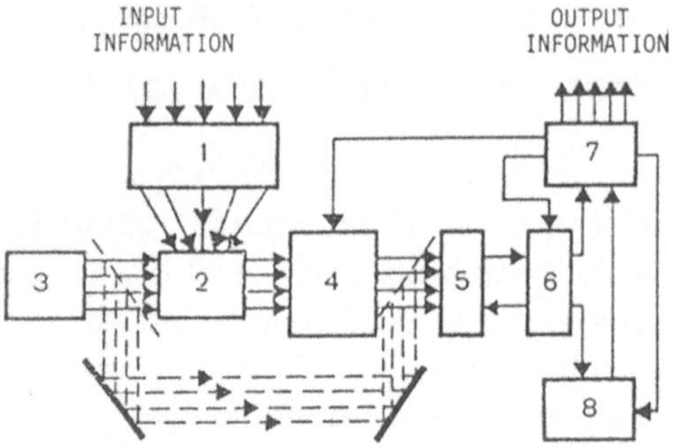

Figure 1.1: Idealized block diagram of electrooptical array.

an optical image is formed of the two-dimensional angular spectrum of the space–time signal being received (the extraction of the noncoordinate information is done through the coherent removal by optical heterodyning—the dashes in Fig. 1.1), or of the one-dimensional angular spectrum with the capability of simultaneous CO processing of the real-time signal (spectral or correlation analysis). In this case, one of the measurements at the output plane of the processor may be used to determine the frequency spectrum, the range or speed of the signals received. Given a more complex signal format (recording forms) at the space–time light modulator, the information at the processor output plane may be more complete. This information is recorded by a (multielement, scanning) photodetector 5 and with the help of an interface 6 is fed into the computer 7 or an analog device 8 for the final processing and decision making. The essence of forming the radiation pattern of an array by coherent optical methods is due primarily to the similarity of the wave nature of optical and microwave (acoustic) fields. A complex excitation function for the array aperture $J(\mathbf{R}_\perp)$ (amplitude-phase distribution) and the radiation pattern $F(\mathbf{K}_\perp)$ is a pair of two-dimensional Fourier transforms [52] (Fig. 1.2a) [4]

$$F(_\perp) = \iint_{-\infty}^{\infty} = J(\mathbf{R}_\perp)\exp(-\mathbf{i}\mathbf{K}_\perp\mathbf{R}_\perp)\mathrm{d}^2\mathbf{R}_\perp = \mathcal{F}\{\dot{J}(\mathbf{R}_\perp)\} \qquad (1.1)$$

where

[4] Vectors are denoted in boldface in the text and by arrows in the figures.

$$\hat{\mathcal{F}}\{\ldots\} = \iint_{-\infty}^{\infty} \ldots \exp(-i\,\mathbf{K}_\perp \mathbf{R}_\perp)\mathrm{d}^2\mathbf{R}_\perp$$

$$= \iint_{-\infty}^{\infty} \ldots \exp[-i(\Omega_X X + \Omega_Y Y)]\mathrm{d}X\,\mathrm{d}Y$$

is the operator of a two-dimensional Fourier transform;

$\mathbf{K}_\perp = \mathbf{n}_X\Omega_X + \mathbf{n}_Y\Omega_Y$ is the vector spatial frequency of the space–time signal;

$\mathbf{n}_X$,$\mathbf{n}_Y$ are the unit vectors of the array coordinates XOY;

$\Omega_X = K\cos\alpha = K\sin\theta\cos\phi$;

$\Omega_Y = K\cos\beta = K\sin\theta\sin\phi$ is the generalized angular variable;

$K = 2\pi/\Lambda = \Omega/c$ is the propagation constant of the radiowave;

Λ is the wavelength;

$\Omega = 2\pi F$ is the angular frequency;

c is the speed of light;
ϕ, θ are the azimuth and the angle of elevation;

$\mathbf{R}_\perp = \mathbf{n}_X X + \mathbf{n}_Y Y$ is the radius vector of the points of the [antenna] aperture;

and $\mathrm{d}^2\mathbf{R}_\perp = \mathrm{d}X\,\mathrm{d}Y$ (the subscript q in Fig. 1.2 denotes the source).

The same pair is formed by the distribution of the complex amplitude of light $\dot{E}(\mathbf{r}_\perp)$ at the forward plane (Π) and $\dot{e}(\mathbf{r}_\perp)$ at the rear focal plane (π)

of the collecting spherical lens (L) [1–6] (Fig. 1.2b):

$$\dot{e}(\mathbf{k}_\perp) = \frac{1}{\lambda f}\hat{\mathcal{F}}\{\dot{E}(\mathbf{r}_\perp)\} \tag{1.2}$$

where

$\mathbf{k}_\perp = \mathbf{n}_x\omega_x + \mathbf{n}_y\omega_y$ is the vector spatial frequency of the optical space–time signal;

$\mathbf{n}_x$ and $\mathbf{n}_y$ are the unit vectors of the processor system xoy;

$\omega_x = kx/f$ and $\omega_y = ky/f$ are the spatial frequencies;

$k = 2\pi\lambda = 2\pi\nu/c$ is the propagation constant of the lightwave;

λ is the wavelength;

c is the speed of light;

f is the focal length of the lens L;

and $\mathbf{r}_\perp = \mathbf{n}_x x + \mathbf{n}_y y$ is the radius vector of points in the plane Π.

Therefore, if we establish the connection $\dot{J}(\mathbf{R}_\perp) \longleftrightarrow \dot{E}(\mathbf{r}_\perp)$, then we obtain the expression $F(\mathbf{K}_\perp) \longleftrightarrow \dot{e}(\mathbf{k}_\perp)$ which expresses the essence of the coherent optical method of generating radiation pattern of the array, an idea which was expressed in [63]. Here we must also note the work in Goodman's work [27] which presents an interesting treatment of electrooptical arrays from a (quasi-) holographic position.

If the space–time signal of an array is treated as a dynamic Fourier microwave hologram (with a real-time basis [13]), and its associated space–time light modulator as a reduced copy (an optical model) of this hologram, then the formation of an optical image of the radio scene (the angular distribution of the emission sources in the far field of the array) may be treated as a reconstruction of the recorded hologram in coherent light. The condition of the geometric similarity of the radio scene and its image, i.e. the condition of an undistorted reproduction of the target object by its image, according

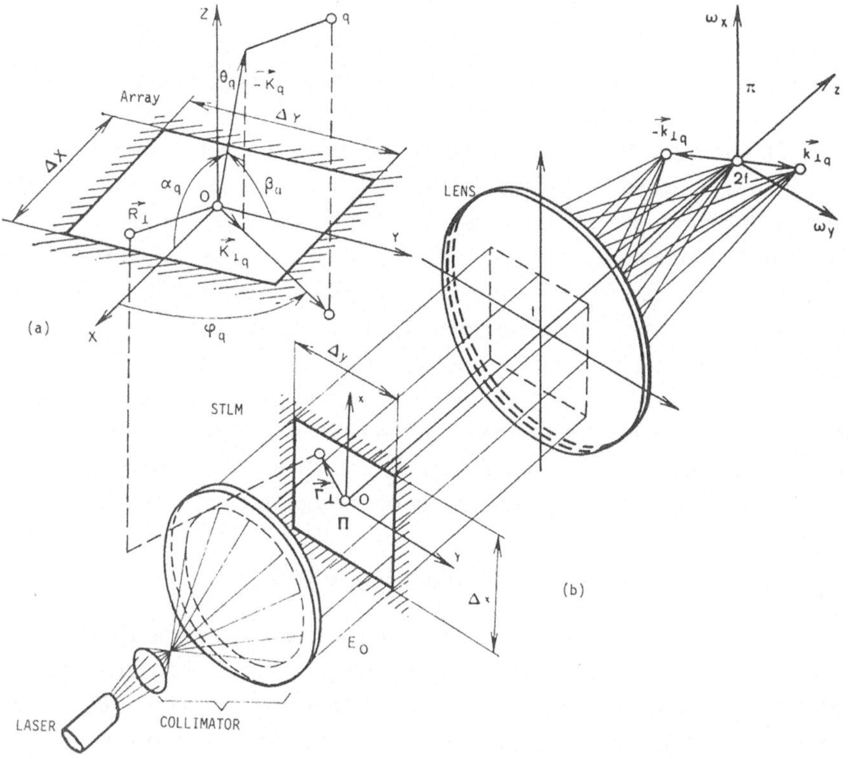

Figure 1.2: (a) Planar aperture (b) a schematic representation of the simplest CO processor.

to the holographic concept, gives rise to the need for the geometrical similarity of the array (microwave hologram) and the space–time light modulator (optical hologram)

$$X = m_x x, \qquad Y = m_y y \tag{1.3}$$

where m_x and m_y ($\sim 10^2 \ldots 10^3$) are scaling factors of the array antenna along the appropriate axes. Inasmuch as the wavelength at restoration is significantly shorter than at recording ($\lambda \ll \Lambda$) while the copy (in the space–time light modulator) is $m_x m_y$ times smaller than the original (at the array) then we observe the holographic reduction [5, 6] in the image of the radio scene along the x and y axes

$$g_x = m_x \lambda / \Lambda, \qquad g_y = m_y \lambda / \Lambda \tag{1.4}$$

whereby the order of magnitude of the holographic reduction is $g_{x,y} \sim 10^{-2} \ldots 10^{-3} \ll 1$.

The holographic approach to the processes in electrooptical array antenna makes it possible to graphically explain still other effects. Thus, the appearance of induced (false) images of the radio scene is the result of the partial loss of phase (amplitude) information when the hologram is made [5, 6], i.e., when the light is modulated in the space–time light modulator by the signals of the array. The formation during restoration of the zeroth diffraction order is explained by the presence in the hologram of a recording reference ("quiescent points" of the space–time light modulator), while the abèrration of the radio scene image in the case of a non-planar but geometrically similar array and a space–time light modulator is explained by the inequality with unity of the coefficients of the holographic reduction $g_{x,y}$.

1.2.2 Types of electrooptical arrays

The practical applications of an electrooptical array significantly depend on the input device (the space–time light modulator) which converts the radar information into a coherent light flux. A CO processor, as a rule, not only generates a continuous light beam in space, but in parallel processes complex time signals (frequency and phase-modulated), the form of which, in the case of active radar, is chosen in conformity to the needs for resolving power and precision of the range measurement and radial velocity [56, 57].

Electrooptical array with a multichannel acoustooptic modulator. The work of Lambert, Arm, and Aimette [63] was one of the first to treat systematically electrooptical arrays which employ a multichannel acoustooptic light modulator (AOM) (Fig. 1.3) as the space–time light modulator. An electrical signal $\mathcal{E}_n(t)$ from the nth receiving element of the antenna array goes to the corresponding piezoelectric transducer on the face of the acoustooptic medium. This excites a traveling acoustic wave resulting in the modulation of the refractive index of the acoustic medium material proportional to the voltage applied. The change in the transmission function of the AOM is described by the expression

$$\dot{T}_n(x, y, t) = T_\circ \mathrm{rect}[(x - ndx)/\delta x] \mathrm{rect}(y/\Delta y) \exp\left[i \frac{\pi}{2\mathcal{E}_{\lambda/2}} \mathcal{E}_n(t - y/v) \right] \tag{1.5}$$

where

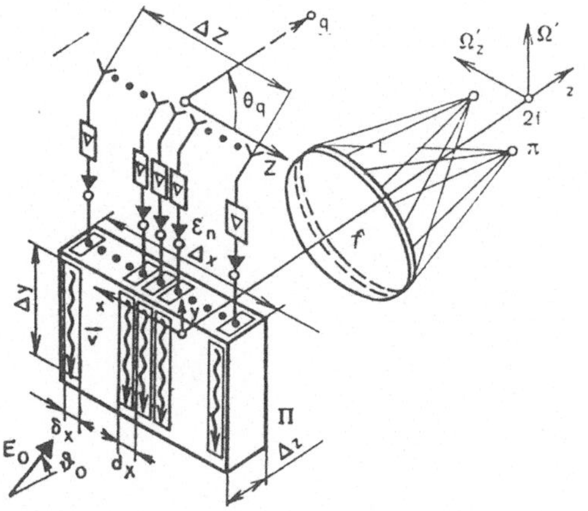

Figure 1.3: Linear electrooptical array with a multichannel AOM.

$T_\circ \leq 1$ is the acoustic medium transmission function when $\mathcal{E}_n(t) = 0$;

rect(y) $= 1$ when $|y| \leq 0.5$ and 0 when $|y| \geq 0.5$;

$\mathcal{E}_{\lambda/2}$ is the halfwave voltage of a channel of the AOM defined by the parameters of the acoustic medium and by the piezoelectric transducer when the phase modulation index is equal to $\pi/2$;

v is the velocity of the acoustic wave in the acoustic medium;

dx is the distance between the channels of the AOM;

δx is the width of the channel;

and Δy is the length of the channel (the remaining notations are clear from Fig. 1.3).

As a result of exposing a multichannel AOM to transmission functions at every channel (Fig. 1.5) with the help of a collimated laser beam with an amplitude $E_\circ$ at the CO processor, an optical signal appears with of the form $\dot{E}(x,y) = E_\circ \dot{T}$, where $\dot{T} = \sum_{n=1}^{N} \dot{T}_n$ (N is the number of channels in

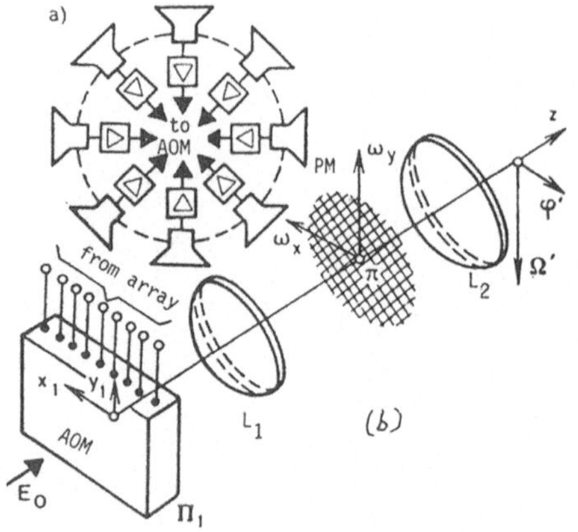

Figure 1.4: Circular electrooptical array with an AOM.

the AOM). This space–time signal is subjected to a Fourier transform with the help of lens L (see Fig. 1.3). In the rear focal plane π of lens L a light distribution is formed proportional to the space–time signal $\mathcal{E}_n(t)$ received by the array. Analyzing the structure of this distribution makes it possible to draw the following conclusions [63], whose mathematical confirmation will be given in section 3.1:

1. All possible angular positions of the beam of an array are produced simultaneously as a continuous wave along one spatial coordinates (Ω', z), therefore a set of targets in space are unambiguously reflected at the output of the processor (see Fig. 1.3, plane π) in the form of diffracted luminous spots.

2. The angular positions of the targets are unambiguously in agreement with the angular coordinates while the angular resolving power of the electrooptical array coincides with that of an equivalent phased array.

3. In the second dimension at the output of the CO processor (Ω' in Fig. 1.3) there is a frequency spectrum of the radio signal shifted to the Doppler frequency.

4. The angular coordinates and the frequencies are interrelated along the orthogonal variables of the processor.

5. Each signal appears at the CO processor output at a time proportional to the distance from the target.

Thus an electrooptical array with one-dimensional direction finding also functions as an optical spectrum analyzer for radio signals. For two-dimensional direction finding Lambert proposed a time-delay multiplexed processor [63]; however, this modification significantly complicates the electronics of the array, decreasing the frequency resolution (see Section 3.3).

A different kind of electrooptical array which optically performs (without time-delay multiplexing) two-dimensional direction finding from radio signals, received by a planar array of a radio-heliograph with elements arranged in a ring as described by McLean [64].

A number of works [65–72] are close to the ideological position of Lambert [63]. The authors of works [73–79] set forth the results of their design work, the manufacturing technology, and their experimental work on liquid and solid-state multielement acoustooptic light modulators, as well as their applications in electrooptic arrays or for solving related problems.

Lewis [80] proposed a coherent optical method for forming the radiation pattern of circular arrays. In essence the CO processor shown in Fig. 1.4 is a correlated spatial filter in the spectral plane of which is placed a passive holographic mask. Besides the one-dimensional direction finding of the azimuth at the processor output plane of the processor, a frequency spectrum is produced of the received radiowave. The work of Lewis [80], expanded and supplemented by Vodovatov, Grinev, and others [81, 82], was directed toward the solution of a certain frequency problem in forming the antenna pattern of a circular array by means of coherent optics.

Electrooptical arrays with electrooptic space–time light modulators with electron-beam addressing. This type of space–time light modulator consists of two electron guns and a crystal target made from electrooptic materials in a vacuum chamber [32, 84–88] (Fig. 1.5). The signal $\mathcal{E}(t)$ after amplification and heterodyning modulates the beam current of a scanning electron gun and the distribution of the charge corresponding to the incoming signal is recorded at the target. The signal from each element of a linear array is recorded as a raster pattern on a separate line of the target. The distribution of the electrical field, which is formed in the crystal when a charged pattern is formed on its surface causes spatial changes in the target's refractive index and amplitude or phase modulation of a collimated beam of light at the processor input, due to the longitudinal electrooptical Pockels effect [89]. The modulation depends on the polarization of the incoming laser beam relative to the principle axes of the index ellipsoid (of the crystallographic axes). The transmission function is

$$\dot{T}_a(x, y) = T_\circ \exp(i\Gamma_\circ/2)\sin[(\Gamma_\circ + \Gamma)/2] \tag{1.6a}$$

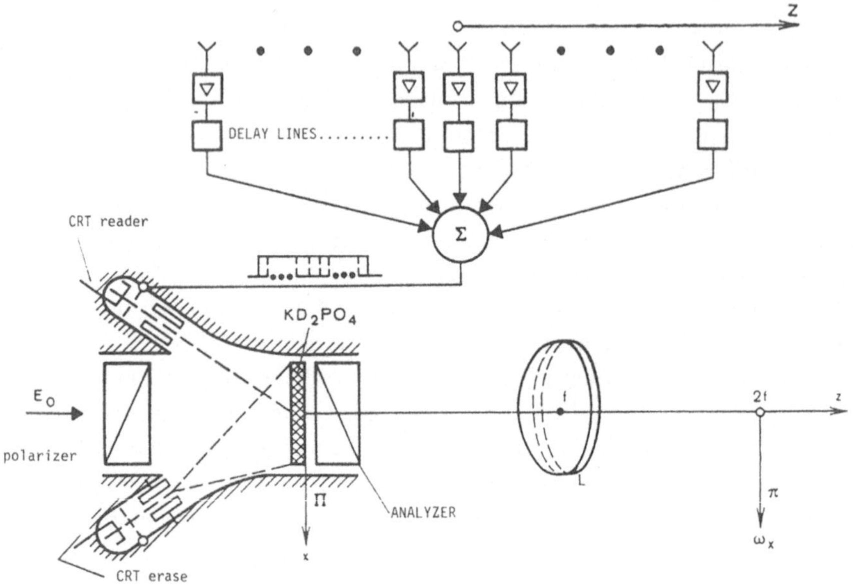

Figure 1.5: A linear electrooptical array with electron-beam addressing.

with amplitude modulation

$$\dot{T}_\phi(x,y) = T_\circ \exp[i(\Gamma_\circ + \Gamma/2)] \tag{1.6b}$$

with phase modulation, where $\Gamma_\circ$ is the phase difference between the ordinary and the extraordinary waves caused by the natural anisotropy of the crystal; $\Gamma(t) = \pi\mathcal{E}(t)/\mathcal{E}_{\lambda/2}$ is the phase difference induced by the voltage $\mathcal{E}(t)$; $\mathcal{E}_{\lambda/2}$ is the halfwave voltage when $\Gamma = \pi$; $T_\circ$ is the maximal transmission coefficient. The recorded charge pattern is erased by a beam from the second electron gun.

Among the different kinds of space–time light modulators the most widely used in CO processors (Fig. 1.5) for processing radar array signals are devices employing monocrystal light valves (chemical formula KD_2PO_4) [22, 84, 85, 88]. Such an electrooptic array works in a reversible cycle (record-read-erase). The signals of a pulse burst Doppler radar with a linear array of 100 elements have been processed in real time. An analogous treatment was performed on signals from a planar array of 70 × 70 elements. In this case, the electrical circuit was simplified by using the format of an open-weave array and the signal spectrum was expanded by employing a 13-bit Barker code.

Electrooptical arrays with a space–time light modulator built on thermoplastic material show promise in synthetic aperture systems and in sonar [13, 90]. The space–time light modulator consists of a quartz base with a clear conductive layer and a thermoplastic layer 5 to 10 μm thick. An electron beam modulated in intensity by the signal scans the surface of the target and reproduces the signal as a charged relief which causes the thermoplastic surface to become deformed. After appropriate treatment the thermoplastic layer preserves the resulting surface image. The signal from each element of the array is recorded on a separate line.

The problems with thermoplastic material in comparison with photographic film are less sensitivity, nonlinearity of phase recording, critical nature of the developing process, and the comparatively small dynamic range.

Electrooptical arrays with a multichannel space–time light modulators addressed by an electrical voltage. A number of works [91-94] describe electrooptical arrays with multichannel space–time light modulators using electrooptical lithium tantalate (niobate) crystals, whose lower surface has a conductive layer and whose upper surface has parallel electrodes according to the number of channels. The signal $\mathcal{E}(t)$ from each array element (amplified and transformed) is fed to its associated electrode and modulates a light beam introduced into the crystal with the help of specialized optical splitting system [95] or masks [92, 96]. By appropriately selecting the section of the crystal, the polarization of the light, and the orientation of the analyzer at the output, it is possible to bring about amplitude (1.6a) or phase (1.6b) modulation of the light by the longitudinal Pockels effect. The optical model of the received spatial signal undergoes a Fourier transformation in the CO processor (Fig. 1.6) as a result of which optically diffracted images of all of the sources in the array's view are formed simultaneously in the π-plane. The time information can be extracted by optical heterodyning [94].

The antenna pattern wase analogously formed with a membrane space–time light modulator deflector [97].

Electrooptical arrays with an optically addressed space–time light modulator. In one of the first electrooptical arrays the input device to the CO processor was a photographic film plate that was advanced on parallel tracks and exposed to an electron gun while simultaneously recording the signals being received by the elements of a linear array. The system performed both one-dimensional direction finding (or two-dimensional, given an orthogonal array), and a correlation analysis of the signals. There have been many proposals for modifying optical processors by recording the signals on pho-

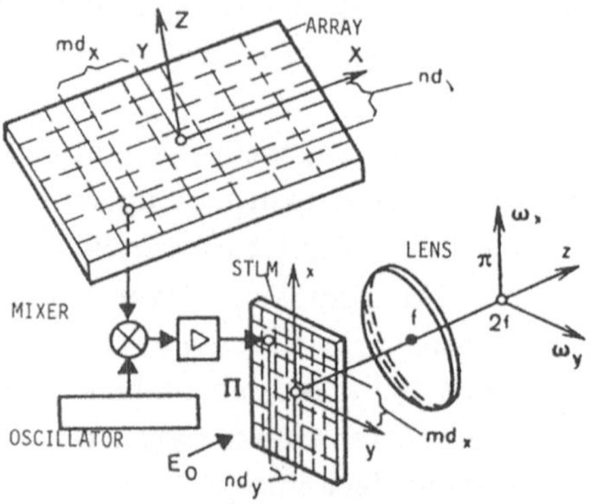

Figure 1.6: A planar electrooptical array with multichannel space–time light modulator addressed by an electrical voltage.

tographic plates, for example, for beam-focusing of a two-dimensional array and correlation analysis by using coded radio impulses [99], and, for simultaneous beam-focusing and pulse compression [100].

Film distortion due to the graininess of an emulsion is one of the most significant sources of noise in a CO processor limiting its dynamic range. Forming the radiation pattern from an array by recording the signal on a reversible analog of film, photosensitive thermoplastic, is treated in [13]. In this case, the recording is made not with an electron beam, but with a spatially-modulated light beam [101]. space–time light modulation with optical addressing by a DKDP crystal light valve is used in a CO processor [102,103]; in this case, the signal recording and reading is performed by coherent light at various wavelengths.

A reversible space–time light modulator which records in parallel from all elements on the carrier (and multichannel optical addressing) makes it possible to eliminate delay lines. There are several types of these space time light modulators. In particular, the transformed signal from each array element modulates the illumination intensity of the corresponding element of the line (matrix) of light diodes whose image is projected and developed on a photosensitive medium with the help of a deflector or by moving the medium itself [104]. In another version the transformed signals $\mathcal{E}_n(t)$ from the array elements 1 (Fig. 1.7) control the corresponding channels of the space–time light modulator 2, for example, an electrooptic one which is illuminated by a collimated light beam with a wavelength λ_{record}. The signal is recorded by scanning the images of channels 2 of the space–time light modulator with the help of a deflector 3 on a photosensitive medium by an optically controlled

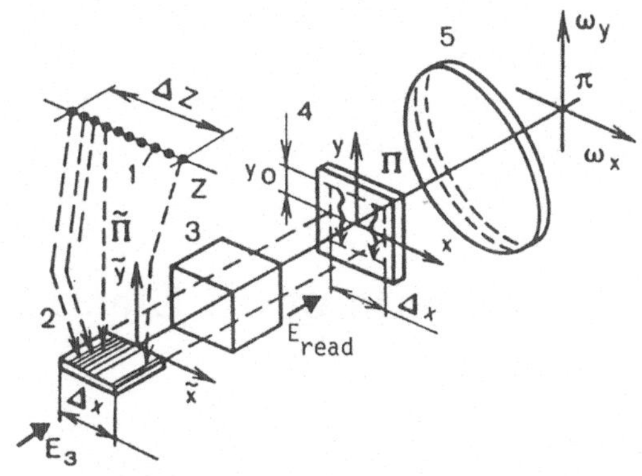

Figure 1.7: A linear electrooptical array with a space–time light modulator having multichannel optical addressing.

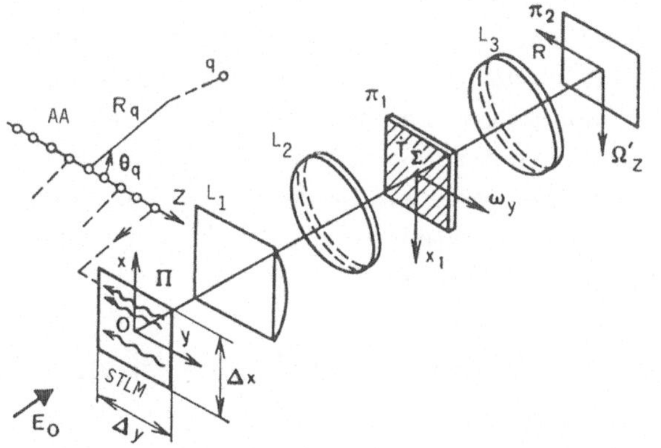

Figure 1.8: A multichannel optical space–time correlator for a linear array.

mask 4, in the capacity of which one can use a Pockels Readout Optical Modulator (PROM), a liquid crystal ligh valve, etc. [105, 106]. Reading is performed by a coherent light beam with a wavelength of λ_{read}.

It is appropriate to note here that the capabilities of optical processors are more fully utilized when, along with the shaping of array beams, one uses the properties of these systems to perform correlation [85, 88, 99, 103]. In particular, a diagram of the multichannel optical correlator for processing signals from a 100-element array receiving sounding signals in the form of a Barker code is shown in Fig. 1.8 [106]. The signals are recorded in raster form (space–time light modulation or with electron addressing is shown in Fig. 1.5) in plane Π. An astigmatic optical lens system L_1, L_2 results in the representation of a raster line along the ox axis and a Fourier transform along the *oy* axis. In plane π_1 there is a multichannel matched filter which has a holographically recorded complex conjugate Fourier transform of the reference function. The coordinates of the correlation maximum at the processor output at the plane π_2 is proportional to the angle of elevation and the distance.

1.3 The Components of an Electrooptical Array Antenna

1.3.1 General information

Electrooptical arrays include a large number of components: miniature amplifiers and receivers, emitters, high quality lenses, diffraction gratings, deflectors, holographic filter masks, etc.

The key components of a CO processor are the space–time light modulator and the output devices (multielement or scanning photodetectors). space–time light modulators are used as the signal input device for optical processors and also as controllable dynamic masks for matched filtering, signal recognition, etc. Below we will briefly discuss only receiving space–time light modulators which operate in coherent light and are reversible for rapid data processing [22, 32, 106–108].

1.3.2 Space–time light modulators in an electrooptical array

They perform spatial (or to be more precise space–time) phase (amplitude) modulation of a coherent light wave in connection with a change in the parameters of the controlling signal at each point. There are many criteria, on the basis of which we can classify space–time light modulators. Thus, space–time light modulators may be classified according to the type of addressing, i.e., the mechanism by which information is transfered to the space–time light modulator; or according to the physical system used to modulate the light (elctrooptical, acoustooptical, etc.); according to the type of modulation and the material used as the modulating medium, and so forth. The following are basic parameters: sensitivity (half-wave voltage), type of addressing, resolution (at 50% modulation) and contrast, dynamic range, linearity, operating frequency, presence of memory, processing speed, diffractional efficiency, optical homogeneity, mechanism of erasing, etc. To compare the different modulators used as input devices we use three criteria:

- the product of the resolution and the linear dimension of the target;

- the number of resolution elements in the image;
- the product of the square of the frequency and the value of the resolution times the linear dimension of the target.

The most successful means for forming the antenna pattern by coherent optical methods are connected with the use of multichannel AOMs [63–83]. The development of high quality multichannel AOMs depends on a proper selection of materials for the acoustic medium and transducer, the development of a specialized manufacturing technology and the necessary system for matching and exciting the piezoelectric transducer. The parameters of a multichannel AOM (the attenuation of the acoustic power within the limits of the working aperture; the operating frequency, a range to 3 dB; and the required power for the excitation signal; and the equivalence of the channels) are determined principally by the nature of the acoustic medium, the properties of the piezoelectric transformation, and by the means for fastening it to the acoustic medium [17, 109–11]. Multichannel AOMs operate in Raman-Nath or Bragg diffraction modes and are carried out with liquid or solid-state acoustic mediums.

Liquid light modulators (an acoustic medium in distilled water) possess a high acoustooptical index. However, due to the great attenuation of the ultrasound waves, the operating frequency does not exceed 40 MHz. Scientists have developed multichannel AOMs with 10 to 50 channels, an acoustic medium length of $\Delta y = 2.5 \ldots 8$ cm, a step of $dx = 0.5 \ldots 1.5$ mm, an operating frequency of 20 ... 40 MHz and a bandwidth of $\delta F = 10 \ldots 40\%$.

At frequencies up to 100 MHz the acoustic medium may be made from fused quartz, special types of glass, and crystals. The disadvantage of quartz is its low acoustooptical index (5 to 10 times lower than glass). Glass has high losses; crystals are expensive and it is difficult to make large-scale acoustic mediums. There are multichannel AOMs with 10 to 64 channels, $\Delta y \leq 7.5$ cm, $dx = 0.5 \ldots 1.5$ cm, an operating frequency of 30 ... 40 MHz and $\Delta F = 30$ to 50%. A modulation index $m = 0.3$, when the intensity of the diffracted light is 3%, is attained when the incoming signal amplitude is 0.5 to 1 V for the liquid space–time light modulator described above and 2 to 2.5 V for a solid state one. The spread of values of the incoming resistances of the modulator channels which occurs in the process of manufacture may be significantly reduced by methods of wide-band matching. The discrimination between channels which are given simultaneous excitation is no less than 20 dB.

At frequencies between 100 and 1,000 MHz multichannel AOMs use quartz and various crystals (lithium niobate, lead molybdate, paratellurite). As a rule, these space–time light modulators operate in the Bragg diffraction mode. There are also fused quartz space–time light modulators with an operating frequency of 200 MHz, N=10 to 30, $\Delta F = 40$ MHz, $\Delta y = 7.5$ cm, $dx = 1.5$ mm, and lead molybdate modulators with an operating frequency of 650 MHz and $\Delta F = 100 \ldots 200$ MHz. Using anisotropic light diffraction in hypersound, it is possible, first of all, to significantly extend the bandwidth (for example, to obtain $\Delta F \approx 500$ MHz at an operating frequency of 1 GHz) by selecting the configuration of the acoustooptical interaction and, second, to decrease significantly the background noise on the photodetector [112].

On the whole, a multichannel AOM permits real-time processing of a large number (50 to 100) of parallel channels, each of which may potentially operate at frequencies of 1 GHz and higher and have a bandwidth of 100 to 300 MHz or more; AOMs are characterized by a low control voltage at a relatively low cost. However, multichannel AOMs have some specific disadvantages: they are difficult to design when there are more than 100 channels; they are not very convenient for processing signals from two-dimensional arrays (even though time multiplexing overcomes this difficulty, it complicates the electronics of the system); they do not process signals larger than 70 to 100 μs (at working frequencies up to 100 MHz) and above 10 μs at high operating frequencies, which limits the frequency resolution from a few to tens of kilohertz, respectively.

Electrooptical space–time light modulators with lithium niobate, lithium tantalate crystals [89] (N = 46 [92], N = 16 [95], N = 32 [93, 94] permit the construction of more than 100 parallel channels with specialized block design techniques (for example, with a system for separating the controlling signals on a polyamide film). They form a two-dimensional structure with a potential bandwidth of more than 1 GHz. Other disadvantages of space–time light modulators are as follows: extracting noncoordinate information is only possible with coherent light (optical heterodyning); a large value for the half-wave voltage $\mathcal{E}_{\lambda/2} \geq 50V$ (there are some methods for lowering $\mathcal{E}_{\lambda/2}$ to a few volts); and the formation of a radiation pattern for the reception of wide-band signals (see Section 2.1). The basic characteristics of space–time light modulators with electron beam and optical addressing, not treated above, are given in Table 1.1 [42].

1.3.3 Output devices

To transform the optical information into electrical signals, the optical signal processor must use multielement photodetectors, photosensitive electron beam devices (vidicons, dissector tubes), or it must scan the angular field of view of the photodetector using various light deflectors. In the latter case, photoelectron multipliers (the bandwidth of the Soviet production model is

80 to 100 MHz), photodiodes, or avalanche photodiodes are often used.

The advantages of optical processors (high speed and parallel data processing) can be fully realized given that the output device transforming the optical signal into an electrical one does not limit the capabilities of the system as a whole. Therefore, for radar problems, the most promising devices for recording optical signals at the CO processor output are multielement photodetectors (linear and matrix), which digitize the optical signals, to effect storage and parallel reading of these signals [113]. Multielement photodetectors are generally manufactured of silicon the spectral characteristics of which cover the range from 0.4 to 1.0 μm; at present, the technology is quite well developed. A photodetector matrix is a set of photosensitive components which together with the commutating elements form the matrix cell. With the help of vertical and horizontal lines the cells are organized into columns and rows so that each cell of the matrix is located at the intersection of the address lines (the reading is, as a rule, by the lines). The photodetector head has one line and much simpler capabilities in comparison with the matrices. The basic parameters of a multielement photodetector are the spatial resolution, the spectral and threshold sensitivity, processing speed (bandwidth), and read-out time. The spatial resolution of the recording system in the appendix is determined by the resolution of the optics used, the sensitivity specifications of the multielement photodetector, and intermodulation distortion, which depend on the interrelationship between the photosensitive elements as well as the technical capabilities.

Sensing the distribution of the light flux corresponding to the slowly changing (in time) spectra of the signals or to the correlation function requires the use of charge-coupled devices (CCD) [37, 114–116], which afford the optimal matching of the parallel optical processor with the subsequent digital system (see Fig. 1.1). The capability of accumulating and storing the charge information in the course of some time uses a CCD as a buffer memory necessary when interfacing information systems of various ratings, while the separation of the reading and accumulation processes makes achieve greater flexibility in solving problems of optical data processing [74, 117, 118]. Using CCD structures increases the resolution of multielement photodetectors to 40 or 60 lines per millimeter, ensures high positional accuracy (≈ 1 μm) of the geometrical dimensions of the elements, and a high uniformity of the photoresponse. There are CCD linear arrays with N = 1,024 photosensitive cells, elements with a dimension of $\delta x_\phi \times \delta y_\phi = 20 \times 20$ μm and a step $d\Phi = 32$ μm, with N=100, $\delta x_\phi \times \delta y_\phi = 20$ μm$\times$ 1 mm and $d_\phi = 25$ μm, and CCD matrices with 100×100, 256×256, and 1024×1024 elements. Information input time for a 256 element CCD linear array is 550 μs, input time for 100 lines of a two-dimensional 100×100 CCD matrix is about 25 ms. These devices, however, are characterized by their lengthy charge accumulation (more than 1 μs) and do not allow parallel information output.

In case it is necessary to record rapidly changing signals at the output of the CO processor (for example, in the case of optical heterodyning in drawing in Fig. 1.6) it is possible to use multielement photodetectors, whose photosensitive element is a photodiode [119-121]. Separate cells of a photodiode matrix have high threshold sensitivity (10^{-10} J) and processing speed ($\tau_\Phi \sim 10 \ldots 25$ ns), and by using avalanche amplification of the photoelectric current up to tenths of a nanosecond. However, their organization into matrices does not allow them to realize the maximum threshold sensitivity due to the presence of large impulse crossover interference in reading and the maximal response speed is limited by the RC-parameters of the matrix.

Table 1.1: Characteristics of Space-time Light Modulators.

Parameter	Thermoplastic STLM	Optical Addressing STLM	PROM STLM	Electron Addressing STLM	Liquid Crystal STLM
3dB Bandwidth, MHz	40 (200)*	–	–	10 (20)	–
Target Dimension, mm	50×50 (100×100)	30×40	30×30 (40×40)	50×50	25×25 (50×50)
Resolution at, 50% Modulation, lines/mm	40 (100)	10 (20)	10 (20)	15 (30)	30 (100)
Threshold Resolution, lines/mm	50 (150)	80	80	30	70 (100)
Number of Resolution Elements	4×10^6 (10^8)	1.2×10^5	9×10^4 (6.4×10^5)	5×10^5	5.6×10^5 (2.2×10^6)
Time of Cycle, s	0.5–1.0	0.01	0.01	0.03	0.03 (10^{-4}–10^{-3})**
Dynamic Range	400:1	100:1	500:1	1000:1	100:1
Longevity	10^5 cycles	3 years	unlimited	3 years	unlimited
Method of Erasing	heat	light	light	electron beam	HF electric field
Duration of Memory	month	1 hr	1 min	1 hr	15 ms***
Erasure time	1 s	0.05 ms	1 ms	0.5 ms	15 ms (0.1 ms)
Recording time	30 ms/frame	50 ns	10 ns	30 ms/frame	10 ns
Sensivity at Maximum Contrast	$1 \mu A$/element	$1 \mu J/cm^2$	$100 \mu J/cm^2$	$50 \mu A$/element	$10 \mu J/cm^2$ (10^{-8} J/cm^2)
Required Voltage, V	7,000	100	1,500–2,000	2,500	10

* Parameters which can be attained are given in parentheses.

** Dual frequency mode of operation.

*** Also possible is a relaxation controlled memory with 2 s time constant.

Chapter 2

Formation of Planar Array Antenna Patterns

This chapter deals with electrooptical planar arrays which form a two-dimensional antenna pattern. Their pattern-forming characteristics (accuracy, power, range) have been treated in a number of works [91, 122–125]. A real array is a system of a finite number of receiving elements; therefore, the space–time radio signal received by the array is a discrete function (known only as points of the receiving elements) and a finite (not equal to zero only within the limits of the array aperture) function of spatial variables. As such, it loses part of the information of the "radio scene" [1] which leads to its incomplete restoration as a result of treatment. The effects of discreteness and finiteness should be studied independently; to do so, this chapter uses the array and the space–time light modulator model in the form of continuously recorded and modulated medium of a limited dimension, the coordinates of which are uniquely connected by some addressing law, for example (1.3). A similar model makes it possible to analyze fully and graphically the basic antenna patterns of planar arrays taking into account specific laws of the displacement of the array elements and the space–time light modulator channels). If we take into account the corresponding radiation pattern of the Huygens element of the recording medium and the radiation pattern of the array receiving element, then this model is equivalent to a real array with an accuracy of digitalization effects which will be studied in Chapter 5.

[1] By *radio scene* we mean in general the multidimensional distribution (by angle, distance, frequency, etc.) of the radio sources undergoing restoration.

2.1 Pattern Control Characteristics of Electrooptical Planar Arrays

Let us assume that a flat continuous recording medium (array model) is placed in the plane XOY (see Fig. 1.2a). Then the space–time signal received by its element with the vector radiation pattern $\mathbf{F}_e$ at the point $\mathbf{R}_\perp$ and at the moment t may be written as [10, 126,127][2]

$$\begin{aligned}
\mathcal{E}(\mathbf{R}_\perp, t) &= \frac{1}{(2\pi)^3}\int_{-\infty}^{\infty} \mathrm{d}\Omega \exp(i\Omega t) \\
&\times \iint_{-\infty}^{\infty} \dot{\mathbf{s}}(\mathbf{K}_\perp, \Omega)\mathcal{F}_e(\mathbf{K}_\perp, \Omega)\exp[i\mathbf{K}_\perp \mathbf{R}_\perp] d^2\mathbf{K}_\perp = \\
&= \hat{\mathbf{F}}_t^{-1}\{\hat{\mathcal{F}}^{-1}\{\dot{s}(\mathbf{K}_\perp \Omega)\}\}
\end{aligned} \tag{2.1}$$

where $\mathcal{F}_t\{\ldots\} = \int_{-\infty}^{\infty} \ldots \exp(-i\Omega t)dt$; "$-1$" is a symbol of the inverse Fourier transformation; $\dot{s}$ is the vector spectrum of rf radiation received at the array from the upper hemisphere which being expressed in angular coordinates θ, ϕ [see eq. (1.1)] is the phase and frequency distribution of the complex vector amplitude of elementary plane waves received at the array from its far field; therefore for $\dot{s}$ we use two equivalent terms: space–time or the frequency-angular spectrum [10]; $\mathrm{d}^2\mathbf{K}_\perp = \mathrm{d}\Omega_X \mathrm{d}\Omega_Y = K^2 \mathrm{d}(\cos\alpha)\mathrm{d}(\cos\beta)$; the remaining symbols are explained in eq. (1.1) and in Fig. 1.2a.

This chapter mainly deals with the formation of the two-dimensional radiation characteristics of the array; therefore the problems of coherent optical processing formulated in its general aspects in Chapter 1 will be limited here to the restoration of spatial information about the radio scene, i.e., to the reproduction of a two-dimensional scalar spectrum $\dot{s} = \mathbf{s}\mathcal{F}_e$ of the space–time signal (2.1), as a function of angular coordinates. Ideally it would be desirable to accomplish this restoration [19] with an infinitely high resolution. In principle, by using the finite angular spectrum ($\dot{s}(K_\perp) = 0$ for $|\mathbf{K}_\perp| > K$ since the sources are located in the far field), it is possible to use such algorithms. However, the realization of similar restoration algorithms is complicated because, due to the incorrectness of the problem of the coherent optical processing at the processor output, there is is an unavoidable

[2] $\mathcal{E}(\mathbf{R}_\perp, t), (V)^{0.5}/\mathrm{M}$ is the scalar analog of the vector radiation intensity [52, p. 156] which in its direction and phase corresponds to the vector of the electric field falling on the array at point $\mathbf{R}_\perp$ while its magnitude is equal to the square root of the magnitude of Poynting's vector.

increase in the noise of the array receiver-amplifier unit, the space–time light modulator, the processor proper, the recording device (not to mention that all these algorithms are realized by Fourier optic means [3]). For technical reasons it is better to accomplish the restoration of the space–time spectrum with a finite resolution corresponding, for example, to the diffractional limit (array resolution). By taking into account Equation (1.1), this algorithm is evidently a Fourier transform. The space–time signal (2.1) in correspondence with Fig. 1.1 controls the transmission function (transmittance) T of the space–time light modulator. Between the coordinates of the array elements $\mathbf{R}_\perp$ and their corresponding space–time light modulator channels $\mathbf{r}_\perp$ there must be a single-valued addressing function

$$\mathbf{R}_\perp = \widehat{m}\mathbf{r}_\perp \tag{2.2}$$

where

$$\widehat{m} = \begin{bmatrix} m_x & m_{yx} \\ m_{xy} & m_y \end{bmatrix}$$

is an affine tensor. The transform (2.2), a particular instance of which is (1.3), considers practically all arrangements of the array elements and the space–time light modulator channels (for example, the case of an array with a triangular or even a trapezoidal network of elements, addressed in the space–time light modulator from a rectangular network of channels and vice versa [13]).

The transmission function T of the space–time light modulator in the general case is non-linear (see section 1.2.2), but since the real space–time signal (2.1) is small ($\mathcal{E} \ll \mathcal{E}_{\lambda/2}$) and, as a rule, increases to a value which excludes a non-linear distortion, this function may be then approximated only by two terms of its power series expansion[3]

$$T(J\mathcal{E}) \approx T^{(0)} + T^{(1)}J\mathcal{E} \tag{2.3}$$

where J is the amplitude phase distribution of the array caused by natural (for example, by edge effects [52–54]), or by artificial (by weighing or apodization) means. Function (2.3) of a three-dimensional space–time signal (2.1) may be treated as an optical model of a dynamic microwave hologram

[3]The nonlinear space–time light modulator effects which appear as subsequent members of a series are treated in chapter 7.

with a time delay recorded on a recording medium (array model). If uses amplitude modulation of light of the type (1.6a), then

$$T^{(0)} = T_{\circ} \sin\left(\frac{\Gamma_{\circ}}{2}\right), \qquad T^{(1)} = T_{\circ} \frac{\pi}{2\mathcal{E}_{\lambda/2}} \cos\left(\frac{\Gamma_{\circ}}{2}\right) \tag{2.4a}$$

If, however, one uses phase modulation (1.6b), then

$$T^{(0)} = T_{\circ} \exp(\Gamma_{\circ}), \qquad T^{(1)} = T_{\circ} \frac{\pi}{2\mathcal{E}_{\lambda/2}} \exp\left[i\left(\Gamma_{\circ} + \frac{\pi}{2}\right)\right] \tag{2.4b}$$

When this hologram is illuminated with a collimated monochromatic beam of light $E_{\circ} \exp(-i2\pi\nu t)$ with amplitude $E_{\circ}$ and frequency $\nu_{\circ}$ in a Fourier processor (see Fig. 1.2b) an optical signal $\dot{E} = \dot{E}^{(0)} + \dot{E}^{(1)}$ is read where

$$\left.\begin{aligned} E^{(0)}(\mathbf{r}_{\perp}, t) &= T^{(0)} E_{\circ} \exp(-i2\pi\nu t) \\ E^{(1)}(\mathbf{r}_{\perp}, t) &= T^{(1)} E_{\circ} J(\widehat{m}\mathbf{r}_{\perp}) \mathcal{E}(\widehat{m}\mathbf{r}_{\perp}, t) \exp(-i2\pi\nu t) \end{aligned}\right\} \tag{2.5}$$

In a linear approximation the components $\dot{E}^{(0)}$ and $\dot{E}^{(1)}$ may be treated separately. The former generates a parasitic effect at the output of the CO processor in the form of a zero order of diffraction, whose contribution to the degeneration of the electrooptical array characteristics are discussed in Chapter 7.

Useful information about the radio scene is carried by the optical space–time signal $\dot{E}^{(1)}$ in (2.5). Let us define its frequency spectrum. According to the modulation theorem [2, 3]

$$\dot{E}^{(1)}_{\omega} = \hat{\mathcal{F}}_t\{\dot{E}^{(1)}(\mathbf{r}_{\perp}, t)\} = T^{(1)} E_{\circ} J(\widehat{m}\mathbf{r}_{\perp}) \dot{\mathcal{E}}_{\omega - 2\pi\nu}(\widehat{m}\mathbf{r}_{\perp}) \tag{2.6}$$

where according to (2.1)

$$\dot{\mathcal{E}}_{\Omega}(\mathbf{R}_{\perp}) = \hat{\mathcal{F}}_t\{\mathcal{E}(\mathbf{R}_{\perp}, t)\} = \hat{\mathcal{F}}^{-1}\{\dot{s}(\mathbf{K}_{\perp}, \Omega)\}$$

In keeping with the above remarks about the need to restore the radio scene with the Fourier algorithm, let us modify Equation (2.6) by Equation (1.2) which describes the transformation accomplished by the collecting spherical lens (see Fig. 1.2b) at the fixed light frequency ν. In the rear focal plane π (the spectral plane) one may observe the following distribution of the light field at a given frequency:

$$\begin{aligned}\dot{e}_{\omega}^{(1)}(\mathbf{k}_{\perp}) &= \frac{1}{\lambda f}\hat{\mathcal{F}}\{\dot{E}_{\omega}^{(1)}(\mathbf{r}_{\perp})\} \\ &= e_{\circ}\iint_{-\infty}^{+\infty}\dot{s}(\mathbf{K}_{\perp},\omega-2\pi\nu)F(\widehat{m}^{-1}\mathbf{k}_{\perp}-\mathbf{K}_{\perp})\mathrm{d}^{2}\mathbf{K} \\ &= e_{\circ}\dot{s}(\widehat{m}^{-1}\mathbf{k}_{\perp},\omega-2\pi\nu)\otimes\otimes F(\widehat{m}^{-1}\mathbf{k}_{\perp}) \end{aligned} \tag{2.7}$$

where $e_{\circ} = T^{(1)}E_{\circ}/(2\pi)||\widehat{m}||\lambda f$; $||\widehat{m}||$det $\widehat{m}$; $\otimes\otimes$ is the symbol for a two-dimensional convolution.

The frequency spectrum (2.7) of the optic signal at the output plane π of the processor (see Fig. 1.2b) may be represented as

$$\dot{e}_{\omega}^{(1)} = \dot{e}_{\omega}^{(-1)} + \dot{e}_{\omega}^{(+1)} \tag{2.8}$$

where $\dot{e}_{\omega}^{(\pm1)} = \dot{e}_{2\pi\nu\pm|\Omega|}^{(1)}$ will be called the ±1 order of diffraction.

Since the modulation is non-linear and the method of complex amplitude is not applicable, it is necessary make the approximation that the $\mathcal{E} \approx \Re e\{\mathcal{E}\}$. Since the signal affects the space–time light modulator and, consequently, the Hermitian conjugate of its space–time spectrum [5, 10]

$$\dot{s}(\mathbf{K}_{\perp},\Omega) \equiv \overset{*}{\dot{s}}(-\mathbf{K}_{\perp},-\Omega) \tag{2.9}$$

($*$ denotes the complex conjugate), it is possible to establish that

$$\dot{e}_{2\pi\nu+|\Omega|}^{(+1)}(\mathbf{k}_{\perp}) = \overset{*}{\dot{e}}{}_{2\pi\nu-|\Omega|}^{(-1)}(-\mathbf{k}_{\perp}) \tag{2.10}$$

Thus, in using a space–time light modulator with transmittance (2.3), i.e., in dual band light modulation [89], the radio scene is reproduced as equal complex-conjugate images, centrally symmetrical relative to the processor axis and on the symmetrical ones relative to frequency, ν (see Fig. 1.2b). This circumstance can lead to an ambiguity in the coordinates of objects and will be treated in Section 5.2.

In the majority of cases the frequency-angular spectrum satisfies the delta correlation condition [126, 127]

$$\overline{\dot{s}(\mathbf{K}_\perp, \Omega) \overset{*}{s} (\mathbf{K}'_\perp, \Omega)} = |\dot{s}(\mathbf{K}_\perp, \Omega)|^2 \delta(\mathbf{K}_\perp - \mathbf{K}'_\perp) \tag{2.11}$$

where the overbar designates the averaging by realization and $\delta(\mathbf{K}_\perp)$ is the Dirac delta function.

Property (2.11) points to the absence of spatial radiation coherence (correlation), arriving at the array from different directions. This circumstance, as a rule, occurs in detection and ranging in the far field (the exception being the case of active detection of extended objects with mooth reflecting surfaces, for example, the sea, ice coverings). Therefore, by virtue of (2.1), the average frequency spectrum of the light flux (2.7) is

$$W_\omega^{(1)}(\mathbf{k}_\perp) = \overline{\dot{e}_\omega^{(1)}(\mathbf{k}_\perp) \overset{*}{e}_\omega (\mathbf{k}_\perp)} =$$

$$e_\circ^2 |\dot{s}(\widehat{m}^{-1}\mathbf{k}_\perp, \omega - 2\pi\nu)|^2 \otimes \otimes |F(\widehat{m}^{-1}k_\perp)|^2 \tag{2.12}$$

The integral operation (2.12), as Goodman pointed out [3], describes the process of forming images in an ideal non-coherent (i.e., linear in intensity) diffraction-limited image forming system in which point $\mathbf{K}_\perp$ at the electrooptical array input (see Fig. 1.2a) corresponds to the point

$$\mathbf{k}_\perp = \pm \widehat{m} \mathbf{K}_\perp \tag{2.13}$$

at the processor output (see Fig. 1.2b). Thus, according to (2.2), the radio scene and its optical image are as similar to each other as the space–time light modulator and its array are affine-like. If, for example, the array elements are situated in the angle of an oblique (rectangular) grid, while the space–time light modulator signal is situated in the angle of a rectangular (oblique angle) grid, then the processor output signal must be read off by a special oblique angular system of coordinates corresponding to equation (2.13).

The image-forming kernel of the integral operation (2.12) with an accuracy of a constant factor $e_\circ^2$ and in an appropriate scale $\widehat{m}$ coincides with the array factor power pattern F^2. Specifically, in the case of an array with a rectangular aperture $J(\mathbf{R}_\perp) = \mathrm{rect}(X/\Delta X)\mathrm{rect}(Y/\Delta Y)$, where $\mathrm{rect}(x) = 1$ at $|x| \leq 0.5$ and $\mathrm{rect}(x) = 0$, if $|x| > 0.5$, the approximation of array factor for a continuous aperture is

$$F(\mathbf{K}_\perp) = \hat{\mathcal{F}}\{J(\mathbf{R}_\perp)\} = \hat{\mathcal{F}}_X\{\text{rect}(X/\Delta X)$$

$$\times \hat{\mathcal{F}}_Y\{\text{rect}(Y/\Delta Y) = \Delta X \Delta Y \text{sinc}(\Omega_X/\delta\Omega_X)\text{sinc}(\Omega_Y/\delta\Omega_Y) \quad (2.14)$$

where $\text{sinc}(x) = \sin(\pi x)/\pi x$, $\delta\Omega_{X,Y} = 2\pi/\Delta X, Y$.

As we know, an array radiation pattern is equal to the product of the radiation pattern of an element of the array (in this chapter the Huygens element $\mathcal{F}_e$) by the array factor F (F_{AA} in Chapter 5) of a system of isotropic radiators situated at positions of the centers of elements in the array aperture. What has been said is true only in the case of independently acting array elements. A discussion of the effects caused by the mutual influence of the array receiving elements is presented in Chapter 5. In the future, however, it will be better to call the array pattern the *array factor* For F_{AA}, since the form of the diffraction spots at the processor output reflecting the point source targets coincides with this factor.

2.2 Accuracy, Power and Range Properties of Electrooptical Arrays

2.2.1 Accuracy

Assume that the matrix $\widehat{m}$ in (2.2) is diagonal: $m_{x,y} = m_{y,x} = 0$, then using (1.1) and (1.2) equation (2.13) becomes

$$\left.\begin{aligned} x &= \sin\theta\cos\varphi\, f\, m_x\Omega/(2\pi\nu \pm \Omega) \\ y &= \sin\theta\sin\varphi\, f\, m_y\Omega/(2\pi\nu \pm \Omega) \end{aligned}\right\} \quad (2.15)$$

As one can see, the mapping of the angular coordinates in the output plane of the processor results in chromatic aberration. An analogous phenomenon, known as aperture dispersion [51], is characteristic of phased arrays (but time delay is not). According to (2.15), with coherent optical processing weakens this phenomena somewhat for the +1 order of diffraction and increases it for the −1 order of diffraction, the more pronounced, the smaller is Λ. Due to the practically complete equality of these orders we will analyze one of them, for example, +1, assuming that the other is taken care of by one of the methods in Section 5.2.

Aberration and other array properties are conveniently analyzed by studying the electrooptical array reaction on a space–time spectrum of the form

$$|\dot{s}(\mathbf{K}_\perp, \Omega)|^2 = |\dot{s}_\phi(\mathbf{K}_\perp, \Omega)|^2 + \frac{1}{K^2}\sum_{q=1}^{Q} |\dot{s}_q(\Omega)|^2 \delta(\vec{\rho} - \vec{\rho}_q) \tag{2.16}$$

where $\vec{\rho} = \mathbf{K}_\perp / K = \mathbf{n}_X \cos\alpha + \mathbf{n}_Y \cos\beta$. The space–time spectrum (2.16) describes the superposition of sky noise $\dot{s}_\phi$ of artificial or natural origin and Q point transmitters with a frequency spectrum $\dot{s}_q(\Omega)$ and coordinates ρ_q. In view of the linearity of the operations (2.12) relative to the frequency spectrum and the sifting property of the δ-function, we obtain

$$W_\omega^{(+1)}(\mathbf{k}_\perp) = W_\omega^{(+1)}(\mathbf{k}_\perp)_\phi + \sum_{q=1}^{Q} W_\omega^{(+1)}(\mathbf{k}_\perp)_q \tag{2.17}$$

where

$$W_\omega^{(+1)}(\mathbf{k}_\perp)_\phi = e_\circ^2 |\dot{s}_\phi(\widehat{m}^{-1}\mathbf{k}_\perp, |\omega - 2\pi\nu|)|^2 \otimes \otimes |F(\widehat{m}^{-1}\mathbf{k}_\perp)|^2 \tag{2.18}$$

$$W_\omega^{(+1)}(\mathbf{k}_\perp)_q = e_\circ^2 |\dot{s}_q(|\omega - 2\pi\nu|) \cdot F(\widehat{m}^{-1}\mathbf{k}_\perp - K\vec{\rho}_q)|^2 \tag{2.19}$$

As (2.17) shows the point sources q are mapped as correspondingly displaced diffractional spots of light (2.19), coinciding in form with the array factor power pattern in the plane of the direction cosines and with a brightness proportional to the power of the radio waves received.

Let us calibrate the π-plane (see Fig. 1.2b) of the processor directly in the direction cosines

$$\mathbf{k}_\perp = \widehat{m}\frac{\Omega_\circ}{c}\vec{\rho}' \tag{2.20}$$

where $\Omega_\circ$ is the radian frequency to be calibrated; $\vec{\rho}' = \mathbf{n}_X \cos\alpha' + \mathbf{n}_Y \cos\beta'$ is the value of the vector $\vec{\rho}$ measured on a graduated scale. Using Parseval's equation, we can calculate the average intensity of the light flux (2.19)

$$I_q^{(+1)}(\vec{\rho}') = \frac{1}{2\pi}\int_{-\infty}^{\infty} W_\omega^{(+1)}(\mathbf{k}_\perp)_q d\omega$$

Expanding vector $\vec{\rho}\,'$ to $\vec{\rho}\,' = \vec{\rho}\,'_q + \vec{\rho}\,''$ where $\vec{\rho}\,'_q || \vec{\rho}_q$, and $\vec{\rho}\,'' \perp \vec{\rho}_q$ we can rewrite this integral as

$$I_q^{(1)}(\vec{\rho}\,') = \frac{e_\circ^2}{2\pi}\left|\dot{s}_q\left(\left|\Omega_\circ \frac{\rho\,'_q}{\rho_q}\right|\right)\right|^2 \otimes\otimes \left|F\left[\left(\Omega_\circ \frac{\rho'_q}{\rho_q}\right)\frac{\vec{\rho}_q}{c} + \frac{\Omega_\circ}{c}\vec{\rho}\,''\right]\right|^2 \qquad (2.21)$$

Expression (2.21) may be interpreted as follows: in the direction $\vec{\rho}\,'_q$ is the frequency spectrum $|\dot{s}_q(|\Omega|)|^2$, and along the orthogonal direction $\vec{\rho}\,''$, the representation coincides with the one for the monochromatic case ($\Omega = \Omega_\circ$), i.e. the optical image of the object coincides in form with the array factor power pattern. Thus, the result of chromatic aberration in the electrooptical array is the "erasing" of the monochromatic image frequency spectrum (2.19) along the elevation coordinates $\rho_q = \sin\theta_q$.

Let the spectrum $\dot{s}(\Omega)$ have a peak at frequency Ω_q, then by reading the coordinates of the object q from Equation (2.21) the maximum of the latter is found at point ($\rho\,'_q = \rho_q\Omega_q/\Omega_\circ$, $\rho'' = 0$) and by taking into account the fact that $\theta = \arcsin\rho$ and $\phi = \arctan(\cos\beta/\cos\alpha)$ [see eq. (1.1)], we obtain the following error in the definition of the angular coordinates of the object

$$\Delta\theta_q \approx -\frac{\Delta\Omega_q}{\Omega_\circ}\tan\theta_q, \qquad \Delta\phi_q = 0 \qquad (2.22)$$

where $\Delta\Omega_q = \Omega_q - \Omega_\circ$.

The same error also occurs in the case of a phased array as a result of only aperture dispersion [51]. However, the latter is also characterized by feeder dispersion of which the electrooptical array is practically free due to the absence of a branched feed system.

Let us finally evaluate the limitations which aperture dispersion places on the electrooptical array parameters. To do this we must pay attention to the fact that the image of the point source is not erased by its frequency spectrum if the width of the latter is less than the resolution of the image-forming kernel. The image-forming kernel in (2.20) is the radiation pattern compressed ρ_q/c times. Therefore, if the width of the radiation pattern is $\delta\Omega_{X,Y}$ then it is valid to use the condition $\Delta\Omega < c/\rho_q\,\delta\Omega_{X,Y}$ which based on the fact that $\delta\Omega_{X,Y} \approx 2\pi/\Delta X, Y$ becomes

$$\Delta F \cdot \Delta X, \qquad Y < c/\sin\theta_{max} \qquad (2.23)$$

where $\theta_{max} = \arcsin \rho_{max}$ is the edge of the scanned sector. As can be seen, the narrower the scanned sector, the weaker the limitation; at the same time the larger the antenna array aperture, the narrower the frequency band and vice versa. Criterion (2.23) guarantees an acceptably small dispersion error (2.22).

2.2.2 Power

By defining the gain of the electrooptical array during reception as the ratio of its excitation from a plane wave to the energy-wise equivalent of the latter (in the limits of the array) from an isotropic radiator, we obtain the corresponding evaluations in both mono- and achromatic cases. In the first case we obviously have

$$D_\circ = W^{(+1)}(\mathbf{k}_{\perp q})_q / W_\omega^{(+1)}(\mathbf{k}_{\perp q})_\phi \tag{2.24}$$

where the numerator is the frequency spectrum of the optical image of the point source (2.19) at a point maximum $\mathbf{k}_{\perp q} = \widehat{m} K \vec{\rho}_q$ and the denominator is the frequency spectrum of the image of an equivalent (using the full solid angle of 4π) uniform radio background (2.18) $4\pi |\dot{s}_\phi|^2 = |\dot{s}_q(\Omega)|^2 / K^2$ at the same point $\mathbf{k}_{\perp q}$. Since

$$\begin{aligned} W_\omega^{(+1)}(\mathbf{k}_{\perp q})_\phi &= e_\circ^2 |\dot{s}_\phi|^2 \iint_{-\infty}^{\infty} |F(\widehat{m}^{-1}\mathbf{k}_\perp)|^2 \mathrm{d}^2(\widehat{m}^{-1}\mathbf{k}_\perp) \\ &= 4\pi^2 e_\circ^2 |\dot{s}_\phi|^2 \iint_{-\infty}^{\infty} |J(\mathbf{R}_\perp)|^2 \mathrm{d}^2\mathbf{R}_\perp \end{aligned}$$

where with respect to the convolution of (2.18), we apply the expansion according moments and Parseval's equation [2], then the equation for gain is

$$D_\circ = \frac{4\pi}{\Lambda^2} \left| \iint_{-\infty}^{\infty} J(\mathbf{R}_\perp) \mathrm{d}^2\mathbf{R}_\perp \right|^2 \Big/ \iint_{-\infty}^{\infty} |J(\mathbf{R}_\perp)|^2 \mathrm{d}^2\mathbf{R}_\perp \tag{2.25}$$

which coincides with the known value of the gain for a planar aperture in the monochromatic case. A dual band input of the space–time signal to the processor produces a pair of matched images (2.8). Therefore, in this case the gain in Equation (2.25) is no more than two time higher than the actual gain (see Chapter 7).

If there are no restrictions (2.23), then we are dealing with a degraded directivity due to aperture dispersion. Equation (2.21) (the electrooptical array reaction to the radiation from a point-source nonmonochromatic object) must be treated as an irregular power pattern of the antenna according to power distorted by dispersion. Therefore, in the case of a nonmonochromatic object, the decrease of directivity in comparison with (2.25) is represented as

$$D/D_\circ = I_q^{(+1)}(\vec{\rho})/I_q^{(+1)}(0)\Big|_{\vec{\rho}_q=0}$$

where the numerator is the value of Equation (2.21) at the maximum $\vec{\rho}\,' = \vec{\rho}\Omega_q/\Omega_\circ$, and the denominator is the same with the absence of $\vec{\rho} = 0$ [see (2.23)]. Therefore, according to (2.19),

$$I_q^{(+1)}(0)\Big|_{\vec{\rho}_q=0} = \frac{e_\circ^2}{2\pi} F^2(0) \int_{-\infty}^{\infty} |\dot{s}_q(\Omega)|^2 d\Omega$$

and, using (2.21) we obtain

$$D/D_\circ = |\dot{s}_q(\Omega)|^2 \otimes |F(\Omega_q\vec{\rho}_q/c)|^2 \Big/ \int_{-\infty}^{\infty} |\dot{s}(\Omega)|^2 d\Omega F^2(0) \tag{2.26}$$

Let us examine a particular case. Let a uniformly excited rectangular array with an amplitude phase distribution $J(\mathbf{R}_\perp) = \mathrm{rect}(X/\Delta X)\mathrm{rect}(Y/\Delta Y)$ receive a rectangular pulse of duration τ_q with $\varphi_q = \arg \vec{\rho}_q = 0$. Then

$$\left.\begin{array}{lcl} |\dot{s}(\Omega)|^2/\int_{-\infty}^{\infty} |\dot{s}(\Omega)|^2 d\Omega &=& \tau_q \mathrm{sinc}^2[(\Omega - \Omega_q)\tau_q/2\pi] \\ F^2(\Omega\vec{\rho}_q/c)/F^2(0) &=& \mathrm{sinc}^2(\Omega\tau_X/c) \end{array}\right\}$$

where $\tau_X \sin\theta\Delta X/c$ is the aperture pulse"filling" time.

Therefore, for (2.26) we have

$$D/D_\circ = \begin{cases} \tau_q/\tau_X\left[1 - \tau_q/(3\tau_X)\right], & \tau_q \le \tau_X \\ 1 - \tau_x/(3\tau_q, & \tau_q > \tau_X \end{cases} \tag{2.27}$$

Considering that dispersion "affects" only the elevation coordinates [see (2.22)], the relative expansion of the radiation pattern may be evaluated

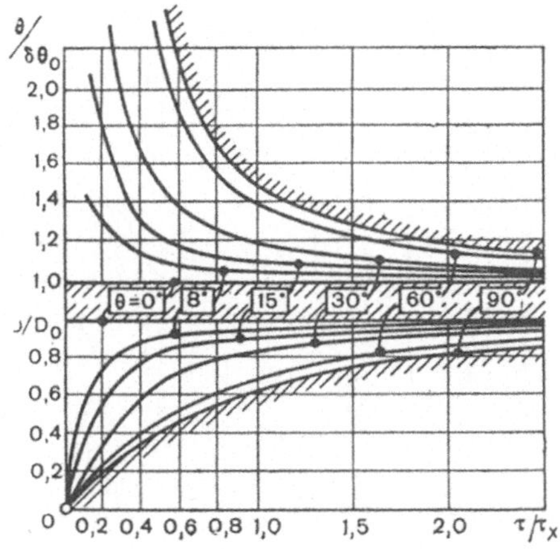

Figure 2.1: Plot of range properties of electrooptic arrays

by the formula [52]

$$\delta\theta/\delta\theta_\circ \approx D_\circ/D \tag{2.28}$$

where $\delta\theta_\circ$ is the radiation pattern width in the monochromatic case. Plots of (2.27) and (2.28) are shown in Fig. 2.1.

These functions illustrate the deterioration of the array directivity with the pulse duration commensurate with the aperture filling time during which there is a violation of the criteria (2.23).

2.2.3 Range properties

Along with the deterioration of the resolution and the directivity, aperture dispersion causes distortion of the radio signals being received. Let us define the excitation of a planar electrooptical array to a single side-band signal, $\dot{\mathcal{E}}_q(t)\exp(i\Omega_q t)$, from a point source, q, with coordinate, ρ_q. According to (2.19) and (2.20), the frequency spectrum of the optical signal at the output plane, π, of the processor

$$\dot{e}_\omega^{(+1)}\left[\widehat{m}\frac{\Omega_\circ}{c}\vec{\rho}\,'\right] = \dot{e}_\omega^{(+1)}(\vec{\rho}\,')_q = e_\circ \dot{s}_q(\omega - 2\pi\nu)F\left(\frac{\Omega_\circ}{c}\vec{\rho}\,' - \frac{\omega - 2\pi\nu}{c}\vec{\rho}_q\right)$$

where $\dot{s}(\Omega) = \hat{\mathcal{F}}_t\{\dot{\mathcal{E}}_q(t)\exp(i\Omega_q t)\}$ is the frequency spectrum of the received radio signal.

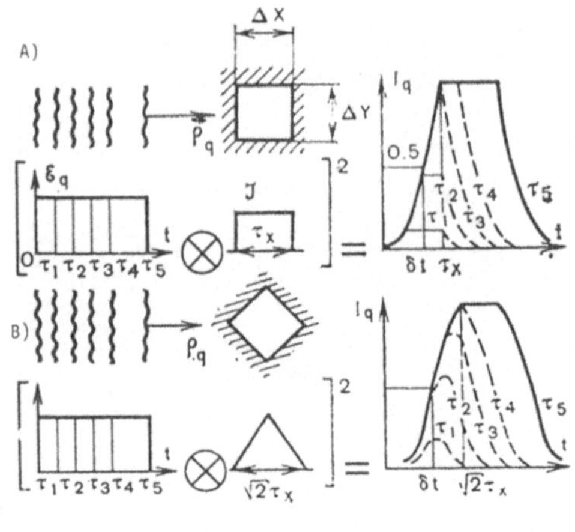

Figure 2.2: Distortion of pulses due to aperture dispersion

The desired excitation is the inverse Fourier transform of $\dot{e}_{\omega}^{(+1)}(\vec{\rho}\,')_q$:

$$\begin{aligned}
\dot{e}_q^{(+1)}(\vec{\rho}\,', t) &= \hat{\mathcal{F}}_t^{-1}\{\dot{e}_{\omega}^{(+1)}(\vec{\rho}\,')_q\} \\
&= e_\circ \hat{\mathcal{F}}_t^{-1}\{\dot{s}_q(\omega - 2\pi\nu)\} \otimes \hat{\mathcal{F}}_t^{-1}\left\{F\left(\frac{\Omega_\circ}{c}\vec{\rho}\,' - \frac{\omega - 2\pi\nu}{c}\vec{\rho}_q\right)\right\} \\
&= c_\circ \dot{\mathcal{E}}_q(t)\exp[i(2\pi\nu + \Omega_q)t] \otimes J_q(\vec{\rho}\,', t)\exp(i2\pi\nu t)
\end{aligned}$$

where

$$J_q(\vec{\rho}\,', t) = \hat{\mathcal{F}}_t^{-1}\left\{F\left(\frac{\Omega_\circ}{c}\vec{\rho}\,' - \frac{\Omega}{c}\vec{\rho}_q\right)\right\}$$

$$= \frac{1}{2\pi}\int_{-\infty}^{\infty} \mathrm{d}\Omega \exp(i\Omega t) \iint_{-\infty}^{\infty} J(\mathbf{R}_\perp)\exp\left[i\left(\frac{\Omega}{c}\vec{\rho}_q - \frac{\Omega_\circ}{c}\vec{\rho}\,'\right)\mathbf{R}_\perp\right]\mathrm{d}^2\mathbf{R}_\perp$$

$$= \frac{c}{\rho_q}\exp\left(i\frac{\Omega_\circ}{\rho_q}\vec{\rho}\,'_q t\right)\iint_{-\infty}^{\infty} J\left(\mathbf{R}'_\perp - \frac{ct}{\rho_q^2}\right)\exp\left(-i\frac{\Omega_\circ}{c}\vec{\rho}\,''\mathbf{R}_\perp\right)\mathrm{d}\mathbf{R}'_\perp$$

where according to (2.21) $\vec{\rho}\,' = \vec{\rho}\,'_q + \vec{\rho}\,''$, $\mathbf{R}'_\perp \perp \vec{\rho}_q$.

Since at a point maximum $\vec{\rho}\,' = \vec{\rho}_q\Omega_q/\Omega_\circ$ of function $\dot{e}_{\omega}^{(+1)}(\vec{\rho}\,')_q$ we have

$$J_q\left(\vec{\rho}_q\frac{\Omega_q}{\Omega_\circ},t\right) = \frac{c}{\rho_q}\exp(i\Omega_\circ t)J_{eqv}(-ct/\rho_q)$$

where

$$J_{eqv}(\mathbf{R}''_\perp) = \int_{-\infty}^{\infty} J(\mathbf{R}'_\perp - \mathbf{R}''_\perp)\mathrm{d}\mathbf{R}'_\perp$$

is the amplitude phase distribution of an equivalent linear array oriented along $\mathbf{R}''_\perp \parallel \vec{\rho}_q$ then,

$$\dot{e}_q^{(+1)}\left(\vec{\rho}_q\frac{\Omega_q}{\Omega_\circ},t\right) = e_\circ\frac{c}{\rho_q}[\dot{\mathcal{E}}_q(t)\otimes J_{eqv}(-ct/\rho_q)\exp[i(2\pi\nu+\Omega_q)t] \qquad (2.29)$$

Thus the electrooptical array excitation to the random radiation of a point source is a signal with a distorted complex envelope and carrier displaced to the frequency of light. If criteria of (2.23) is fulfilled, then the distortion of the envelope may be neglected since, in this case, the function $J_{eqv}(-ct/\rho_q)$ is a very short pulse ($\tau_X \approx 1..10$ ns) in comparison to the envelope. Using coherent heterodyning (see Chapter 7), we can extract from the optical signal (2.29) information about the amplitude and about the phase of the radio signal $\dot{\mathcal{E}}_q(t)\exp(i\Omega_q t)$. In non-coherent photodetection, phase information is lost since, in doing so, the intensity of light is registered

$$\begin{aligned} I_q^{(+1)}(\vec{\rho}_q\Omega_q/\Omega_\circ,t) &= \overline{\dot{e}_q^{(+1)}(\vec{\rho}_q\Omega_q/\Omega_\circ,t)\,\overset{*}{e}{}_q^{(+1)}(\vec{\rho}_q\Omega_q/\Omega_\circ,t)} \\ &= e_\circ^2(c/\rho_q)^2|\dot{\mathcal{E}}_q(t)\otimes J_{eqv}(-ct/\rho_q)|^2 \qquad (2.30)\end{aligned}$$

Let us illustrate the signal distortion of (2.30) when the criteria of (2.23) is violated. Let the envelope be a rectangular pulse $\dot{\mathcal{E}}_q(t) = \mathcal{E}\,\mathrm{rect}[(t-0.5\tau_q)/\tau_q]$ of duration τ_q and amplitude $\mathcal{E}_q$, while an array has a square aperture $\Delta X = \Delta Y$ with a uniform amplitude phase distribution $J(\mathbf{R}_\perp) = \mathrm{rect}(X/\Delta X)\mathrm{rect}(Y/\Delta Y)$. Then, the form of the signal (2.30) at the processor output, when pulses of length $\tau_q = \tau_1,\ldots,\tau_5$ with directions $\phi_q = 0$ and 45° arrive at the array antenna, takes on the form shown in Fig. 2.2a and 2.2b respectively. The front half-build-up-time is $\delta t = 0.71_{\tau_X}$ in the first case and 0.73 τ_X in the second where, $\tau_X = \sin\theta_q \Delta X/c < 10$ ns at $\Delta X < 30$ m. For all practical purposes, a similar signal distortion is not important, therefore, instead of convolution (2.30), it is possible to use only the first term of its expansion by moments [2]:

$$I_q^{(+1)}(\rho_q \Omega_q/\Omega_\circ, t) \approx |e_\circ \sum_\circ \dot{\mathcal{E}}_q(t)|^2 \tag{2.31}$$

where

$$\begin{aligned} \sum_\circ &= (c/\rho_q)\int_{-\infty}^{\infty} \mathcal{J}_{eqv}(-ct/\rho_q)\mathrm{d}t \\ &= \int_{-\infty}^{\infty} \mathcal{J}_{eqv}(\mathbf{R}'')\mathrm{d}\mathbf{R}''_\perp = \iint_\Sigma \mathcal{J}(\mathbf{R}_\perp)\mathrm{d}^2\mathbf{R}_\perp \end{aligned}$$

2.2.4 Transformation of the information block

These results give a useful graphical illustration. Look at Fig. 2.3. The cylinder (Fig. 2.3a) of height $\Delta\Omega$ (the electrooptical array antenna bandwidth) and radius $\rho_{max} = 1$ (dimension of the hemisphere) characterizes the initial information content of the radio scene. The content contains a random realization of the frequency-angular spectrum of the signal being received, especially, the signal from the point source q with a coordinate $\vec{\rho}_q$ and spectrum $\dot{s}_q(\Omega)$. The electrooptical array antenna "interprets" the radio scene in the form of a bipolar space–time spectrum whose informational content is two centrally symmetrical cones with a height $\Delta\Omega$ and a generator passing through the origin of the coordinates since between the points of the contents (Fig. 2.3a) and (Fig. 2.3b) there is a correspondence $\mathbf{K}_\perp = \vec{\rho}\Omega/c$ [see (2.16)]. The lower cone corresponds to the lower hemisphere and for the present does not cause any ambiguity, since it is known that array elements receive radiation from the upper hemisphere. Further (see Fig. 2.3c) the electrooptical array operates as an ideal noncoherent diffraction-limited imaging system (2.12): a transformation of affine coordinates (2.13) is carried out and the frequency spectrum is displaced to the optical carrier $2\pi\nu$. In this case, if there is bipolar light modulation, then both cones are displaced; if, however, there is a monopolar modulation, then one of them (the upper or lower cone depending on the type of monopolar modulation) is displaced. The upper cone of the information content corresponds to the +1 order of diffraction,

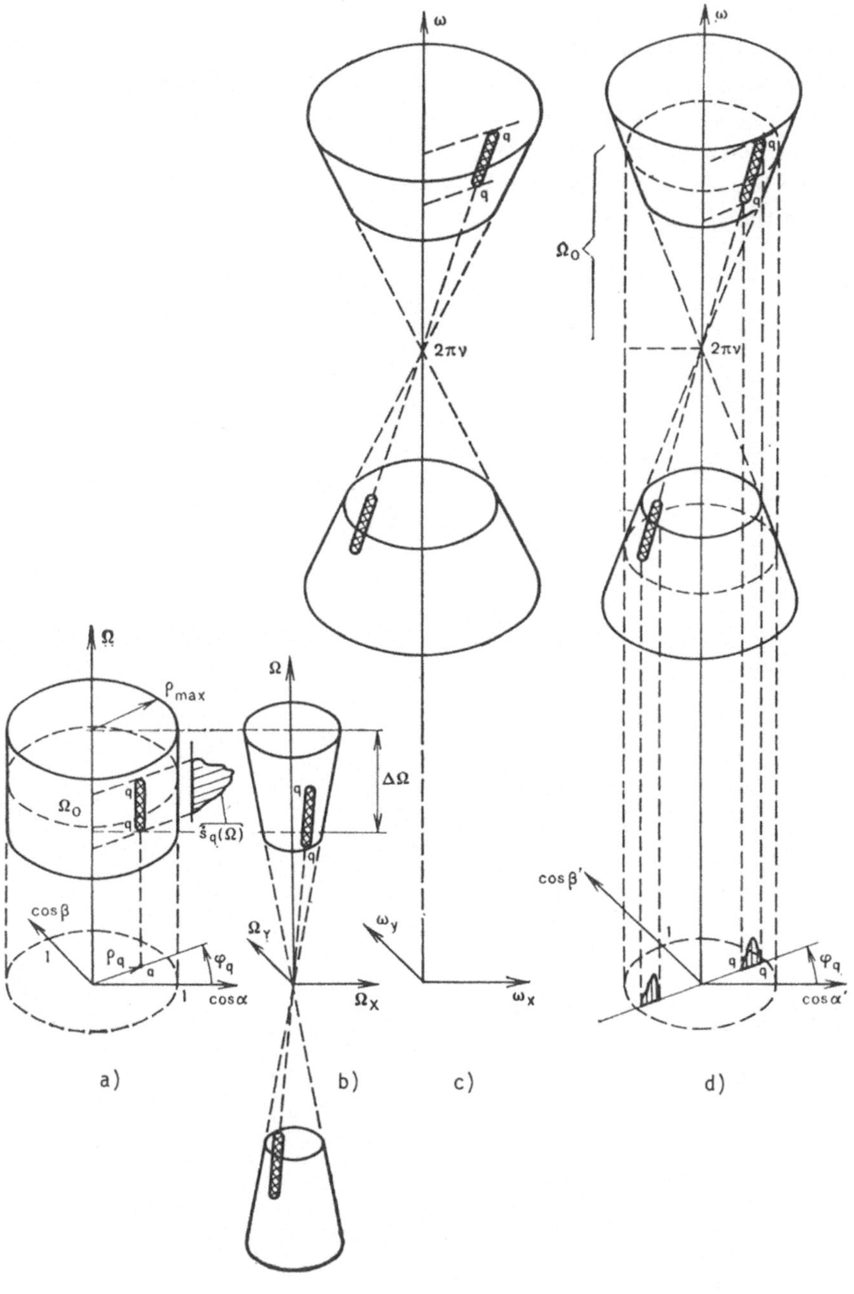

Figure 2.3: Transformation of the information block.

the lower to -1 order [see (2.10)]. The volume of Fig. 2.3d contains the end result of the coherent optical processing of the space–time signal. It is obtained from the preceding transformation (2.20). The object q is reflected in the processor plane λ (see Fig. 1.2b) as a projection of the enclosed volume (Fig. 2.3d) on the scale of the direction cosines. As can be seen, point q on Fig. 2.3a is reproduced as the segment q–q along the radius traced at angle ϕ_q. The broader the spectrum of $\dot{s}_q$, the longer this segment and greater the error of the elevation angle.

Chapter 3

Array Antennas with Space–Time Signal Processing

The range of problems solved by modern radar systems includes angular coordinate determination and spectral analysis (in passive radar), of, for example, the angular coordinates, the distance, speed, and acceleration using correlation processing in the active detection mode (for more details see Section 1.1).

Chapter 1 discussed the merits of optical signal processing involving time processing methods. The potential of coherent optical (CO) processors becomes more apparent when time processing is combined with spatial processing. This chapter basically deals with linear (one-dimensional) electrooptical arrays, which perform one-dimensional panoramic spatial scanning involving time signal processing (by spectral and correlation analysis). We give an overall view of their pattern and spectrum forming properties. We will treat coherent optical (CO) processor input devices using a multichannel acoustoopic modulator (AOM) in various diffraction modes (Raman-Nath and Bragg) and space–time light modulator with electron and optical signal addressing. We will discuss several complex processor input formats that make it possible to process the space–time signals from two-dimensional array antennas, which produce at the processor output information about the two angular coordinates and the signal spectrum. The chapter closes with an example of a CO processor which forms the spatial relational characteristics and the correlation signal processing.

3.1 Pattern and Spectrum Shaping Characteristics of Linear Electrooptical Arrays Using Multichannel Acoustooptical Space–Time Light Modulation

In one-dimensional parallel scanning by a linear array the CO processor does not use one spatial measurement. Scanning of the space–time signal from the array elements using a multichannel AOM supplements the electrooptical array with an optical spectrum analyzer which may subsequently undergo CO processing (for example, correlation analysis and pulse compression). The theory behind optical spectrum analyzers has been studied in sufficient detail [4, 17, 22, 35, 44, 45]; however, electrooptical arrays have certain characteristics that influence their pattern shaping properties.

3.1.1 Coherent optical regeneration of the frequency angular spectrum

In view of the above assumptions about the continuity of the arrangement of the receiving elements and the channels of the AOM, the space–time signal supplied to the CO processor may be represented as (see Fig. 1.3).

$$\mathcal{E}(Z,t) = \hat{\mathcal{F}}_t^{-1}\{\hat{\mathcal{F}}_Z^{-1}\{\dot{s}(\Omega_Z,\Omega)\}\} \tag{3.1}$$

where

$$\hat{\mathcal{F}}_Z\{\ldots\} = \int_{-\infty}^{\infty} \ldots \exp(-i\Omega_Z Z)\mathrm{d}Z$$

and

$$\hat{\mathcal{F}}_t\{\ldots\} = \int_{-\infty}^{\infty} \ldots \exp(-i\Omega_t t)\mathrm{d}t$$

are the operators of a one-dimensional Fourier-transform on Z and t, respectively;

-1 is the symbol of the inverse operator;

$\dot{s}(\Omega_Z, \Omega) = K \int_0^{2\pi} \dot{s}(\mathbf{K}_\perp, \Omega) d\varphi$ is a two-dimensional analog of a three-dimensional scalar frequency angular (or space–time) spectrum of the received emission $\dot{s}(\mathbf{K}_\perp, \Omega)$ from the upper (lower) hemisphere when $\Omega_Z \geq 0 (\Omega_Z < 0)$ [123, 125];

$\Omega_Z = K \cos\theta = \pm\sqrt{K^2 - |\mathbf{K}_\perp|^2}$ is the spatial frequency;

$K = 2\pi/\Lambda$ is the wave number of the radio wave.

Assume that an AOM operating in the Raman-Nath mode ($\vartheta_\circ = 0$) is placed in the object plane Π of a CO processor (see Fig. 1.3). Then in correspondence with (1.5) in the case of continuous arrangement of the AOM channels, the processor reads in the following optical signal (we are considering only its "informational" part):

$$\begin{aligned} E^{(1)}(x, y, t) &= T^{(1)} E_\circ \mathcal{J}(m_x x) \mathcal{J}_t(y) \\ &\times \mathcal{E}[m_x x; t + (y - y_\circ)/v] \exp(-i 2\pi \nu t) \end{aligned} \tag{3.2}$$

where

$\mathcal{J}(Z) = \mathcal{J}(m_x x)$ is a one-dimensional amplitude phase distribution;

$m_x = \Delta Z / \Delta X$ is the degree of modulation;

$\mathcal{J}_t(y)$ is the weighting function of the AOM channel;

$y_\circ = \Delta y / 2$ is the coordinate of the bar of the piezoelectric transducer;

v is the speed of sound in the medium (the latter is supposed to be nondispersive i.e., $v \equiv \text{const}$).

The frequency spectrum of the optical signal (3.2) is

$$\begin{aligned} \dot{E}_\omega^{(1)}(x, y) &= \hat{\mathcal{F}}_t\{E^{(1)}(x, y, t)\} \\ &= T^{(1)} E_\circ \mathcal{J}(m_x x) \mathcal{J}_t(y) \mathcal{E}_\Omega(m_x x) \exp\left[i\frac{\Omega}{v}(y - y_\circ)\right] \end{aligned} \tag{3.3}$$

where $\mathcal{E}_\Omega(Z) = \hat{\mathcal{F}}_t\{\mathcal{E}(Z, t)\} = \hat{\mathcal{F}}_Z^{-1}\{\dot{s}(\Omega_Z, \Omega)\}$ is the frequency spectrum of the radio signal at the array (3.1); $\omega = 2\pi\nu \pm \Omega$.

At the rear focal plane π of the Fourier processor (see Fig. 1.3) there is formed a distribution of the light field at a given frequency

$$\dot{e}_{\omega}^{(1)}(\omega_x, \omega_y) = \frac{1}{\lambda f}\hat{\mathcal{F}}\{\dot{E}_{\omega}^{(1)}(x,y)\} = \frac{T^{(1)}E_{\circ}}{\lambda f}\hat{\mathcal{F}}_x\{J(m_x x)\mathcal{E}_{\Omega}(m_x x)\}$$

$$\times\hat{\mathcal{F}}_y\left\{J_t(y)\exp\left[i\frac{\Omega}{v}(y-y_{\circ})\right]\right\}$$

$$= \frac{T^{(1)}E_{\circ}}{2\pi\lambda f m_x}\left[\dot{s}\left(\frac{\omega_x}{m_x},\Omega\right)\otimes F\left(\frac{\omega_x}{m_x}\right)\right]F_t(\omega_y - \Omega/v)\exp\left(-\frac{iy_{\circ}}{v}\Omega\right) \quad (3.4)$$

where $F(\Omega_Z) = \hat{\mathcal{F}}_Z\{J(Z)\} = m_x\hat{\mathcal{F}}\{J(m_x x)\} = F(\omega_x/m_x)$ is the directivity of a linear array; $F_t(\omega_y) = \hat{\mathcal{F}}_y\{J(y)\}$ is the image forming kernel of the optical spectrum analyzer.

Thus, at the processor spectral plane π one can observe a light flux with the intensity

$$I^{(1)}(\omega_x,\omega_y) = \frac{1}{2\pi}\int_{-\infty}^{\infty}\overline{\dot{e}_{\omega}^{(1)}(\omega_x,\omega_y)\,\overset{*(1)}{e_{\omega}}(\omega_x,\omega_y)d\omega}$$

$$= I_{\circ}\left|F\left(\frac{\omega_x}{m_x}\right)\right|^2\otimes\left|\dot{s}\left(\frac{\omega_x}{m_x},v\omega_y\right)\right|^2\otimes|F_t(\omega_y)|^2 \quad (3.5)$$

where we use Parseval's equation and take into account the property of δ-correlation of the frequency-angular spectrum (2.11).

As is apparent from (3.5), electrooptical array supplemented by a spectrum analyzer is equivalent to an ideal non-coherent diffractionally limited image-forming system which performs an affine transformation as

$$\omega_x = m_x\Omega_Z, \qquad \omega_y = \Omega/v \quad (3.6)$$

Thus, a linear electrooptical array with a multichannel AOM performs a simultaneous panoramic scan by frequency and by spatial coordinate; in the latter case the transformation (3.6) is analogous to the transformation in a two-dimensional electrooptical array (2.13).

A multichannel AOM in the Raman-Nath diffraction mode (Fig. 3.1, $\varphi = 0$) performs a bipolar modulation of light [89]. Therefore, just like (2.8) the optical image (3.5) in the processor spectral plane may be represented

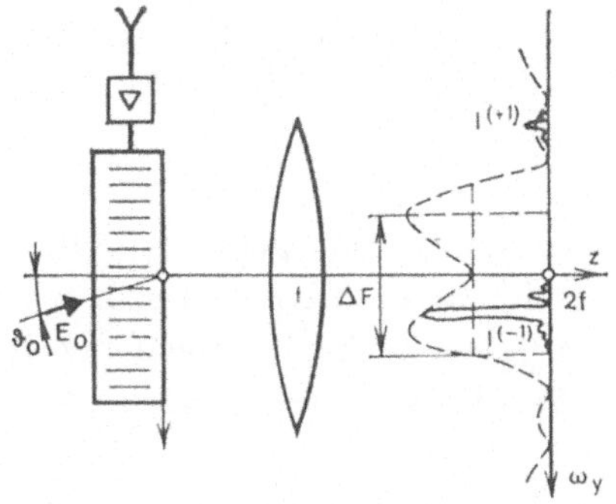

Figure 3.1: The Bragg effect.

as a superposition

$$I^{(1)}(\omega_x, \omega_y) = I^{(+1)}(\omega_x, \omega_y) + I^{(1)}(\omega_x, \omega_y) \tag{3.7}$$

where

$$I^{(\pm 1)}(\omega_x, \omega_y) = I_\circ \left| F\left(\frac{\omega_x}{m_x}\right)\right|^2 \otimes \left|\dot{s}\left(\pm\frac{\omega_x}{m_x}, \pm|v\omega_y|\right)\right|^2 \otimes |F_t(\omega_y)|^2$$

is a $\pm 1^{\mathrm{st}}$ order diffraction. Since a two-dimensional frequency-angular spectrum as its three-dimensional prototype [see the explanation to (3.1)] satisfies the Hermitian condition (2.9) $\dot{s}(\Omega_Z, \Omega) \equiv \overset{*}{\dot{s}}(-\Omega_Z, -\Omega)$, then

$$I^{(+1)}(\omega_x, \omega_y) \equiv I^{(-1)}(-\omega_x, -\omega_y) \tag{3.8}$$

However, in this case, the latter circumstance does not cause any ambiguity, since $\pm 1^{\mathrm{st}}$ order diffractions are formed in various hemispheres (see Fig. 1.3, $\omega_y \gtrless 0$).

3.1.2 Precision and power characteristics

As in our analysis of planar electrooptical array in Section 2.2, we will give a specific example of (3.5) for the frequency angular power spectrum of the received radio emission as

$$|\dot{s}(\Omega_Z,\Omega)|^2 = \frac{1}{K}\sum_{q=1}^{Q} |\dot{s}_q|^2 \delta\left(\frac{\Omega_Z - \Omega_{Z_q}}{K}\right)\delta(\Omega - \Omega_q) \tag{3.9}$$

The frequency angular power spectrum (3.9) describes the superposition of harmonic radiation at the frequencies Ω_q with complex amplitudes $\dot{s}_q$ from Q point sources in the far field of the array. In view of the linearity of the operations in (3.5) relative to the power (3.9)

$$I^{(1)}(\omega_x,\omega_y) = \sum_{q=1}^{Q} I_q^{(1)}(\omega_x,\omega_y) \tag{3.10}$$

where

$$I_q^{(1)}(\omega_x,\omega_y) = I_o|\dot{s}_q F(m_x^{-1}\omega_x - \Omega_{Z_q})F_t(\omega_y - \Omega_q/v)|^2 \tag{3.11}$$

As is apparent from (3.10), the point source q emitting harmonic signals of frequency Ω_q are reflected at the processor output (see Fig. 1.3) as correspondingly displaced diffractional light spots (3.11), whose distribution and intensity coincides in form with the directivity factor of the array in power $|F|^2$ along the dimension ω_x and with image forming kernel of the spectrum analyzer $|F_t|^2$ along the dimension ω_y.

In the case of a linear array with amplitude frequency modulation $J(Z) = \text{rect}(Z/\Delta Z)$ and acoustic line AOM with a weighting function $J_t(y) = \text{rect}(y/\Delta y)\exp[-\kappa(y - \Delta y/2)]$ (Δy is the length of the acoustic line; κ is the coefficient of attenuation of an acoustic wave in the acoustic line of an AOM) the radiation pattern and the image-forming kernel in (3.11) is defined by the following:

$$|F(m_x^{-1}x - \Omega_{Z_q})|^2 = \left\{\frac{\Delta Z \sin\left[\pi\left(\omega_x/m_x - \frac{\Omega_q}{c}\cos\theta_q\right)\right]}{\pi(\omega_x/m_x - \Omega_q/c\cos\theta_q)}\right\}^2 \tag{3.12}$$

$$|F_t(\omega_y - \Omega_q/v)|^2$$

$$= 4\exp(-\kappa\Delta y)\frac{\sinh^2(\kappa\Delta y/2) + \sin^2[(\omega_y - \Omega_q/v)/2\Delta y]}{\kappa^2 + (\omega_y - \Omega_q/v)^2} \tag{3.13}$$

whereby, when $\kappa \to 0$, $|F_t(\omega_y)|^2$ strives toward a function of the form (3.12). The maximum of the light intensity distribution in the diffraction spectrum (3.11), for example,of the $+1^{st}$ order, in correspondence with (3.12), (3.13) and (1.2) has the coordinates (see Fig. 1.3)

$$x_q = m_x f \frac{\Omega_q}{2\pi\nu} \cos\theta_q, \qquad u_q = \frac{c}{v} f \frac{\Omega_q}{2\pi\nu} = \frac{f}{kv}\dot{\Omega}_q \tag{3.14}$$

Equations (3.12) and (3.13) show that the angular coordinate resolution of an electrooptical array is no worse that with an equally-sized phased array ($\delta\Omega_Z = 2\pi/\Delta Z$), while the resolution of the optical spectrum analyzer $\delta\Omega = v\delta\omega_y \geq 2\pi v/\Delta y = 2\pi/\Delta\tau$ ($\Delta\tau$ is the acoustic line filling time).

These expressions describe the operation of a Fourier processor as a paraxial approximation ($\gamma \sim 10^{-2}\ldots 10^{-1}$ is the maximum field angle of the paraxial region [8,16]; therefore, $\omega_y = \Omega/v \leq \gamma k$ or $v \geq c\lambda/\gamma\Lambda$. For all practical purposes, $\lambda/\gamma\Lambda \approx 10^{-3}\ldots 10^{-2}$, consequently $v > 3\cdot 10^4 m/s$. Since for real acoustic lines of an AOM $v < 5\cdot 10^3 m/s$, then the spectrum of the radio signal must be displaced to a region of lower intermediate frequencies $\Omega_\pi = 2\pi F_\pi = \Omega_h$ (Ω_h is the frequency of the heterodyne of the electrooptical receiver). From these equations it follows that $\Delta F < F_\pi < \gamma v/\lambda \approx 100\ldots 1{,}000$ MHz (for an AOM with a solid state acoustic line). With a heterodyne displacement, the frequency correlation (3.5) should be written

$$I^{(1)}(\omega_x,\omega_y) = I_\circ \left|F\left(\frac{\omega_x}{m_x}\right)\right|^2 \otimes \left|\dot{s}\left(\frac{\omega_x}{m_x},\Omega_h - v\omega_y\right)\right|^2 \otimes |F_t(\omega_y)|^2 \tag{3.15}$$

Let us determine the directivity of a linear electrooptical array at reception using the approach illustrated in Section 2.2.2. Then

$$D = \left.\frac{I_q^{(1)}(\omega_x,\omega_y)}{I_\phi^{(1)}(\omega_x,\omega_y)}\right|_{\omega_x/m_x=(\Omega_h-v\omega_y)\theta_q/c}$$

$$= \frac{s|F(0)|^2}{\int_{-K}^{K}|F(\Omega_Z)|^2 d\Omega_Z} \tag{3.16}$$

where $I_\phi^{(1)}$ is the array reaction to the isotropic radiation equivalent in power to a planar wave within the limits of an array from source q.

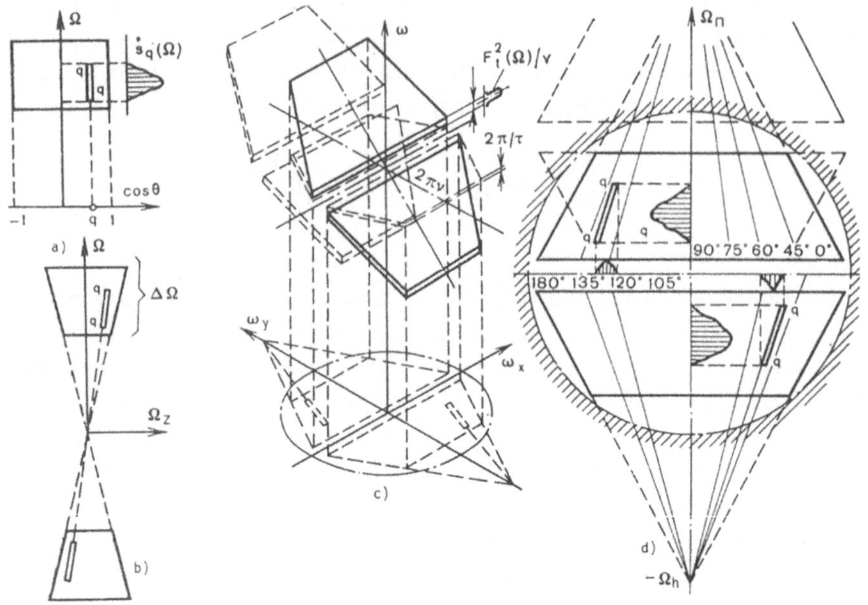

Figure 3.2: Transformation of the information block.

The directivity (3.16) coincides with the known expression for directivity of a linear array [52] even in the non-monochromatic case ($\Omega \neq \text{const}$). In this case, in contrast to a two-dimensional electrooptical array (see Section 2.2), the actual directivity of a linear electrooptical array also coincides with (3.16) (it is not two times smaller), for the conjugate images (3.7) illuminate each other. The presence of an optical spectrum analyzer in the electrooptical array has still another advantage. Thus, equation (3.16) occurs along the lines $\omega_x/m_x = (\Omega_h - v\omega_y)\cos\theta_q/c$ on which the values of elevation coordinates are unchanged. This circumstance makes is possible to avoid the dispersion error of the type (2.22) at the expense of a suitable calibration of the plane π of the processor (see Fig. 3.2) even when condition (2.23) is violated.

3.1.3 Comparison of Raman-Nath and Bragg diffraction in coherent optical processors

If the reading beam E_o is incident on a multichannel AOM at an angle ϑ_o to the axis Z (see Fig. 1.3 and 3.1), then the diffraction pattern (3.15) or (3.5) is displaced to the subcarrier spatial frequency $\omega_{y_o} = k\sin\vartheta_o$ and is considered the envelope [89,110]

$$\mathrm{sinc}^2\left[\frac{\Delta z}{n_\circ\lambda}(\cos\vartheta - \cos\vartheta_\circ)\right] \approx \mathrm{sinc}^2\left[\frac{\Delta z}{2n_\circ\lambda k^2}(\omega_y^2 - \omega_{y_\circ}^2)\right]$$

where $\Delta z/n_\circ\lambda$ is the optical "thickness" of the AOM channel (Δz is the geometrical thickness, $n_\circ$ is the refraction index of the acoustic line) and taking into account that $\cos\vartheta = \sqrt{1-\sin^2\vartheta} = \sqrt{1-\omega_y^2/k^2} \approx 1-\omega_y^2/2k$. Therefore, instead of (3.15) for Bragg diffraction we have

$$I_{BR}^{(1)}(\omega_x,\omega_y) = \mathrm{sinc}^2\left[\frac{\Delta z}{n_\circ\lambda 2k^2}(\omega_y^2 - \omega_{y_\circ}^2)\right] I^{(1)}(\omega_x,\omega_y - \omega_{y_\circ}) \qquad (3.17)$$

Note that the coefficient $I^{(0)}$ in (3.15) in this linear approximation [see (2.3)] is the same for both kinds of diffraction (Raman-Nath and Bragg)[1] since in the first case $T^{(1)} =\approx J_1(\pi/2\mathcal{E}/\mathcal{E}_{\lambda/2})$ (J_1 is a first-order Bessel function), and in the second case $T^{(1)} \approx \sin[(\pi/4)\mathcal{E}/\mathcal{E}_{\lambda/2}]$.

In defining $\vartheta_\circ$ in the paraxial region $\omega_y \leq \gamma k$ only one of the diffractional orders (for example, -1 in Fig. 3.1) is formed. The service band of an optical processor with a multichannel AOM is determined by the bandwidth of a piezoelectric transducer with an appropriate electrical circuit and a band of efficient opto-acoustic interaction [44]. For Bragg diffraction modulators the optimal choice for the band of efficient opto-acoustic interaction is in the alignment of the edges of the paraxial region $\omega_y = \gamma k$ with the spatial frequency $\omega_{y_\circ} = k\sin\vartheta_\circ = \gamma k$, hence, by taking into account the smallness of $\gamma(10^{-2}\ldots 10^{-1})$, we obtain $\vartheta_\circ \approx \gamma$. In this case, the frequency band of a spectrum analyzer is almost two times wider than in the case of Raman-Nath diffraction since here the paraxial region is completely occupied only by a one-order diffraction (-1 here).

However, in this AOM diffraction mode the previously mentioned envelope determines the inequality of the optical spectrum analyzer frequency characteristics. In order to reduce this circumstance, it is necessary to limit the inequality of the envelope, for example, by 0.5 from its maximum. Since the envelope has a minimum at $\omega_y = 0$, and the maximums are reached at $\omega_y = \pm\omega_{y_\circ}$, then the suggested requirement is fulfilled if $\mathrm{sinc}^2(\Delta\omega_{y_\circ}^2/n_\circ\lambda 2k^2) \geq 0.5$ or

$$\Delta z/n_\circ\lambda \leq 2\kappa_{0.5}/\gamma^2 \qquad (3.18)$$

[1]The nonlinear AOM effects which occur as successive members of the series (2.3) is treated in chapter 7.

where $\mathrm{sinc}^2\kappa_{0.5} = 0.5$.

Thus, in the Bragg diffraction mode the the range of efficient optic-acoustic interaction $\Delta F < F_n < 2\gamma v/\lambda \approx 2,000$ MHz, which in essence characterizes the potential optical spectrum analyzer range, which, for all practical purposes, is limited by the piezoelectric transducer bandwidth and the circuit diagram of the AOM.

Given these processor and AOM parameters the latter is, for all practical purposes, equivalent to a monopolar space–time light modulator and possesses a very high diffractional efficiency since in this case one of the diffraction orders (+1) is beyond the range of efficient acoustooptical interaction and reduced by the envelope to no less than 13 dB (see Fig. 3.1, the level of the first side lobe).

3.1.4 Transformation of the information block

These results are graphically shown in Fig. 3.2. The rectangle (Fig. 3.2a) with height $\Delta\Omega$ and width 2 is a two-dimensional analog of the initial information content of possible states of the radio scene (see Fig. 1.3a) containing a random realization of the frequency-angular spectrum of the emission, especially the emission from the point source q with coordinate $\boldsymbol{\theta}_q$ and spectrum $\dot{s}_q(\Omega)$. The system "perceives" the radio scene as a frequency-angular spectrum whose information block is illustrated in Fig. 3.2b. As a result of heterodyning and CO processing in a multichannel space–time light modulator (Raman-Nath diffraction mode), the entire information block is transferred to an information "layer" (Fig. 3.2c). The thickness of the layer is determined by the resolution of the optical spectrum analyzer. Projection of the "layer" onto the plane $\omega_x o \omega_y$ represents two isosceles trapezoids which are close to each other and therefore fit into the paraxial region (i.e., inside the circle outlined in dashes). The same dashes show the information "layer" of the AOM in the Bragg diffraction mode which is displaced along the axis ω_y to the left. By carefully selecting the AOM diffraction mode in the paraxial region it is possible to obtain only one trapezoid. This makes it possible to almost double the range of the spectrum analyzer and the diffraction efficiency of the AOM in comparison to that of the Raman-Nath mode. Fig. 3.2d shows the output plane of the processor with calibration to remove the elevation dispersion error (2.22). As can be seen, $\pm 1^{\mathrm{st}}$ orders are in different hemispheres (upper and lower) and do not interfere and do not cause any ambiguity if $\Omega_h > \Omega$. In Bragg diffraction only one order is seen (-1st).

3.2 Linear Electrooptical Array with Electron Beam and Optical Signal Addressing Input Devices

Besides a multichannel AOM, a space–time light modulator with electron and optical signal addressing may be used for a spatial sweep of a time dependent space–time signal received at the array. Data about electrooptical array with similar space–time light modulator inputs (see Figs. 1.5 and 1.7) is given in Section 1.2.2. In essence, these systems differ in the recording of the array space–time signal on a carrier (photographic film or plate, thermoplastic film or electrooptical crystal) by an optical or electron beam, unfolded consecutively (in the presence of delay lines, see Fig. 1.5) or in parallel (for example, with the help of a matrix of light-emitting diodes of a linear multichannel space–time light modulator addressed by electrical voltage—see Fig. 1.7). Therefore, an analysis of an electrooptical array with such a space–time light modulator may be done uniformly while bearing in mind that the space–time light modulator selected does not influence the basic pattern-forming properties of the processor (directive gain, resolution, etc.), but determines the operating range, the dynamic range, the sensitivity, the cycle of information updating, lifespan, and other signal parameters.

3.2.1 Pattern- and spectrum-forming properties

As before, we will use expression (3.1) as the initial space–time signal supplied to the CO processor. This input operates either continually or frame-by-frame (record-read-erase) with a time cycle $T_c = T_{rec} + T_{read} + T_{erase}$. Thus, in the interval $t_n \leq t \leq t_n + T_{read}$ ($t_n = t_\circ + nT_{erase}$, $t_\circ$ is the beginning of reading of the null cycle, n is the number of the current cycle) the processor reads from the surface Π of the combined space–time light modulator (see Fig. 1.7) an optical signal with the amplitude [128]

$$E^{(1)}(x,y) = T^{(1)} E_{read} J(m_x x) J_t(y) \tilde{\mathcal{E}} \left(m_x x; t_n + \frac{y - y_\circ}{v_p} \right) \tag{3.19}$$

where $T^{(1)}$ is expanded in (2.3); E_{read} is the amplitude of the read beam, $J(Z)$, $J_t(y)$ is the weighting function (amplitude phase distribution) in the corresponding dimensions; $Z = m_x x$ is the addressing law ($m_x = \Delta Z / \Delta x$ is the scale of modeling the array of length ΔZ with the help of a space–time light modulator of length ΔZ); v_p is the signal scanning rate in plane Π (by a deflector or electron beam tube); $y = v_p t - y_\circ$; $\tilde{\mathcal{E}}$ is space–time signal image $\mathcal{E}(Z,t)$ at plane Π (see Fig. 1.7), which with a finite dimension δ_y of the recording beam along the sweep is defined by the equation

$$\tilde{E}(m_x x, t) = \int_{-\infty}^{\infty} \mathcal{E}(m_x x, t')\text{rect}[(t-t')/\delta_t]dt' \tag{3.20}$$

where $\delta_t = \delta_y / v_p$.

Substituting (3.20) into (1.2), we find that in the output plane π of a Fourier-processor (see Figs. 1.5 and 1.7) an optical image of the space–time spectrum is formed with a complex amplitude

$$\dot{e}(\omega_x, \omega_y) = \frac{1}{\lambda_{read} f} \mathcal{F}\{E^{(1)}(x, y)\}$$

$$\frac{T^{(1)} E_{read}}{(2\pi)^2 \lambda_{read} f} \mathcal{F}\{J(m_x x) J_t(y)\}$$

$$\otimes \otimes \mathcal{F}\{\tilde{\mathcal{E}}[m_x x; t_n + (y - y_\circ)/v_p]\}$$

$$= e_t F\left(\frac{\omega_x}{m_x}\right) \otimes \left\{\dot{s}\left(\frac{\omega_x v_p \omega_y}{m_x}\right) \text{sinc}(\omega_y \delta_y / 2\pi) \exp[i(v_p t_n - y_\circ)\omega_y]\right\} \otimes F_t(\omega_y)$$

(3.21)

where $e_t = T^{(1)} E_{read} v_p \delta y / (2\pi)^2 \lambda_{read} f m_x$; $F(\omega_x / m_x) = F(\Omega_Z) = \mathcal{F}_Z\{J(Z)\} = m_x \mathcal{F}_x\{J(m_x x)\}$ is the radiation pattern of a linear array; $\delta_y \,\text{sinc}(\omega_y \delta_y / 2\pi) = \mathcal{F}_y\{\text{rect}(y/\delta y)\}$; $F_t(\omega_y) = \mathcal{F}_y\{J_t(y)\}$ and, consequently, we take into account the convolution theorem, factorization of the two-dimensional Fourier transform, equations (3.1)-(3.5), and the shift and similarity theorems.

Note that expression (3.21) cannot be put into the form of (3.5) due to the time uniformity of the optical image (3.21) in the reading interval. Therefore, this is equivalent to a coherent diffractionally-limited image-forming system [3], linear in intensity inasmuch as T_c significantly exceeds the constant time of the photodetector. However, this does not give the systems under consideration any noticeable advantage with respect to resolution, gain and other very important electrooptical array parameters in comparison with non-coherent image-forming systems [3].

Thus, from (3.21) it follows that linear arrays with a space–time light modulator CO processor using electron and optical addressing just as when using multichannel AOM (see Section 3.1), one simultaneously performs a panoramic sweep along one spatial coordinate with a highly-directional beam of the radiation pattern $F(\Omega_Z)$ and along a frequency with an image-forming kernel $F_t(\omega_y)$. In this case, the frequency-angular spectrum of the incoming signals $\dot{s}(\Omega_Z, \Omega)$, just as before is reflected at the processor output (in plane π, see Figs. 1.5 and 1.7) on a scale (3.1). However, in restoring the signal's frequency spectrum there is a distortion (not peculiar to multichannel AOM) of the envelope function of the kind $\mathrm{sinc}(\omega_y \delta_y / 2\lambda)$, due to the finite dimension of the recording beam.

3.2.2 Requirements for space–time light modulators

Let us evaluate the last factor for which we will limit the inequality of the envelope by a level $\sqrt{0.5}$ of its maximum value. Inasmuch as $\mathrm{sinc}\kappa_{0.5} = \sqrt{0.5}$ for $\kappa_{0.5} \approx 0.44$, then taking into account (3.6) the maximum frequency (range ΔF) $F_{max} = \Omega_{max}/2\pi = \Delta F$ of the frequency-angular spectrum $\dot{s}(\Omega_Z, \Omega)$ of the incoming signals is defined by the expression

$$\Delta F = F_{max} = 0.44 v_p / \delta_y \tag{3.22}$$

The value $F \leq F_{max}$ agrees with the physically obvious requirement $\delta_t = \delta_y / v_p < 1$. Specifically when $\delta_y = 5$ μm; 100 μm and $v_p = 10^{-1}$ m/s, value $F_{max} \approx 9$ kHz and 0.44 kHz, respectively [2, 3], but when $\delta_y = 50$ μm and $v_p = 1{,}500$ m/s, $F_{max} \approx 12$ MHz [85].

For processing signals with a base $\Delta T \Delta F (\Delta T, \Delta F$ is the length and the width of the signals) with the help of an optical processor with a number of elements of the space–time light modulator along the "time" coordinate $\Delta y R_y$ (Δy is the dimension of the signal sweep, R_y is the resolution of the material in lines/mm) the condition $\Delta T \Delta F \leq \Delta y R_y$ must be fulfilled. Since $\Delta T = \Delta y / v_p$ then

$$R_y \gg F_{max}/v_p = \Delta F v_p \tag{3.23}$$

The inequality (3.23) determines the requirement for the resolution of the material. In particular when $F_{max} = 10$ MHz, $v_p = 15 \cdot 10^5$ mm/s, $R_y \geq 7$ mm^{-1}and when $F_{max} = 9$ kHz; $v_p = 10^3$ mm/s, $R_y \geq 9$ mm^{-1}

In accordance with Section 3.1.2., the resolution of an optical spectrum analyzer is in the best case

$$\delta F = \delta\Omega / 2\pi = v_p / \Delta y \tag{3.24}$$

A comparison of (3.23) and (3.24) shows that the attainment of the maximum bandwidth ΔF is inconsistent with an increase in the resolution δF of the spectrum analyzer. To enhance the maximum bandwidth at a given spectrum analyzer resolution, in accordance with (3.22), (3.23), and (3.24), it is necessary to maximize the number of the space–time light modulator resolution elements $(\Delta y R_y)$:

$$\Delta F/\delta F = \Delta y R_y \tag{3.25}$$

In particular, with a target dimension $\Delta y = 50$ mm of the space–time light modulator and a resolution $R_y = 15\ \text{mm}^{-1}$ (see Table 1.1), $\Delta F/\delta F = 750$.

As already noted, the merit of this space–time light modulator is the possibility of varying the sweep rate v_p, which gives it a certain flexibility (as opposed to an AOM) with a compromise between the resolution (3.24) and the operating frequency band (3.22) of an electrooptical array.

Let us also elaborate the requirements for the stability of the sweep rate v_p of the time signal on the carrier. In the first approximation, it is possible to be limited by the linear law of the rate of change $v_p = v_p(t) = v_\circ a_\circ t$ ($v_\circ$ is the initial rate, $a_\circ$ is the acceleration). Evidently a signal sweep with variable speed $v_p(t)$ results in nonlinear distortion. Actually, the current position of the recording beam at time t is

$$y(t) - y_\circ = \int_\circ^t v_p(t')dt' = v_\circ t + \frac{a_\circ}{2}t^2$$

hence, from the solution of the last quadratic equation, time

$$t = \sqrt{v_\circ^2 + 2a_\circ(y - y_\circ)}/a_\circ - v_\circ/a_\circ \approx (y - y_\circ)/v_\circ - a_\circ(y - y_\circ)^2 2v_\circ^3 \tag{3.26}$$

Substituting (3.26) into (3.19) instead of the argument $t = (y - y_\circ)/v_p$ ($t_n = 0$) and being limited by two members of the expansion of the signal $\tilde{\mathcal{E}}$ into the power series:

$$\tilde{\mathcal{E}}[m_x x, (y - y_\circ)/v_\circ - a_\circ(y - y_\circ)^2/2v_\circ^3]$$

$$\approx \tilde{\mathcal{E}}[m_x x(y - y_\circ)/v_\circ] - \frac{a_\circ(y - y_\circ)^2}{2v_\circ^3}\frac{\partial\tilde{\mathcal{E}}(m_x x, t)}{\partial t}\bigg|_{t=(y-y_\circ)/v_\circ}$$

instead of (3.21) with regard to the Fourier transform properties of an arbitrary function [2: 72], we obtain

$$\dot{\tilde{e}}^{(1)}(\omega_x, \omega_y) = \dot{e}^{(1)}(\omega_x, \omega_y) - ie_t F$$

$$\times(\omega_x/m_x) \otimes_{\omega_x} [\mathrm{sinc}(\omega_y \delta_y/2\pi)\dot{s}(\omega_x/m_x, v_\circ \omega_y)$$

$$\times v_\circ \omega_y \exp(-iy_\circ \omega_y)] \otimes_{\omega_y} \frac{a_\circ}{2v_\circ^3} frac \partial^2 F_t(\omega_y) \partial \omega_y^2 \qquad (3.27)$$

Equation (3.27) shows that the instability of the scanning rate produces noise in the image (3.21) at the processor output plane π (see Fig. 1.7) by an additive component which must be treated as parasitic since its image-forming kernel $\partial^2 F_t(\omega_y)/\partial\omega^2 y$ is not the same as that of (3.21) and consequently distorts the useful image. To reduce this distortion it is necessary to limit the relative level of the parasitic noise by the value of the processor's dynamic range (DR).

$$\left| v_\circ \omega_y \frac{a_\circ}{2v_\circ^3} \frac{\partial^2 F_t(\omega_y)}{\partial \omega_y^2} / F_t(\omega_y) \right|^2 \leq 1/DR \qquad (3.28)$$

In keeping with the upper limit estimates of the values limited along the spectrum and its second derivative [2: 195], and taking into account that $v_\circ \omega_y \leq \Omega_{max}$, we obtain

$$a_\circ \leq (2\pi)^2 8\sqrt{5} v_\circ^3 / \sqrt{(DR)} \Omega_{max} \Delta y^2 \qquad (3.29)$$

Thus the instability of the scanning rate must not exceed

$$\frac{\Delta v_p}{v_\circ} \cdot 100\% \approx \frac{a_\circ \Delta \tau}{v_\circ} \cdot 100\% = \frac{a_\circ \Delta_y}{v_\circ^2} \cdot 100\%$$

$$\leq \frac{(2\pi)^2 8\sqrt{5} v_\circ^3}{\sqrt{(DR)} \Omega_{max} \Delta y^2} \frac{\Delta y}{v_\circ^2} \cdot 100\%$$

$$\approx \frac{10^2 \delta F}{\sqrt{(DR)} F_{max}} \cdot 100\% \approx 10^4 / \sqrt{(DR)} \Delta y R_y, \% \qquad (3.30)$$

where (3.25) and (3.24) are used. In particular, when the dynamic range

$DR = 10^4$ and the number of resolution elements is $\Delta y R_y = 10^3$ the scanning rate instability must be less than 0.1%.

In conclusion we will point out the sensitivity requirement E_ϕ, J/cm^2 of the space–time light modulator photoconductor which describes the minimum energy flux density for the light recording necessary to ensure the required contrast for the signal in the processor. Obviously, this value is derived from

$$E_\phi \leq P_{rec}\eta_{rec}\tau_{rec}/\Delta x \Delta y = P_{rec}\eta_{rec}/\Delta x v_p \tag{3.31}$$

where P_{rec} is the power of the recording laser in watts; η_{rec} is the coefficient which accounts for the light lost in the optical processor during the recording. In particular, when $P_{rec} = 25 \cdot 10^{-3}$ W which corresponds to a real "blue" laser ($\lambda_{rec} = 0.4416\ \mu$m), $\eta_{rec} = 0.5$ (taking into account the loss in a linear multichannel space–time light modulator 2 and deflector 3 of Fig. 1.7; $\Delta x =$ 5 cm, $v_p = 10$ cm/s, the sensitivity $E_\phi \leq 2.5 \cdot 10^{-3}$J/cm^2 which may ensure with the help of PROM structures ($E_\phi = 10^{-4} \ldots 10^{-5}$J/cm^2) and a liquid crystal photoconductor ($E_\phi = 10^{-5} \ldots 10^{-7}$J/cm^2).

A complete description of the possibilities of a CO processor with this hybrid input requires a full treatment of the device which make up the space–time light modulator with a multichannel optical addressing [105,107]. In particular, it is necessary to know the frequency contrast characteristics of the device in which is turned the time signal (the experiment Section 9.2 uses a PROM) for determining the reproduction of the results of signal processing at the processor output at the given dynamic range, the diffraction efficiency (the ratio of the power of light, which has diffracted to a useful order of diffraction to the intensity of the light incident on the device), and the power and time characteristics. These contradictory requirements may be overcome by compromises. Thus, the higher spatial frequency of the signal being recorded F/v_p, the greater the exposure (sensitivity) necessary to produce the maximum diffraction efficiency. However, the exposure value is limited by the maximum frequency of the signal, by the scanning speed, and by the dimension δ_y. With an increase in exposure, the resolution at a given contrast level grows, while the diffraction efficiency falls. The power-time characteristics of the device impose limitations on the time interval between frames, i.e., on the work cycle.

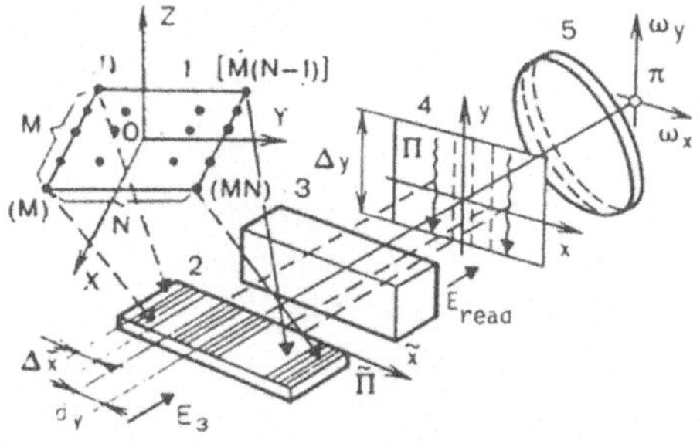

Figure 3.3: Block diagram of CO processor with a complex signal recording format.

3.3 Coherent Optical Processors for Planar Array Antennas with Space–Time Light Modulator with a Complex Law for Addressing the Input Signal

The loss of information about the second spatial coordinate in a number of cases is a defect of the processors discussed above with an input format (3.2) or (3.19). When a space–time light modulator discussed in this chapter is used it is possible to perform also a two-dimensional parallel scanning of space with the help of a two-dimensional array antenna at the cost of complicating the law of input signal addressing. For this Lambert [63] and Casasent [85] use, for example, the method of time multiplexing (see Fig. 1.5) making it possible to enhance the two-dimensional parallel scan of space at the expense of a significantly complicated electronic circuit for the electrooptical array antenna (introduction of lines of delay and adders). There is another solution for this problem: one may complicate the input signal [104, 129], which, however, in the case of a space–time light modulator with electron addressing also complicates the electronic part of the electrooptical array antenna and in the case of a multichannel AOM limits the number of elements of a two-dimensional array by the maximally possible number of AOM channels (approximately 100). Therefore, we will show below the possibilities for using a space–time light modulator with multichannel optical addressing to overcome these problems [130, 131].

3.3.1 Complex input format

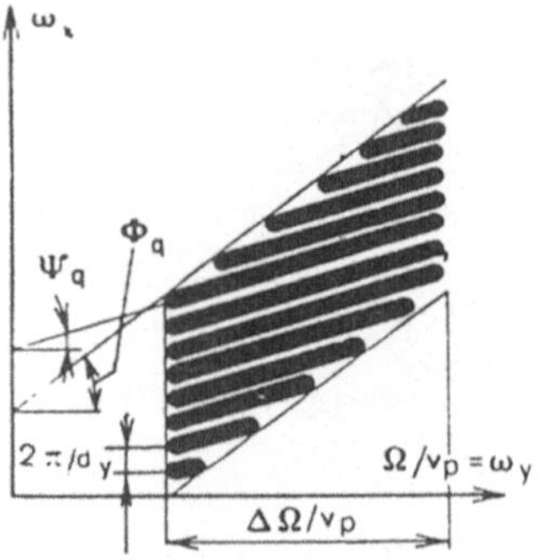

Figure 3.4: Reading information.

This case corresponds to the block diagram of the CO processor shown in Fig. 3.3. Signals from the elements of a two-dimensional array 1 consisting of M rows and N columns control the channels of the linear space–time light modulator 2, for example, a multichannel electrooptical modulator (see Chapter 1), and then with the help of deflector 3 is unwrapped in space on the carrier 4. According to Prokhorov [13] for an array element with the coordinates (X, Y) there is a space–time light modulator channel 2 with the coordinates

$$\tilde{x} = X/m_x + Y/\widetilde{m}_y \tag{3.32}$$

where $m_x = \Delta X/\Delta x$ is the address range in a column of antenna elements (ΔX is the dimension of the array along the X axis; Δx is the length of a group of M space–time light modulator channels corresponding to an array column; d_Y is the distance between array channels; $\widetilde{m} = d_Y/d_y$ is the scale of addressing between the element columns; d_y is the distance between groups of space–time light modulator channels). By contrast, a space–time light modulator channel with coordinate x has an array element with the coordinates

$$X = m_x\tilde{x} - \widetilde{m}_y[\tilde{x}/d_y], \qquad Y = \widetilde{m}_y[\tilde{x}/d_y]dy \tag{3.33}$$

where [...] is a symbol for the integral part of a number.

As is apparent from (3.33), addressing (3.32) designates that the recording of array signals on photoconductor 4 is combined into N subgroups (according to the number of columns), each of which contain M channels (according to the number of rows). Within the subgroups addressing is analogous to the above-mentioned case of a linear array, and adjacent subgroups are recorded with a step d_y and corresponds to adjacent array columns.

Therefore, in this case two-dimensional information is compacted along one spatial dimension which makes it possible to free the second one for spectral or correlation analysis.

Using the recording for a two-dimensional array signal (2.5) in which by substituting X, Y into (3.33) and taking into account (3.19), we find that the processor reads from plane Π (see Fig. 3.3) the optical signal

$$E^{(1)}(x,y) = T^{(1)} E_{read} J_X(m_x x - \widetilde{m}_y[x/d_y]d_y) J_Y(\widetilde{m}_y[x/d_y]d_y)$$
$$\times J_t(y)\mathcal{E}\left(m_x x - \widetilde{m}_y[x/d_y]d_y, \widetilde{m}_y[x/d_y]d_y; t_n + \frac{y - y_\circ}{v_p}\right) \qquad (3.34)$$

where the sign $\sim$ above the x and $\mathcal{E}$ is omitted because in the first case the reading is done from the plane Π; and, in the second case, inasmuch as the the dimension finiteness effect of the recording spot is not treated here, $J(X,Y) = J_X(X)J_Y(Y)$.

As a result of a two-dimensional Fourier-transform (1.2) with the help of lens 5 at the processor output plane π we find the following spatial distribution of the light field (see (3.21)):

$$\dot{e}^{(1)}(x,y)(\omega_x, \omega_y) = \frac{1}{\lambda_{read} f} \hat{\mathcal{F}}\{E^{(1)}(x,y)\} =$$

$$\frac{T^{(1)} E_{read}}{(2\pi)^2 \lambda_{read} f} \hat{\mathcal{F}}\{J_X(m_x x - \widetilde{m}_y[x/d_y]d_y) J_Y(\widetilde{m}_y[x/d_y]d_y) J_t(y)\}$$

$$\otimes \otimes \hat{\mathcal{F}}\left\{\mathcal{E}\left(m_x x - \widetilde{m}_y[x/d_y]d_y, \widetilde{m}_y[x/d_y]d_y; t_n + \frac{y - y_\circ}{v_p}\right)\right\}$$

$$= \frac{T^{(1)} E_{read}}{(2\pi)^2 \lambda_{read} f} \sum_{[x/d_y]=-\infty}^{\infty} \exp(-i\omega_x \widetilde{m}_y[x/d_y]d_y)$$

$$\times \hat{\mathcal{F}}_x\{J_X(m_x x)\} J_Y(\widetilde{m}_y[x/d_y]d_y) \hat{\mathcal{F}}_y\{J_t(y)\}$$

$$\otimes \sum_{[x/d_y]=-\infty}^{\infty} \exp(-i\omega_x \widetilde{m}_y[x/d_y]d_y)$$

$$\mathcal{F}\left\{\mathcal{E}\left(m_x x, \widetilde{m}_y[x/d_y]d_y; t_n + \frac{y - y_\circ}{v_p}\right)\right\}$$

$$= \tilde{e}_t \cdot F_X\left(\frac{\omega_x}{m_x}\right) F_Y\left(\frac{\omega_x^Y}{\widetilde{m}_y}\right)$$

$$\otimes_{\omega_x} \otimes_{\omega_x^Y} \left\{\dot{s}\left(\frac{\omega_x}{m_x}, \frac{\omega_x^Y}{\widetilde{m}_y}, v_p\omega_y\right) \exp[i(v_p t_n - y_\circ)\omega_y]\right\}\Bigg|_{\omega_x^Y = \omega_x} \otimes F_t(\omega_y) \tag{3.35}$$

Here $\tilde{e}_t = T^{(1)} E_{read} v_p/(2\pi)^2 \lambda_{read} f m_x \widetilde{m}_y$; $F_X(\Omega_X) = \mathcal{F}_X\{J_X(X)\}$ is the array radiation pattern at the plane XOZ;

$$F_Y(\Omega_Y) = F_Y\left(\frac{\omega_x^Y}{\widetilde{m}_y}\right) = \sum_{[x/d_y]=-\infty}^{\infty} \exp(-i\omega_x^Y \widetilde{m}_y[x/d_y]d_y) J_Y(\widetilde{m}_y[x/d_y]d_y) \tag{3.36}$$

is the array radiation pattern at plane YOZ (a discrete Fourier transform, see [52]), where $\omega_x^Y = \Omega_y \widetilde{m}_y$ is the spatial frequency (argument) of a discrete Fourier transform along x, whereby ω_x^Y is measured in the same scale that is ω_x, i.e. $\omega_x^Y = kx/f$; at the conclusion of the expression (3.35) we use the convolution, shift, and similarity theorems relative to a Fourier transform [3], and also we have considered the circumstance that a discrete Fourier transform of signal $\mathcal{E}(X, Y = \widetilde{m}_y[x/d_y]d_y, t)$ along $[x/d_y]$ coincides with the space–time spectrum $\dot{s}(\Omega_X, \Omega_Y = \omega_x^Y/\widetilde{m}_y, \Omega)$ due to the finiteness of the latter and the assumption that the array radiation pattern lacks a diffraction maximum of a higher order.

The spatial structure of the light field distribution (3.35) at the processor output plane π as contrasted to, for example, (3.21) makes it possible to determine both the frequency and the two-dimensional angular (spatial spectrum) of the incoming signals; and, the needed generalized angular coordinate of the entities $\Omega_Y = K \sin\theta \sin\varphi = K \cos\alpha$ is determined by a "fine" diffraction pattern of the field described by factor (3.36), and the angular coordinate $\Omega_X = K \sin\theta \cos\varphi = K \cos\beta$ is described by a "crude" pattern described by factor F_x.

A similar diffraction pattern has been formulated thanks to the compression of the three-dimensional input space–time signal into a complex format at the addressing stage at which time one spatial dimension of the array (X coordinate) is represented as a sublattice with a small period dx

and the other (Y coordinate) as an array composed of subarrays with a period $d_y \gg dx$. Because periodic signals have been Fourier transformed to a specified subarray and constituent array there is a "crude" and a "fine" diffraction pattern.

Let us illustrate what we have said for cases involving narrow-band and wide-band signals from point sources q. Assume

$$\dot{s}(\Omega_X,\Omega_Y,\Omega) = \frac{1}{K^2}\dot{s}_q \cdot \delta\left(\frac{\Omega_X - \Omega_{X_q}}{K}\right) \cdot \delta\left(\frac{\Omega_Y - \Omega_{Y_q}}{K}\right) \cdot \delta(\Omega - \Omega_q)$$

i.e., a monochromatic signal at frequency Ω_q with a complex amplitude $\dot{s}_q$ from a point source with the generalized coordinates Ω_{X_q},Ω_{Y_q}. Then in correspondence with (3.35) we obtain

$$e_q^{(1)}(\omega_x,\omega_y) = \tilde{e}_o\dot{s}_q F_X\left(\frac{\omega_x}{m_x} - \Omega_{X_q}\right) F_Y\left(\frac{\omega_x}{\widetilde{m}_x} - \Omega_{Y_q}\right) F_t(\omega_y - \Omega_q/v_q) \quad (3.37)$$

i.e., the spatial coordinates and the values for the frequencies of object q is determined by the following:

$$\begin{aligned} \Omega_{X_q} &= \omega_{xq}/m_x \\ \Omega_{Y_q} &= \omega_{xq}/\widetilde{m}_y - \left[\frac{\omega_{x_q} d_y}{2\pi}\right] \frac{(2\pi/d_y)}{\widetilde{m}_y} \\ \Omega_q &= \omega_{y_q} v_p \end{aligned} \quad (3.38)$$

where [...] is the symbol for the integral part of the number; $2\pi/d_y$ is the period of function (3.36).

Suppose now that the incoming signal is a wideband one, for example

$$\dot{s}(\Omega_X,\Omega_Y,\Omega)$$
$$= \frac{1}{K^2}\dot{s}_q \cdot \delta\left(\frac{\Omega_X - \Omega_{X_q}}{K}\right) \cdot \delta\left(\frac{\Omega_Y - \Omega_{Y_q}}{K}\right) \mathrm{rect}[(\Omega - \Omega_q)/\Delta\Omega],$$

where Ω_q is the central frequency; $\Delta\Omega$ is the signal frequency band q. Substituting the latter expression into (3.35) and taking into account the filtering property of δ-functions, accurate to the first member of the expansion of the convolution of the "wide" function "rect" with the narrow F_t into a series by moments, we obtain

$$e_q^{(1)}(\omega_x, \omega_y) \approx \frac{\tilde{e}_\circ}{2\pi} \dot{s}_q F_X \left(\frac{\omega_x}{m_x} - \frac{\omega_y v_p}{c} \cos \alpha_q \right)$$

$$\times \; F_Y \left(\omega_x \widetilde{m}_y - \frac{\omega_y v_p}{c} \cos \beta_q \right) \mathrm{rect}[(\omega_y v_p - \Omega_q)/\Delta\Omega] \quad (3.39)$$

where

$$\frac{1}{2\pi} = \int_{-\infty}^{\infty} F_t(\omega_y) d\omega_y$$

and c is the signal propagation velocity from the object (light or sound wave). From (3.39) we see that the image of a wideband point-source space–time signal is "blurred" along the straight lines

$$\frac{\omega_x}{\widetilde{m}_y} - \frac{\omega_y v_p}{c} \cos \beta_q = n \frac{2\pi}{d_y} / \widetilde{m}_y, \qquad n = 0, 1, 2, \ldots \quad (3.40)$$

along which the angular coordinate β_q (direction $cos\beta_q$) preserves its value; and, its value determines the slope Ψ_q:

$$\tan \Psi_q = \frac{v_p}{c} \cos \beta_q \widetilde{m}_y = \frac{d_Y}{d_y} \frac{v_p}{c} \cos \beta_q \quad (3.41)$$

The system of parallel lines (3.40) is "cut off" by the factor F_x of a "crude" structure, whose maximum is on a line

$$\frac{\omega_x}{m_x} - \frac{\omega_y v_p}{c} \cos \alpha_q = 0 \quad (3.42)$$

passing at an angle Φ_q:

$$\tan \Phi_q = \frac{v_p \cos \alpha_q}{c} m_x = \frac{\Delta X}{\Delta \tilde{x}} \frac{v_p}{c} \cos \alpha_q \quad (3.43)$$

By measuring angles Φ_q, Ψ_q it is possible to determine two spatial coordinates of an object $\cos \alpha_q, \cos \beta_q$ whereby the signal frequency spectrum is reflected

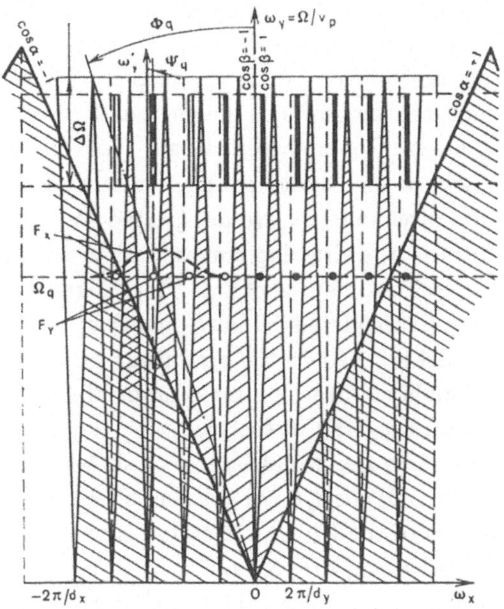

Figure 3.5: Reading information at the processor output plane.

along the axis ω_y of the processor (Fig. 3.4). The requirement for precision in measuring the slope Ψ_q of a "fine" structure of a light field is evidently $m_x/\widetilde{m}_y$ times higher than for $\Phi_q(\widetilde{m}_y \ll m_x)$.

Depending on the width of the signal spectrum $\Delta\Omega$ Nakhmanson [129] recommends that the angular coordinates be determined by (3.38), or (3.41) and (3.43), whereby the feasibility of using the latter correlations is determined by the requirement that the width of the image of the frequency spectrum of a "fine" structure exceed the spatial period of the latter ($\Delta\Omega/v_p \gg 2\pi/d_y$). However, it is inconvenient to use this technique because it requires continuous control in fulfilling the given criteria. Therefore, it is better to use a single system for reading in the data. Fig. 3.5 shows an example of the calibration of the processor output plane of diffraction pattern (3.37), received from a monochromatic point-source light q with a frequency Ω_q and with director cones: $\cos\alpha_q = -0.75$ and $\cos\beta_q = +0.5$. The "fine" structure of a diffraction pattern which is described by a periodic factor F_Y with a period $2\pi/d_y$ and shifted along axis ω_y by a value $\widetilde{m}_y = \frac{\Omega_q}{c}\cos\beta_q$, is "weighted" by the "crude" structure (at the processor output we see only unblocked diffraction maximums of factor F_Y). The latter is described by factor F_X, shifted along axis ω_x by a value $\widetilde{m}_x = \frac{\Omega_q}{c}\cos\alpha_q$. Locating the coordinates $\cos\alpha_q$, $\cos\beta_q$ and the frequency Δ_q in correspondence with (3.38), (3.41), and (3.43) is achieved along angles Φ_q, Ψ_q, and $\omega_q v_p$ shown in Fig. 3.5. In this case angle Φ_q is read from angle ω_y up to a position on the maximum of the "crude" structure; the limits of the change Φ_q are designated by lines $\cos\alpha = \pm 1$. Angle Ψ_q is measured from axis ω'_y (parallel to ω_y) of the local coordinate system where the "spot" of the structure falls with the greatest intensity. The limits of change Ψ_q are designated by $\cos\beta = \pm 1$. The area with the broken lines in the drawing correspond to the space–time regions in which the diffraction maximums of the "fine" structure do not fall (this circumstance must be considered in implementing an information input). When $\Omega > \Omega_{max}$ (see Fig. 3.5) in the scanning region ($|\cos\alpha, \cos\beta| \leq 1$) there are images of higher diffraction maximums of the array that result in the nonuniqueness and the drop in the gain.

Fig. 3.5 shows also the broad-band signal (band $\Delta\Omega$) with the same angular coordinates ($\cos\alpha_q \cos\beta_q$), the reading of which is done analogously. The diffraction pattern of the wide-band signal is in essence a variant of Fig. 3.4 on a scale extended along the axis ω_x.

In conclusion we must note that the shortcomings of this processor are in the device for picking up information from its output plane λ in order to identify the "fine" and the "crude" structure of a light field and in the greater number of channels needed in the linear space–time light modulator which is constrained both by the manufacturing technology and by the optics requirements.

3.3.2 Two-dimensional input format

This case corresponds to the block diagram of the CO processor shown in Fig. 3.6. Signals $\mathcal{E}(X,Y,t)$ received by elements 1 of a two-dimensional array with the coordinates X, Y control the corresponding channels of a two-dimensional space–time light modulator 2 with the coordinates $\tilde{x}$, $\tilde{y}$ at the processor plane $\tilde{\Pi}$. Then with the help of a deflector 3 the format of the signal received at plane $\tilde{\Pi}$ is deflected onto a photosensitive carrier 4 in plane Π. The aperture interval Δy of the temporal dependency of the signal in space determines the processor's frequency resolution and is not greater than the distance d_y between space–time light modulator channels.[2]

Taking into account the addressing law (1.3) of array element 1 to space–time light modulator 2 and also the circumstance that $\Delta y \leq d_y$ by

[2]In realizing a two-dimensional input format with the help of a multichannel AOM [63], the role of a two-dimensional space–time light modulator and deflector 3 is carried out by a linear multichannel AOM supplemented by the corresponding delay lines and by an adder (see Chapter 1).

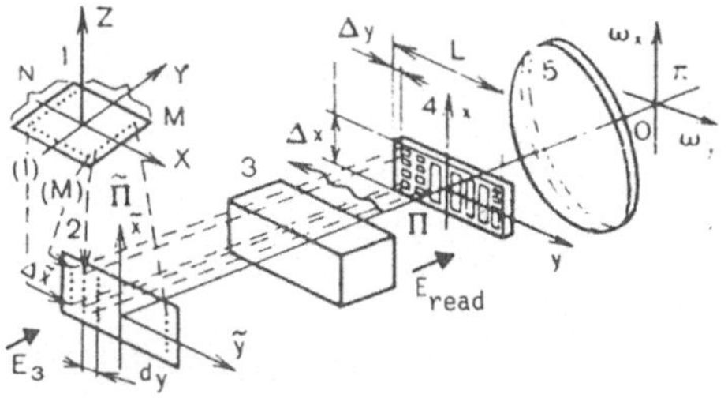

Figure 3.6: Block diagram of CO processor with a two-dimensional signal recording format.

analogy to (3.33) for the input plane Π of a Fourier processor (with 5 lens) we find the following relation between the coordinates X, Y of the space–time signal received by the array and the coordinates x, y of its optical model reproduced at the plane Π:

$$X = m_x x, \qquad Y = m_y \lfloor y/d_y \rfloor d_y \tag{3.44}$$

where $\lfloor ... \rfloor$ is the symbol for the integral part of the number.

When analyzing a two-dimensional signal input format recorded at plane Π by a reading beam of coherent light E_{read} of wave length λ_{read} the Fourier processor is fed an optical signal whose information part, by analogy with (3.19) and (3.34), is represented as

$$\begin{aligned} E^{(1)}(x,y) &= T^{(1)} E_{read} J_X(m_x x) J_Y(m_y \lfloor y/d_y \rfloor d_y) \cdot J_t(y - \lfloor y/d_y \rfloor d_y) \\ &\times \; \mathcal{E}\left(m_x, m_y \lfloor y/d_y \rfloor d_y;\, t_n + \frac{y - \lfloor y/d_y \rfloor d_y}{v_p}\right) \end{aligned} \tag{3.45}$$

where it is assumed that in this case $y_\circ = \lfloor y/d_y \rfloor d_y$ (see Fig. 1.7).

As a result of a two-dimensional Fourier transform (1.2) with the help of lens 5 the following spatial distribution of the light field is formed at the processor output plane π [cf. (3.35)]:

$$\dot{e}^{(1)}(\omega_x, \omega_y) = \frac{T^{(1)} E_{read}}{(2\pi)^2 \lambda_{read} f} \hat{\mathcal{F}}\{J_X(m_x x) J_Y(m_y \lfloor y/d_y \rfloor d_y) J_t(y - \lfloor y/d_y \rfloor d_y)\}$$

$$\otimes \otimes\, \hat{\mathcal{F}}\left\{\mathcal{E}\left(m_x x, m_y \lfloor y/d_y \rfloor d_y; t_n + \frac{y - \lfloor y/d_y \rfloor d_y}{v_p}\right)\right\}$$

$$= \frac{T^{(1)} E_{read}}{(2\pi)^2 \lambda_{read} f} \sum_{[y/d_y]d_y=-\infty}^{\infty} \exp(-i\omega_y m_y [y/d_y] d_y)$$

$$\times \hat{\mathcal{F}}_x\{J_X(m_x x)\} J_Y(m_y [y/d_y] d_y) \times$$

$$\times \hat{\mathcal{F}}_y\{J_t(y)\} \otimes \otimes \sum_{[y/d_y]=-\infty}^{\infty} \exp(-i\omega_y m_y [y/d_y] d_y)$$

$$\times \hat{\mathcal{F}}\{\mathcal{E}(m_x x, m_y [y/d_y] d_y; t + y/v_p)\}$$

$$\tilde{\tilde{e}}_t E_X(\omega_x/m_x) F_Y(\omega_y/m_y)$$

$$\otimes_{\omega_x} \otimes_{\omega_y} \{\dot{s}(\omega_x/m_x, \omega_y/m_y, v_p \omega'_y) \exp(i v_p t_n \omega_y)\} \otimes_{\omega'_y} F_t(\omega'_y)\Big|_{\omega'_y=\omega_y} \quad (3.46)$$

Here $\tilde{\tilde{e}}_t = T^{(1)} E_{read} v_p / (2\pi)^2 \lambda_{read} f m_x m_y$; $F_x(\Omega_x) = \hat{\mathcal{F}}\{Jx(X)\}$ is the array radiation pattern at plane XOZ

$$F_Y(\Omega_Y) = F_Y(\omega_y/m_y)$$

$$= \sum_{[y/d_y]=-\infty}^{\infty} \exp(-i\omega_y m_y [y/d_y] d_y) J_Y(m_y [y/d_y] d_y) \quad (3.47)$$

is the array radiation pattern at plane YOZ [cf. (3.36)].

Thus the spatial structure of a light field (3.46) at the processor output plane π (see Fig. 3.6) as well as the distribution (3.35) makes it possible to determine both the frequency and the two-dimensional angular (spatial) spectrum of the incoming signals. In the case of point-source narrow-band (monochromatic) point sources its generalized angular coordinates and frequency may be calculated from the following correlations by analogy with (3.38)

$$\begin{aligned} \Omega_{x_q} &= \omega_{x_q}/m_x, \\ \Omega_{Y_q} &= \omega_{x_q}/m_y - \left[\frac{\omega_{y_q} d_y}{2\pi}\right] 2\pi d_y/m_y, \\ \Omega_q &= \omega'_{y_q} v_p \end{aligned} \quad (3.48)$$

where Ω_{x_q}, Ω_{y_q} are the coordinates of the maximum of the optical image of object q; Ω'_{y_q} is the coordinate of the maximum of the "crude" diffraction structure, defined as the envelope of the "fine" structure (3.47) [see Fig. 3.5].

If the incoming signal is a wideband one, then to determine the coordinates and spectrum of object q we can use (3.48) as before if the criteria (2.24) are met. In the opposite case there will be an error (2.23) in the determination of the angular coordinates and a drop in the directive gain (2.27).

The problem with this kind of a processor is in its limited frequency resolution because the aperture dimension Δy, which determines its resolution ($\delta F = v_p/\Delta y$), is N times (N is the number of elements in the array along axis Y) less than the overall dimension of the space–time light modulator (due to the fact that $\Delta_q y \leq L/N$).

3.4 Linear Arrays with Correlation Signal Processing

As we noted in Chapter 1, the potential for optical processors performing one-dimensional parallel spatial processing and removal of the time (non-coordinate) information is most fully realized when one uses the properties of these systems to do correlation analysis of time data.

To increase the Doppler shift or range resolution and to improve the signal-to-noise ratio in an active radar system it is necessary to use complex (linear-frequency modulated, phase-manipulated, etc.) coded signals that possess clearly expressed correlation properties [56, 57, 132]. The product of the signal duration and the bandwidth ($\Delta T \Delta F$) in modern radar systems reaches $10^4 \ldots 10^6$. For more efficient use of the signal's energy, and also for greater precision in determining the range and speed of objects it is necessary to use up to 20 kinds of probing signals with a different structure. Electrooptical arrays that combine one-dimensional panoramic scanning and spectral signal analysis that make it possible to separate the Doppler shift are treated in Sections 3.1–3.3 and 4.2. A radar system with an increased accuracy for determining the Doppler frequency shift or range (in optical value intervals), as a rule, uses the principle of correlation processing. Thus, if it is necessary to increase the resolving power in determining the Doppler frequency shift, such processing is done in the spectral region; in determining the range, however, correlation analysis is done in the time field. Let us examine processing that uses an electrooptical array as an example to represent a multichannel optical array signal correlator that scans space by one coordinate and range [22, 88]. Suppose that a radar unit emits a sounding signal $\mathcal{E}_\Sigma(t) = \Re\{\dot{\mathcal{E}}_\Sigma(t) \times \exp(i\Omega_\circ t)\}$; ($\dot{\mathcal{E}}_\Sigma(t) = | \dot{\mathcal{E}}_\Sigma(t) | \exp[i\Phi(t)]$ is the complex amplitude). Then, at the output of the receiving element of a linear array with coordinate Z (see Fig. 1.8) we find the signal

$$e_q(\tilde{Z},t) = \Re e\left\{\dot{\mathcal{E}}_q\left(t - \frac{2R_q - Z\cos\theta_q}{c}\right)\exp[i(\Omega_\circ - \Omega_d)t]\right\} \tag{3.49}$$

where $\dot{\mathcal{E}}$ is the complex amplitude of the incoming signal proportional to $\dot{\mathcal{E}}_\Sigma$ accurate to a coefficient which takes into account the reflecting properties of an object q and the attenuation of the signal on the path; R_q, θ_q is the range and the angle of elevation of object q; $\Omega_q = -\frac{2}{\Lambda} dR_q/dt$ is the Doppler frequency shift ($\Lambda = 2\pi c/\Omega$); signal (3.49) is taken as the true one because it controls the non-linear space–time light modulator input.

A multichannel space–time light modulator performs in plane Π (see Fig. 1.8) a spatial sweep of the time-based dependency of the signals (3.49) along the measurement y of all array elements addressed at the space–time light modulator according to the law $Z = m_x x$. When the signal of a one-dimensional Fourier transform along the Oy axis is read into the processor, and the image is transferred on the Ox axis in a 1:1 scale ($x = x_1$) with the help of an astigmatic lens system λ_1, λ_2 in a plane π_1 analogous to (3.4) and (3.21), the following distribution of the light field is formed:

$$\begin{aligned}\dot{e}(1)(x_1,\omega_y) &= \frac{1}{\sqrt{\lambda_{read} f_2}}\hat{\mathcal{F}}\left\{T^{(1)}E_{read}J(m_x x_1)J_t(y)e_q\left[m_x x_1, \frac{y-y_\circ}{v_p}\right]\right\}\\ &= \frac{T^{(1)}E_{read}v_p}{2\pi\sqrt{\lambda_{read} f_2}}J(m_x x_1)\left[E_t(\omega_y)\otimes\hat{\mathcal{F}}\left\{e_q\left[m_x x_1, \frac{y-y_\circ}{v_p}\right]\right\}\right]\end{aligned} \tag{3.50}$$

Here λ_{read}, E_{read}, J, J_t, v_p, and E_t are expanded in (3.4) and (3.21), while t_n appearing in (3.21) is taken as equal to zero because this leads only to the displacement of the beginning of the distance reading at the output plane π_2, f_2 is the focal length of lens Λ_2:

$$\begin{aligned}&\hat{\mathcal{F}}_y\left\{e_q\left(m_x x_1, \frac{y-y_\circ}{v_p}\right)\right\}\\ &= \dot{s}_q(v_p\omega_y - \Omega_d)\exp\left[-i\left(\frac{2R_q - m_x x_1\cos\theta_q}{c}v_p - \frac{y_\circ}{v_p}\right)\omega_y\right]\end{aligned} \tag{3.51}$$

where $\dot{s}_q(\Omega) = \hat{\mathcal{F}}_t\{\Re e[\dot{\mathcal{E}}_q(t)\exp(i\Omega_\circ t]\}$.

As evident from (3.51) information about the Doppler frequency shift may be extracted at the intermediate plane π_1. To extract at the processor output plane π_2 the "angle-distance" information from the plane π_1 it is necessary to use a matched mask $\dot{T}_\Sigma$ whose transmission function is complexly related to a Fourier transform of the reference signal $e_\Sigma(t)$:

$$\begin{aligned}\dot{T}_\sigma(x_1,\omega_y) &= T_\sigma \cdot \hat{\mathcal{F}}_y\{\Re e[\overset{*}{\mathcal{E}}_\sigma(-y/v_p)\exp(i\Omega_\circ y/v_p)]\}\exp(i\omega_y y_\sigma) \\ &= T_\sigma \hat{\mathcal{F}}_y\left\{e_\sigma\left(\frac{-y+y_\sigma}{v_p}\right)\right\} \qquad (3.52)\end{aligned}$$

where T_σ is a coefficient of proportionality; $y_\sigma = f_2 \sin\alpha_\sigma$ (α_σ is the angle of incidence of the reference wave to the mask when the holographic method is used to record the latter). After a two-dimensional Fourier transform of the light field (3.50) using lens Λ_3 the weighted/suspended mask (3.52) at the processor output plane π_2 may be used to observe the optical signal

$$\begin{aligned}\dot{e}_\pi^{(1)}(\omega_x,y_2) &= \hat{\mathcal{F}}_x\hat{\mathcal{F}}_y^{-1}\{\dot{e}^{(1)}(x_1,\omega_y)\dot{T}_\sigma(x_1,\omega_y)\} \\ &= e_\pi F\left(\frac{\omega_x}{m_x}-\frac{\Omega_\circ}{c}\cos\theta_q\right) \\ &\times \left[e_\sigma\left(\frac{-y_2+y_\sigma}{v_p}\right)\otimes e_q\left(0,\frac{y_2-y_\circ}{v_p}-\frac{2R_q}{c}\right)\mathcal{J}_t(y)\right] \\ &= e_\pi F\left(\frac{\omega_x}{m_x}-\frac{\Omega_\circ}{c}\cos\theta_q\right)[e_\sigma(t)*e_q(0,t)]\Big|_{t=(y_2-y_\circ-y_\sigma)/v_p-2R_q/c}\end{aligned} \qquad (3.53)$$

where $e_\pi = (T_\Sigma T^{(1)}E_{read}v_p)/(2\pi\sqrt{\lambda_{read}f_2})$; "$*$" is the symbol for the correlation operation.

Fig. 3.7 shows an example of an optical image (3.53) at the processor plane π_2 (see Fig. 1.8). Its maximum is at a point with coordinates $\omega_x = m_x\Omega_\circ/c\cos\theta$ [cf. (3.6)] and $y_2 = y_\circ + y_\Sigma + 2Rq/cv_p$. Thus, along with an increased signal-to-noise ratio, which is achieved by pulse compression $e_\Sigma(t) * e_q(0,t)$, this processor makes it possible to simultaneously extract information about the angle of elevation θ_q and the distance R_q of the source.

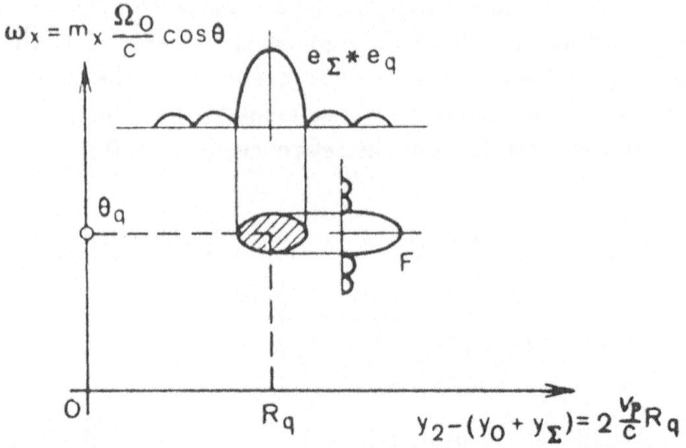

Figure 3.7: Output plane of correlation processor.

Chapter 4

Formation of the Antenna Pattern of Nonplanar Array Antennas

Nonplanar (convex, conformal) arrays which consist of radiating elements on a nonplanar, geometrical or conductive surface (for example, a cylindrical or spherical one) allow wide angle scanning without (or, for all practical purposes, without) a decrease in directivity [133]. The formation of such highly directivite radiation patterns (or reception) with the aid of such arrays is an extremely complex problem. It is even more difficult to perform a panoramic sweep analogous to those treated in Chapters 2 and 3.

This chapter demonstrates the most likely possibilities of forming an array antenna pattern, which are characteristic of coherent optical methods. Below we systematically explain the algorithms for processing space–time signals received by an arbitrary nonplanar array and we describe coherent optical processors for them. We will examine in detail piecewise-planar, cylindrical, and circular arrays (for possible spectral analysis of a space–time signal) made possible by coherent optical and holographic methods. We will discuss analog (holographic) and digital (with the help of precision photoplotter) methods for implementing conjugate masks which play a key role in these processors. We will show the general cases of these processors for arbitrary axial symmetric arrays when three-dimensional holographic masks are used. In this chapter, as in the previous ones, we use an approximation of continuous aperture, i.e., an array and a space–time light modulator are represented as a continuous recording and modulating medium of a finite dimension. The discreteness effects of the aperture are studied in detail in Chapter 5.

4.1 The Restoration Algorithm

Coherent optical beam formation by nonplanar arrays has been treated earlier [80–82]. In essence, these works treated only a particular case of a circular array for forming a one-dimensional frequency-angular spectrum of the signal which is not possible to embody in a more general case of an arbitrary nonplanar array. Below is an approach to the construction of such coherent optical processors.

Let the surface of an arbitrary nonplanar array be represented as (Fig. 4.1a)

$$\mathbf{R} = \mathbf{n}_X X + \mathbf{n}_Y Y + \mathbf{n}_Z Z \tag{4.1}$$

where $Z(X,Y) = Z(\mathbf{R}_\perp)$ is a continuous or piecewise continuous function (the radius of the curve exceeds λ) which is further assumed to be either unique (the surface is open) or ambiguous (the surface is closed).

Let it further be assumed that the radiation pattern of the array elements $\mathcal{F}_e(\mathbf{K},\mathbf{R})$ at any point on the surface (4.1), including coupling, is obtained by, for example, direct measurement or solution of an external electrodynamic problem [133]. The radio emission from the far field of the array will be characterized by a frequency-angular spectrum $\dot{S}(\mathbf{K})(\mathbf{K} = -\mathbf{n}_X cos\alpha - \mathbf{n}_Y \mathbf{K} cos\beta - \mathbf{n}_Z cos\theta = -\mathbf{K}_\perp - \mathbf{n}_Z \Omega_Z$ is the wave vector, see (1.1), while $\mathbf{S} \perp \mathbf{K}$) by which we will mean a three-dimensional frequency distribution $\Omega = c|\mathbf{K}|$ and two generalized angular variables ς, ξ (for example θ and ϕ or cos α or cos β, or cos θ and ϕ, etc) of vector complex amplitudes of elementary planar monochromatic waves incident on the array in the direction of vector $\mathbf{K}$.

4.1.1 The coherent optical processing algorithm

Radio-frequency emissions from the far field with a frequency-angular spectrum $\dot{S}(\mathbf{K})$ excite the array elements with an element pattern $\mathcal{F}_e(\mathbf{K},\mathbf{R})$ [See (2.1)], to produce a space–time signal with the following frequency spectrum at points $\mathbf{R}$:

$$\dot{\mathcal{E}}_\Omega(\mathbf{R}) = \hat{\mathcal{F}}_t\{\mathcal{E}(\mathbf{R},t)\} = \int\int_{4\pi} \mathcal{F}_e(\mathbf{K},\mathbf{R})\dot{\mathbf{S}}(\mathbf{K}) \exp[-i\,\mathbf{K}\mathbf{R}]\mathrm{d}^2\mathbf{K} \tag{4.2}$$

where 4π (in steradians) is the full solid angle, $d^2\mathbf{K} = d\varsigma d\xi = \sin\theta d\theta d\varphi$ is the elementary solid angle, $\mathcal{F}_e\mathbf{S}$ is the scalar product.

The purpose of processing the space–time signal is to restore the frequency-angular spectrum $\mathbf{S}(\mathbf{K})$ which, being expressed in variables ς, ξ,

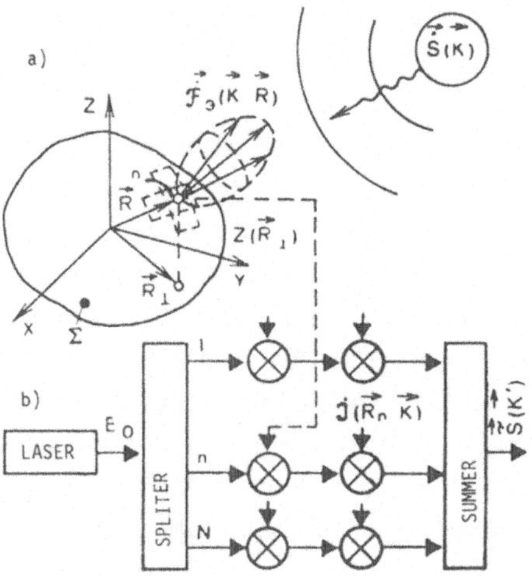

Figure 4.1: (a) Nonplanar array surface; (b) correlated coherent optical processor.

Ω, characterizes the phase and frequency distribution of the sources (target area). From a mathematical point of view, this is equivalent to the solution of a Fredholm integral equation of the first order (4.2) characteristic of antenna synthesis problems [134, 135]. In principle, a precise solution (restoring the frequency-angular spectrum with infinite resolution) is possible. However, it is unstable due to the incorrect formulation of the problem (4.2), which makes the search for such solutions by coherent optical methods impractical. Therefore, as in Chapter 2, we will be limited by the approximate restoration of the spectrum $\dot{S}(\mathbf{K})$ with a finite resolution corresponding to the resolving power of the array, i.e.

$$\tilde{\dot{S}}(\mathbf{K}') = \int\int_{4\pi} \dot{\mathbf{S}}(\mathbf{K})\mathbf{F}(\mathbf{K},\mathbf{K}')\mathrm{d}^2\mathbf{K} \tag{4.3}$$

Here

$$\mathbf{F}(\mathbf{K},\mathbf{K}') = \int\int_{\Sigma} \mathcal{F}_e(\mathbf{K},\mathbf{R})\dot{J}(\mathbf{R},\mathbf{K}')\exp[-i\,\mathbf{K}\mathbf{R}]\mathrm{d}^2\mathbf{R} \tag{4.4}$$

is the array pattern (in the Huygens-Kirchhoff approximation) with very high gain in the direction $\mathbf{K}'$; $J(\mathbf{R},\mathbf{K}') = |\dot{J}(\mathbf{R},\mathbf{K}') = |\exp[i\mathbf{K}'\mathbf{R}|$ is the optimum (for example, directive gain, sidelobe levels, etc.) amplitude-phase distribution of the excitation of the array elements at points $\mathbf{R}$ corresponding to the gain in the direction $\mathbf{K}'$; $\mathrm{d}^2\mathbf{R}$ is an element on the array surface Σ

(see Fig. 4.1a). The approximate solution of (4.3) is an extension of Equation (2.7) for the output response of a planar electrooptical array applied to a nonplanar surface (4.1). As before, it describes a continuous beam of a highly-directional radiation pattern at reception which is modulated by the frequency-angular spectrum $\dot{\mathbf{S}}$, formed during a parallel space sweep. In contrast to (2.7), however, the superposition integral (4.3) is not a convolution since the form of the radiation pattern of a nonplanar array (4.4) generally depends on the reception direction $\mathbf{K}'$ and is dependent on the shift.

Thus the task of coherent optical processing amounts to the following: a known input (4.2) yields an output in the form of an approximate solution (4.3). Let us determine the form of the transformation (the processing algorithm) necessary for such an operation. Substituting (4.4) into (4.3) we will change the order of integration and take into account (4.2). Thus,

$$\begin{aligned}\tilde{\dot{S}}(\mathbf{K}') &= \iint_{4\pi} \mathrm{d}^2\mathbf{K}\dot{\mathbf{S}}(\mathbf{K}) \cdot \iint_{\Sigma} \mathcal{F}_e(\mathbf{K},\mathbf{R})\dot{J}(\mathbf{R},\mathbf{K}) \exp[-i\,\mathbf{K}\mathbf{R}]\mathrm{d}^2\mathbf{R} \\ &= \iint_{\Sigma} \mathrm{d}^2\mathbf{R}J(\mathbf{R},\mathbf{K}') \iint_{4\pi} \mathcal{F}_e(\mathbf{K},\mathbf{R})\dot{\mathbf{S}}(\mathbf{K}) \exp[-i\,\mathbf{K}\mathbf{R}]\mathrm{d}^2\mathbf{K} \\ &= \iint_{\Sigma} \dot{\mathcal{E}}_{\Omega}(\mathbf{R})\dot{J}(\mathbf{R},\mathbf{K}')\mathrm{d}^2\mathbf{R} \qquad (4.5)\end{aligned}$$

The desired coherent optical processor for a nonplanar array signal must execute the processing algorithm in terms of the following integral operator:

$$\hat{L}\{\ldots\} = \iint_{\Sigma} \ldots \dot{J}(\mathbf{R},\mathbf{K}')\mathrm{d}^2\mathbf{R} \qquad (4.6)$$

where $\{\ldots\}$ is the object of the operator action, i.e., the space–time signal (4.2); $\mathbf{K}' \in 4\pi$.

4.1.2 Characteristics of the algorithm

The formation of the antenna pattern of a nonplanar array according to the algorithm (4.6) gives a resolution, directive gain, and polarization efficiency equal to or better than an equivalent nonplanar phased array. Actually, in the general case, the vector radiation pattern (4.4) is represented as [52]

$$\mathbf{F}(\mathbf{K},\mathbf{K}') = \mathbf{p}(\mathbf{K},\mathbf{K}')F(\mathbf{K},\mathbf{K}') \qquad (4.7)$$

where

$$\mathbf{p}(\mathbf{K},\mathbf{K}') = \mathbf{n}_{\parallel}(\mathbf{K})\,\dot{a}(\mathbf{K},\mathbf{K}') + \mathbf{n}_{\perp}(\mathbf{K})\,\dot{b}(\mathbf{K},\mathbf{K}')$$

is the individual polarization vector; $\mathbf{n}_{\parallel}$ and $\mathbf{n}_{\perp}$ are the basic individual co- and cross-polarization vectors, respectively, $(\mathbf{n}_{\parallel}\mathbf{n}_{\perp} = 0)$; $|\dot{a}(\mathbf{K},\mathbf{K}')|^2 = 1 - |\dot{b}(\mathbf{K},\mathbf{K}')|^2$ is the polarization efficiency of the array in the direction $\mathbf{K}$ with the maximum gain along $\mathbf{K}'$. The radiation pattern (4.7) may be represented as

$$\mathbf{F}(\mathbf{K},\mathbf{K}') = \mathbf{n}_{\parallel}(\mathbf{K})\,F_{\parallel}(\mathbf{K},\mathbf{K}') + \mathbf{n}_{\perp}F_{\perp}(\mathbf{K},\mathbf{K}') \tag{4.8}$$

where $F_{\parallel} = \dot{a}F$ and $F_{\perp} = \dot{b}F$ are the scalar radiation patterns of the array on the co- and cross-polarization ones, respectively. Taking into account (4.8) in (4.3) and representing the vector frequency-phase spectrum as $\dot{S} = \mathbf{n}_{\parallel}\dot{S} + \mathbf{n}_{\perp}\dot{S}_{\perp}$ ($\dot{S}_{\parallel}, \dot{S}_{\perp}$ are the components $\dot{S}$ along the axes $\mathbf{n}_{\parallel}, \mathbf{n}_{\perp}$), we obtain

$$\tilde{\dot{S}}(\mathbf{K}') = \tilde{\dot{S}}_{\parallel}(\mathbf{K}') + \tilde{\dot{S}}_{\perp}(\mathbf{K}') \tag{4.9}$$

Here

$$\tilde{\dot{S}}_{\parallel,\perp}(\mathbf{K}') = \iint_{4\pi} \dot{S}_{\parallel,\perp}(\mathbf{K}')\,F_{\parallel,\perp}(\mathbf{K},\mathbf{K}')\mathrm{d}^2\mathbf{K} \tag{4.10}$$

is the corresponding component of the frequency-angular spectrum.

Thus, the approximate solution (4.3) is a superposition (4.9) of the images (4.10) of the target area when received at the co- and cross-polarizations, i.e. the coherent optical processor with the processing algorithm (4.6) preserves the polarization properties of the nonplanar array.

Using the property of δ-correlation coefficient [125, 127] [see also (2.11)] we obtain

$$\overline{\dot{\mathbf{S}}(\mathbf{K})\,\overset{*}{\mathbf{S}}(\mathbf{K}')} = |\dot{\mathbf{S}}(\mathbf{K})|^2\delta\left(\frac{\mathbf{K}-\mathbf{K}'}{K}\right) \tag{4.11}$$

Equation (4.9) may be represented as

$$\overline{|\tilde{\dot{S}}(\mathbf{K}')|^2} = \overline{|\tilde{\dot{S}}_{\parallel}(\mathbf{K}')|^2} + \overline{|\tilde{\dot{S}}_{\perp}(\mathbf{K}')|^2} \tag{4.12}$$

where

$$\overline{|\tilde{\dot{S}}(\mathbf{K}')|^2} = \iint_{4\pi} |\dot{S}_{\parallel,\perp}(\mathbf{K})\ F_{\parallel,\perp}(\mathbf{K},\mathbf{K}')|^2 \mathrm{d}^2\mathbf{K} \tag{4.13}$$

Equation (4.12) treats the behavior of a "space-array-processor" system as the superposition of two independent (partial) noncoherent diffraction limited image-forming systems operating in orthogonal polarizations which represent sources with the coordinates defined by the vector $\mathbf{K}$ as a light image at the processor output with the maxima at points

$$\mathbf{K}' = \mathbf{K} \tag{4.14}$$

Inasmuch as the image-forming kernel of these diffraction limited image-forming systems coincide with the corresponding radiation pattern of the array in power, then the resolution of the directive gain (both partial and full) and the polarization efficiency of a convex electrooptical array with the processing algorithm (4.6) coincides with the same equivalent nonplanar phased array antenna [133].

If the space–time signal (4.2) is passed to a coherent optical processor with the help of an amplitude or phase (i.e., double-band) space–time light modulator, then representing it as $\dot{\mathcal{J}}_\Omega(\mathbf{R}) = \dot{\mathcal{J}}_{|\Omega|}(\mathbf{R}) + \dot{\mathcal{J}}_{-|\Omega|}(\mathbf{R})$ at the processor output with the algorithm (4.6) produces the following image of the target area:

$$\hat{L}\{\dot{\mathcal{E}}_\Omega(\mathbf{R})\} = \hat{L}\{\dot{\mathcal{E}}_{|\Omega|}(\mathbf{R})\} + \hat{L}\{\dot{\mathcal{E}}_{-|\Omega|}(\mathbf{R})\} = \tilde{\dot{S}}^{(+1)}(\mathbf{K}') + \tilde{\dot{S}}^{(-1)}(\mathbf{K}') \tag{4.15}$$

where $\tilde{\dot{S}}^{(\pm1)}$ is what is called a $\pm 1^{\mathrm{st}}$ order of diffraction. Since the real space–time signal is real $\mathcal{E} \equiv \Re e\{\mathcal{E}\}$, then its frequency spectrum (4.2) is characterized by the Hermitian property of [10]

$$\dot{\mathcal{E}}_\Omega(\mathbf{R}) \equiv \overset{*}{\dot{\mathcal{E}}}_{-\Omega}(\mathbf{R}) \tag{4.16}$$

Using (4.16) and (4.5), we will determine that a -1st order of diffraction is described by the expression

$$\tilde{\dot{S}}^{(-1)}(\mathbf{K}') = \iint_{4\pi} \overset{*}{\dot{\mathbf{S}}}(\mathbf{K})\mathbf{F}^{(-1)}(\mathbf{K},\mathbf{K}')\mathrm{d}^2\mathbf{K} \tag{4.17}$$

where in contrast to (4.4)

$$\mathbf{F}^{(-1)}(\mathbf{K},\mathbf{K}') = \iint_{\Sigma} \overset{*}{\mathcal{F}}_e (\mathbf{K},\mathbf{R}) \mathcal{J}(\mathbf{R},\mathbf{K}') \exp[i\mathbf{K}\mathbf{R}] \mathrm{d}^2\mathbf{R} \tag{4.18}$$

is the -1st order imagining kernel. An image-forming kernel of the $+1^{\text{st}}$ order agrees with (4.4). In comparing (4.4) and (4.18) we discover that their phases are opposite. Since the direction $\mathbf{K}' = -\mathbf{K}$ corresponds to the backscatter radiation of all the elements on an "exposed" side of the array, then the imaging kernel (4.18) does not have an acceptable directivity, while the image (4.17) is defocused ("blurred") and for all practical purposes creates a uniform exposure of the image (4.3). Thus, nonplanar electrooptical arrays with a double band space–time light modulators are free from the ambiguity characteristic of planar arrays. However, their actual directive gain is somewhat less than the directive gain of a strictly nonplanar array but no less than 50% less.

Note that along with algorithm (4.6) there is an equivalent variant

$$\overset{*}{\hat{L}} \{\ldots\} = \iint_{\Sigma} \ldots \overset{*}{\mathcal{J}} (\mathbf{R},\mathbf{K}') \mathrm{d}^2\mathbf{R} \tag{4.19}$$

which provides for the formation of an acceptable (in terms of resolution, directive gain, and polarizing efficiency) image of the target area in the form of a -1st order of diffraction

$$\tilde{\dot{S}}^{(-1)}(\mathbf{K}') = \iint_{4\pi} \overset{*}{\dot{S}} (\mathbf{K}) \overset{*}{\dot{F}} (\mathbf{K},\mathbf{K}') \mathrm{d}^2\mathbf{K} \tag{4.20}$$

and a parasitic exposure in the form of a $+1^{\text{st}}$ order of diffraction

$$\tilde{\dot{S}}^{(-1)}(\mathbf{K}') = \iint_{4\pi} \dot{S}(\mathbf{K}) \overset{*}{\dot{F}} (\mathbf{K},\mathbf{K}') \mathrm{d}^2\mathbf{K} \tag{4.21}$$

Let the output of the processor be calibrated to the frequency $\Omega_\circ$ in some angular variables ς, ξ. Then (4.14) may be written as

$$\mathbf{K}(\varsigma', \xi'; \Omega_\circ) = \mathbf{K}(\varsigma, \xi; \Omega) \tag{4.22}$$

where ς', ξ' are angular variables measured along a graduated scale. Since, in the general case, when $\Omega \neq \Omega_\circ$, then $\varsigma' \neq \varsigma, \xi' \neq \xi$, and (4.22) may be treated as a parametric equation of the curve

$$\varsigma' = \varsigma'(\varsigma, \xi; \Omega/\Omega_\circ), \qquad \xi' = \xi'(\varsigma, \xi; \Omega/\Omega_\circ) \tag{4.23}$$

given in an implicit form. In the nonmonochromatic case $[\dot{S}_q(\Omega) \neq \delta(\Omega - \Omega_\circ)]$, along this curve we can observe a blurring (by the spectrum) of the monochromatic image of the point source q. The actual form of the curve (4.23) is determined by the geometry of the array and by the structure of the coherent optical processor [in the case of a planar electrooptical array see (2.22)]. However, in the general case, it is also possible to formulate the criteria that guarantee a small dispersion error in comparison with array resolution. Actually it is valid to apply criteria (2.33) to any local planar section $d^2\mathbf{R}$ of a smooth surface (4.1), having the length: $\Delta F|d\mathbf{R}|\sin\theta < c$ (θ_R is the greatest scan angle at point $\mathbf{R}$). If we extend all of $d\mathbf{R}$ along the longest contour t of a planar equivalent aperture of the surface (4.1), then the latter inequality becomes

$$\int_t \Delta F|d\mathbf{R}|\sin\theta_R = |\Delta F \Delta D < c$$

where ΔD is the largest diameter of the surface (4.1). Thus, the dispersion effects may be disregarded when

$$\Delta F \Delta D < c \tag{4.24}$$

which is somewhat stricter than (2.23). Recall that when applied to a phased array antenna and a multibeam array, these criteria guarantee only aperture dispersion (and not feed dispersion).

4.2 Implementing the Processing Algorithm with Coherent Optics

As we know, the problem of structural and parametric synthesis of (electromechanical and electrooptical) devices with a known transfer function does not have a unique solution [138]. To implement the algorithm (4.6) (or (4.19)) by using coherent optics one can proceed in two ways: without taking into account, or by taking into account the surface peculiarities of the array (4.1) [136, 137]. The first approach is widely used and tested in radio engineering;

it consists of using coherent optics to reconstruct the corresponding mathematical operations that describe the processing algorithm (signals from the array elements are multiplied by the value of the current amplitude phase distribution and added). Fig. 4.1b gives a structural diagram of one possible coherent optical processor that reproduces algorithm (4.6). (It is approximate for an array model in the form of a continuous recording medium and exact for an array with a finite number of elements). The processor consists of a laser beam splitter E_0 at N independent processor channels (N is the number of array elements), which multiply the signals from the array elements by the appropriate amplitude phase distribution values with the help of a controllable and a passive mask and also with the help of an adder.

A more constructive approach would be the second one: to reconstruct the obtained processing algorithm by one (or a combination of) of the optically implemented transformations (Fresnel, Fourier, Bessel, Hilbert, convolution, etc. [2–6]) with due regard to the array geometry. However, the class of surfaces (4.1) that permit such a reconstruction (within the framework of a cylindrical, spherical, or conic optics) is limited. These surfaces include, in particular, polyhedrons and rotation bodies which are of the greatest interest from the point of view of practical applications [51, 133].

4.2.1 Piecewise planar arrays

Large nonplanar arrays generally consist of a set of planar subarray modules shaped in the form of a square, octagon, and other polygons (Fig. 4.2). The basic difficulty in controlling highly directional radiation (or reception) with the help of these structures is the electrical connections of the borders of the subarray modules without significant power losses [133]. It is comparatively simple, even in parallel scanning to solve this problem by coherent optical methods.

Let the surface (4.1) consist of squares σ_n (Fig. 4.3a) $\Sigma = \sum_n \sigma_n$. Then the operator (4.6) may be represented as

$$\{\ldots\} = \sum_n \hat{L}_n\{\ldots\} \tag{4.25}$$

Here

$$\begin{aligned}\hat{L}\{\ldots\} &= \iint_{\sigma_n} \ldots \jmath(\mathbf{R},\mathbf{K}')\mathrm{d}^2\mathbf{R} \\ &= \exp(\imath\mathbf{K}\mathbf{R}_n)\iint_{-\infty}^{\infty} \ldots J_n\left(\mathbf{R}_\perp^{(n)}\right)\exp\left(-\imath\,\mathbf{K}_\perp^{(n)}\mathbf{R}_\perp^{(n)}\right)\mathrm{d}^2\mathbf{R}_\perp^{(n)}\end{aligned}$$

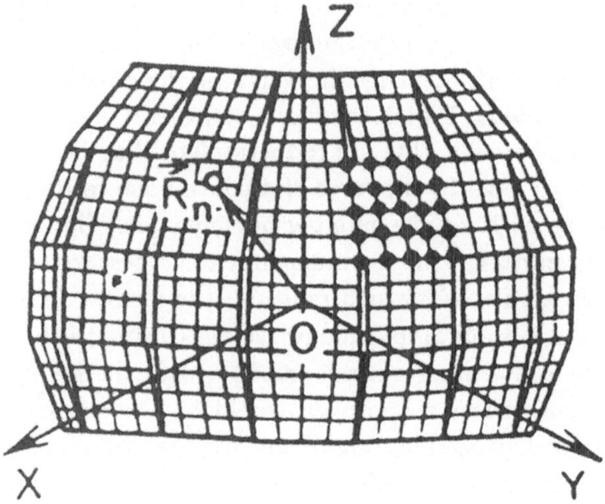

Figure 4.2: A modular implementation of a nonplanar array.

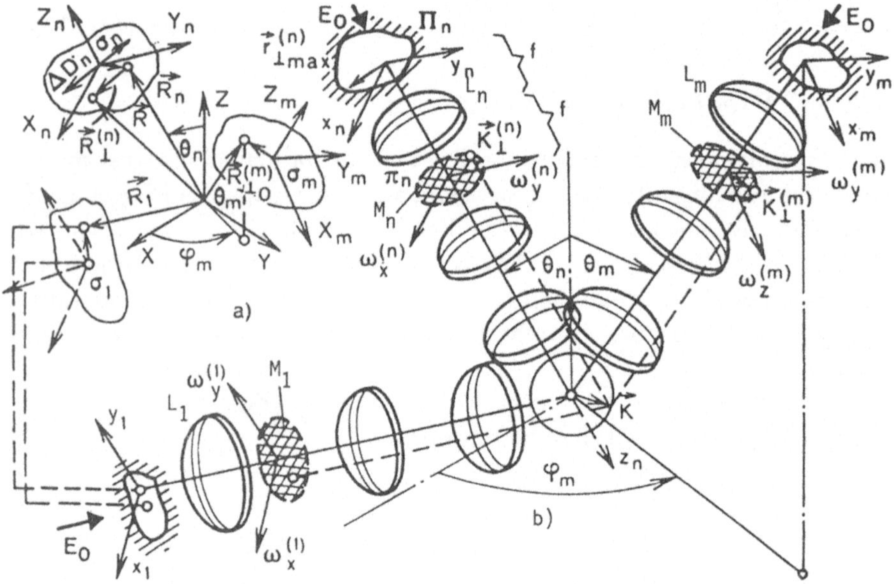

Figure 4.3: Structure of a coherent optical processor for a piecewise-planar array.

$$= \exp(-i\Omega_Z^{(n)} R_\perp^{(n)})\hat{\mathcal{F}}\left\{\ldots J_n\left(\mathbf{R}_\perp^{(n)}\right)\right\} \tag{4.26}$$

where $\mathbf{R} = \mathbf{R}_n + \mathbf{R}_\perp^{(n)}$ (see Fig. 4.3a); $\mathbf{R}_n$ is the normal to σ_n; $\mathbf{K} = \mathbf{K}_\perp^{(n)} - \mathbf{n}_Z^{(n)}\Omega_Z^{(n)}$ ($\mathbf{K}_\perp^{(n)}$ and $\Omega_Z^{(n)}$ are spatial frequencies in the coordinate system σ_n, $\mathbf{n}_Z^{(n)}$ is the normal to σ_n); $J_n(\mathbf{R})$ is the amplitude phase distribution of the excitation σ_n. As we can see from (4.25) and (4.26), the processing algorithm (4.6) in this case amounts to a Fourier transform of space–time signals received by the subarrays σ_n with the weighting function $J_n(\mathbf{R})$ multiplied by $T - n = \exp(-i\Omega_Z^{(n)} R_n)$ and addition of the results. The first two operations may be performed by the Fourier processors of the planar electrooptical array with phase masks T_n (Fig. 4.3b). Adding (joining borders) is complicated because no spectral surface $\omega_x^{(n)}$, $\omega_y^{(n)}$ can serve as a common output surface. In order to establish the presence and view of the latter, we will use the identity [see the explanation to (4.26)]

$$\mathbf{K}_\perp^{(m)} + \mathbf{n}_Z^{(m)}\Omega_Z^{(m)} \equiv \mathbf{K}_\perp^{(n)} + \mathbf{n}_Z^{(n)}\Omega_Z^{(n)} = -\mathbf{K} \tag{4.27}$$

which is valid for any arrays σ_m and σ_n. From (4.27) it follows that the addition of the operators $\hat{L}_m$ and $\hat{L}_n$ is possible only on the radius K with the center at the point of intersection of the optical axes of all the Fourier processors. In ordinary units of length this radius is [see (1.2)–(1.4) and (2.15)]

$$r = fm \qquad K/k = fg \tag{4.28}$$

where $m = m_x = m_y$, and $g = g_x = g_y$. However, when the output signal of a coherent optical processor is removed defocusing occurs, which is caused by the displacement of the output points at $\Delta z_n \leq r$ from the focal plane . Therefore the radius (4.28) should be limited by the condition of the approach of a "shadow"[9, 10] - $\Delta z \leq r \leq 0.1 \cdot 2\pi 2k/(kr_{\perp max}/f)_2$, where $(kr_{\perp max}/f)$ is the maximum spatial frequency in the spectral planes $\omega_x^{(n)}$, $\omega_y^{(N)}$ ($r_{\perp max}$ is the greatest radius of the space–time light modulator). The latter inequality may be simplified:

$$\Delta D/\Lambda \leq 0.8 f/\Delta d \tag{4.29}$$

where $\Delta D = m\Delta d$ is the largest diameter of the subarray. Condition (4.29) provides an approximate equation $\hat{L} \approx \exp(-ik\Delta z_n)\hat{L}_n$ with a phase error no greater than $2\pi/10$. The supplemental phase shift that occurs here

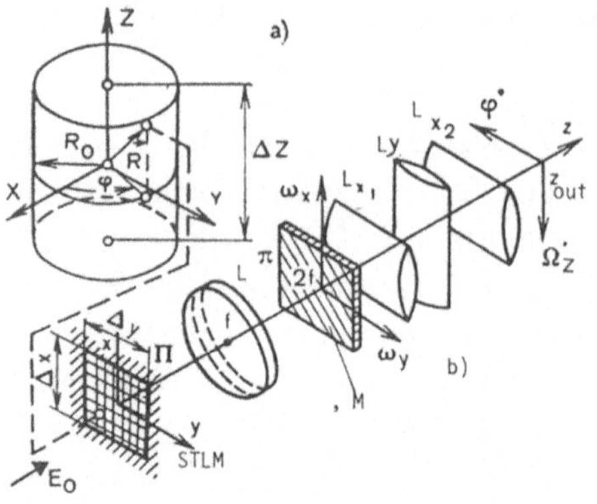

Figure 4.4: (a) Cylindrical array; (b) its coherent optical processor.

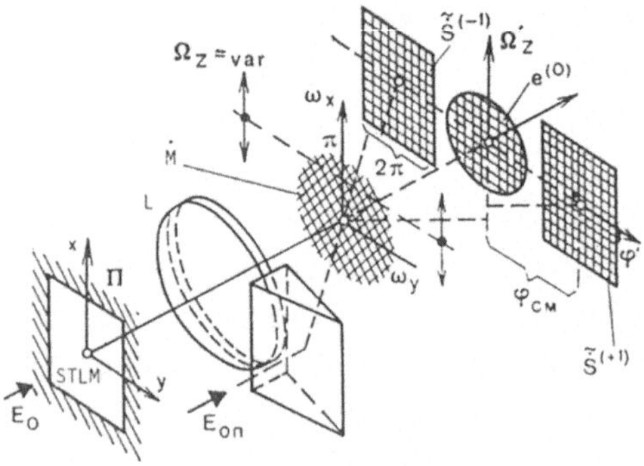

Figure 4.5: Recording a mask.

$\exp(-ik\Delta_n)$ must be compensated by the masks T_n which in terms of the previous phase shifts must have the transmission $T_n = \exp[i\Omega_Z^{(n)}(mf - R_n)$. As can be seen, when $R_1 = \ldots = R_n = mf$, which is valid for a rectilinear polygon (see Fig. 4.2) masks are not needed. The basic shortcoming of this processor is the difficulty in extracting the information.

4.2.2 Array antennas on the surface of a circular cylinder

Lewis [80] proposed a method for shaping the radiation pattern of an array by coherent optics, a method which restores only the fixed elevation cross section (for example, $\theta = \pi/2$) of the angular spectrum within the limits of $0 \leq \varphi \leq 2\pi$. If $0 \neq \pi/2$, then the gain and azimuth resolution will decrease. Below we show a coherent optical processor for a cylindrical array which produces a two-dimensional angular spectrum at the full solid angle and which does not simultaneously decrease the resolution and gain [136, 137, 139].

In this case the surface equation (4.1) has the form

$$\mathbf{R} = \mathbf{n}_X R_\circ \cos\varphi + \mathbf{n}_Y \sin\varphi + \mathbf{n}_Z Z \tag{4.30}$$

where $R_\circ$ is the radius of the cylinder, while φ and Z are the cylindrical coordinates (Fig. 4.4a). In terms of the surface symmetry (4.30) the amplitude phase distribution which belongs in the operator (4.6), may be represented as

$$\begin{aligned}
\dot{J}(\mathbf{R}, \mathbf{K}') &= J_\varphi(\varphi' - \varphi) J_Z(Z) \\
&\times \exp\{-i[K R_\circ(\sin\theta' \cos\varphi' \cos\varphi + \sin\theta' \sin\varphi \sin\varphi) + \Omega'_Z Z]\} \\
&= J_\varphi(\varphi' - \varphi) J_Z(Z) \exp[-i\sqrt{K^2 - \Omega'^2_Z} R_\circ \cos(\varphi' - \varphi)] \exp(i\Omega'_Z Z) \\
&= \dot{J}_\varphi(\varphi' - \varphi, \Omega'_Z) J_Z(Z) \exp(-i\Omega'_Z Z)
\end{aligned} \tag{4.31}$$

Here

$$\begin{aligned}
K' &= -(\mathbf{n}_X K \sin\theta' \cos\varphi' + \mathbf{n}_Y K \sin\theta' \sin\varphi' + \mathbf{n}_Z K \cos\theta) \\
&= -[\sqrt{K^2 + \Omega'^2_Z}(\mathbf{n}_X \cos\varphi' + \mathbf{n}_Y \sin\varphi') + \mathbf{n}_Z \Omega'_Z]; \\
\dot{J}_\varphi(\varphi, \Omega_Z) &= J_\varphi(\varphi) \exp(-i\sqrt{K^2 + \Omega'^2_Z} R_\circ \cos\varphi)
\end{aligned} \tag{4.32}$$

where $J_\varphi(\varphi)$ and $J_Z(Z)$ is the partial amplitude phase distribution, while $J_\varphi = J_\varphi \mathrm{rect}(\varphi)/pi)$ ("the shadowed" half of the cylinder in shaping the radiation pattern, for all practical purposes, does not participate), $J_Z(Z) = J_Z(Z)\mathrm{rect}(Z/\Delta Z)$ (see Fig. 4.4a). Substituting (4.31) into (4.6) we find

$$\begin{aligned}
\hat{L}\{\ldots\} &= R_\circ \int_{-\infty}^{\infty} d\varphi \dot{J}_\varphi(\varphi' - \varphi, \Omega'_Z) \int_{-\infty}^{\infty} \ldots J_Z(Z) \exp(-i\Omega'_Z Z) dZ \\
&= R_\circ \hat{\mathcal{F}}_Z\{\ldots J_Z(Z)\} \otimes_\varphi \dot{J}(\varphi', \Omega'_Z)
\end{aligned}$$

where $\otimes_\varphi$ is the convolution along φ. Applying the convolution theorem we finally obtain

$$\hat{L}\{\ldots\} = R_\circ \hat{\mathcal{F}}_\varphi^{-1}\{\hat{\mathcal{F}}\{\ldots J_Z(Z)\}\dot{T}(\Omega_\varphi, \Omega'_Z)\} \tag{4.33}$$

where

$$\dot{T}(\Omega_\varphi, \Omega_Z) = \hat{\mathcal{F}}_\varphi\{\dot{J}_\varphi(\varphi, \Omega_Z)\} \tag{4.34}$$

As is apparent, in the case of a cylindrical array, Equation (4.6) amounts to the product of operators particular to the optical system. Fig. 4.4b shows a coherent optical processor based on algorithm (4.33) where between the receiving elements of the array and the channels of the space–time light modulator in the object plane, Π, the addressing law is fixed as

$$Z = m_x x, \qquad \varphi = m_y y \tag{4.35}$$

In accordance with (4.33), the processor consists of a spherical lens L which performs a two-dimensional Fourier transformation of a geometrically similar flat scan $\Delta_x \Delta_y$ of the space–time signal (4.2) of a cylindrical array (see Fig. 4.4a) with weighting $J_Z(Z)$, a passive amplitude-phase mask (4.34) in the spectral plane, Π, recorded on the scale

$$\omega_x = m_x \Omega_Z, \qquad \omega_y = m_y \Omega_\varphi \tag{4.36}$$

(Ω_φ is the supplemental spatial frequency), and an astigmatic system containing a Fourier processor at the azimuth measurement (the cylindrical lens L_y) and an imaging system in the elevation measurement (cylindrical lens L_{x_1} and L_{x_1}). In accordance with (4.35) and (4.36), the objects' coordinates are mapped at the processor output by the following scale in compliance with Equation (1.2)

$$\Omega'_Z = \omega_x / m_x = kx / f m_x, \qquad \varphi' = m_y y \tag{4.37}$$

and the case of the optical lens enlargement (reduction) by an astigmatic system is considered to be evident. The mask (4.34) may be obtained either

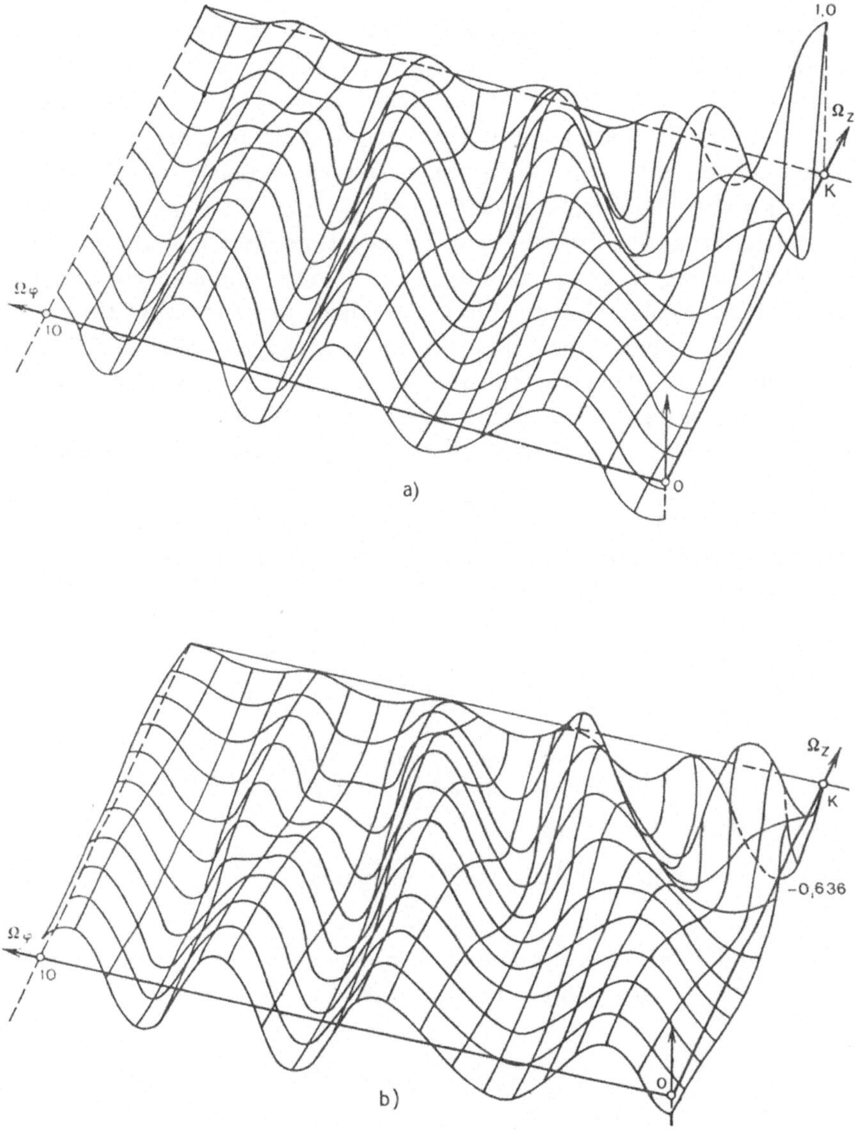

Figure 4.6: Transmission function of a mask: (a) real; (b) imaginary.

by computer design and subsequent microphotography or by a holographic technique [3, 6, 142]. The latter (called the Vander Lugt method) records complex masks with an amplitude transmittance function on photographic materials. Its advantage is that one does not need to know the transmittance function. No special equipment is required for making a mask since it can be made directly with the help of a coherent optical processor modified according to the Vander Lugt method (Fig. 4.5). This not only excludes instrument errors but directly uses a space–time light modulator which is controllable by an array signal (4.2) (or by signals from a specialized multichannel array simulator) instead of a specialized mask (4.32) at the plane Π (see Fig. 4.4). When a mask like this is used, the ±1 order of diffractions are displaced $\pm\varphi_{cm}$ from the coherent optical processor axis (see Fig. 4.5), which eliminates their exposure by a zero order of diffraction (given the appropriate choice φ_{cm} is the incidence angle of the reference beam split by the prism).

The structure of the mask (4.34) is determined by the specific amplitude phase distribution (4.30). In particular, given the equal excitation of the array elements on the azimuth $J_\varphi(\varphi) = \mathrm{rect}(\varphi/\pi)$ we obtain

$$\begin{aligned} \dot{T}(\Omega_\varphi, \Omega_Z) &= \int_{-\pi/2}^{\pi/2} \exp\left[i\left(\sqrt{K^2 - \Omega_Z^2} R_\circ \cos\varphi + \Omega_\varphi \varphi\right)\right] d\varphi \\ &= \pi \exp\left(i\frac{\pi}{2}\Omega_\varphi\right)\left[A_{-\Omega_\varphi}\left(\sqrt{K^2 - \Omega_Z^2} R_\circ\right)\right. \\ &\quad \left. + iV_{-\Omega_\varphi}\left(\sqrt{K^2 - \Omega_Z^2} R_\circ\right)\right] \end{aligned} \tag{4.38}$$

where A and V are Anger's and Weber's function [144, 145]. Quadrants $(0 \leq \Omega_\varphi \leq 10,\ 0 \leq \Omega_Z \leq K)$ of the surfaces

$$\Re e\left\{\dot{T} \exp\left(-i\frac{\pi}{2}\Omega_\varphi\right)/\pi\right\}$$

and

$$\Im m\left\{\dot{T} \exp\left(-i\frac{\pi}{2}\Omega_\varphi\right)/\pi\right\}$$

when $KR_\circ = 10$ (the case of a small $KR_\circ$ is sufficiently demonstrated since as $KR_\circ \to \infty$ the functions A and V asymptotically approach the harmonic [144, 145]) are shown in Figs. 4.6a and b, respectively.

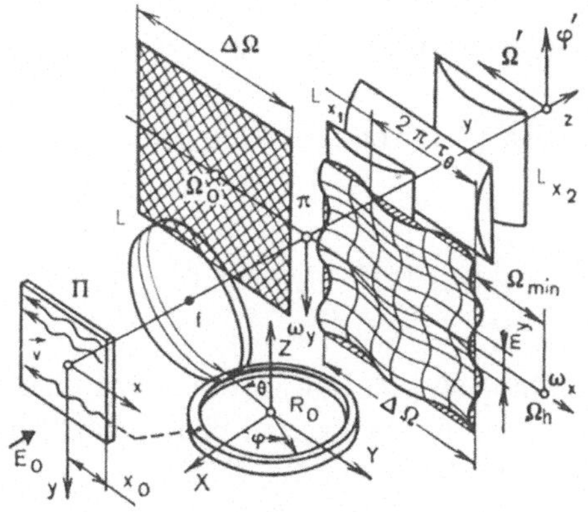

Figure 4.7: A circular array and its coherent optical processor.

4.2.3 Circular arrays

If the elements of an array are arranged in a circle (see Fig. 4.7) then the input device may be a space–time light modulator which makes a spatial sweep of the time signals. This permits the use of a circular array for a simultaneous panoramic azimuth and frequency scan [80-83], in addition to a correlation analysis [105]. The results obtained in these works were developed by Grinev and Voronin [137] who presented the structure of a complex coordinate mask for a coherent optical processor in a circular electrooptical array and evaluated the image defocusing when the elevation differed from the given mask, and studied the possibility of reproducing such a mask by holography and microphotography.

The specific details about the operation of coherent optical processors which employ various space–time light modulators that perform spatial signal sweep are explained in Sections 3.1 and 3.2 which show that they (the details) are not decisive with respect to the pattern-forming properties of the electrooptical array. Let us look at the circular electrooptical array using as an example a multichannel acoustooptic modulator (AOM) as the space–time light modulator.

Let the space–time signal of the circular array be addressed by the AOM channels according to the law (4.5) $\varphi = m_y y$. Then the processing algorithm (4.6) agrees with (4.33) in the azimuth measurement φ, but in the orthogonal measurement it differs in that at the processor output instead of the elevation coordinates Ω'_Z the frequency $\Omega' = \Omega'_Z c/cos\theta_\circ$ ($\theta_\circ$ is the elevation at which the CO processor is set) must be reproduced in the specific scale. Let us convince ourselves that the processing algorithm (4.33) modified in this way, enhances the pattern- and spectrum-forming properties of a circular array no worse than an equivalent circular array and an optical spectrum analyzer.

On analogy with (3.2) the following optical signal (Fig. 4.7) is read into the processor

$$E^{(1)}(x,y,t) = T^{(1)} E_\circ J_t(x) J(y) \mathcal{E}[m_y y; t + (x - x_\circ)/v] \exp(-i2\pi\nu t) \quad (4.39)$$

where $J_t(x)$ is the weighting function of the AOM channel; $J(y)$ is the amplitude phase distribution of the excited channels $[J(y) = \mathrm{rect}(y/\Delta y)$ is in correspondence with algorithm (4.33)]; $\mathcal{E}(\varphi, t)$ is the space–time signal at the output of the array elements [see (4.2)]; $x_\circ = 0.5\Delta x$ is the coordinate of the bar of the piezoelectric transducers.

The frequency spectrum of the optical signal (4.39)

$$\begin{aligned} \dot{E}^{(1)}_\omega(x,y) &= \hat{\mathcal{F}}_t\{E^{(1)}(x,y,t)\} \\ &= T^{(1)} E_\circ J_t(x) J(y) \dot{\mathcal{E}}_\Omega(m_y y) \exp\left[i\frac{\Omega}{v}(x - x_\circ)\right] \quad (4.40) \end{aligned}$$

where $\dot{\mathcal{E}}_\Omega(\varphi) = \hat{\mathcal{F}}_t\{\mathcal{E}(\varphi, t)\}$ is the signal's frequency spectrum at the array (4.2); $\omega = 2\pi\nu \pm \Omega$.

Substituting (4.40) into (4.33) and taking into account that due to "degeneration" of the cylinder to a circle $J_Z(Z) = J_Z(0) = 1$ and $\Omega'_Z = \Omega/c \cos\theta_\circ$. Then

$$\dot{e}^{(1)}_\omega(\varphi', \Omega') = \hat{L}\{\dot{E}^{(1)}_\omega(x,y)\}$$

$$= R_\circ \hat{\mathcal{F}}^{-1}_\varphi \left\{\hat{\mathcal{F}}\left\{T^{(1)} E_\circ J_t(x) \mathrm{rect}(y/\Delta y) \dot{\mathcal{E}}_\Omega(m_y y) \exp\left[i\frac{\Omega}{v}(x - x_\circ)\right]\right\}\right.$$

$$\left.\times \dot{T}\left(\Omega_\varphi, \frac{\Omega'}{c}\cos\theta_\circ\right)\right\}$$

$$\frac{R_\circ T^{(1)} E_\circ}{2\pi\lambda f m_y} \hat{\mathcal{F}}_x \left\{J_t(x) \exp\left[i\frac{\Omega}{v}(x - x_\circ)\right]\right\}$$

$$\times \hat{\mathcal{F}}^{-1}_\varphi \left\{\hat{\mathcal{F}}_\varphi\{\dot{\mathcal{E}}_\Omega(\varphi)\} \dot{T}\left(\Omega_\varphi, \frac{\Omega'}{c}\cos\theta_\circ\right)\right\}$$

$$= \frac{T^{(1)}E_\circ}{2\pi\lambda f m_y}[\hat{L}_\varphi\{\dot{\mathcal{E}}_\Omega(\varphi)\}]\left[F_t\left(\frac{\Omega'-\Omega}{v}\right)\exp\left(-i\frac{x_\circ}{v}\Omega\right)\right] \tag{4.41}$$

Here

$$\hat{L}_\varphi\{\ldots\} = R_\circ \hat{\mathcal{F}}_\varphi^{-1}\left\{\hat{\mathcal{F}}_\varphi\{\ldots\}\dot{T}\left(\Omega_\varphi, \frac{\Omega'}{c}\cos\theta_\circ\right)\right\} \tag{4.42}$$

is the "azimuth" processing algorithm adjusted for the frequency Ω' and the fixed elevation $\theta_\circ$; $F_t(\omega_x) = \hat{\mathcal{F}}_x\{\mathcal{J}_t(x)\}$ is the imaging kernel of the optical spectrum analyzer which in the case of an equal weighting function $\mathcal{J}_t(x) = \text{rect}(x/\Delta x)$ is $\text{sinc}(\Omega'/\delta\Omega)$ ($\Omega' = v\omega_x$ is the frequency read from the processor output; $\delta\Omega = 2\pi v/\Delta x$ is the resolution of the spectrum analyzer).

As can be seen, the expression (4.41) is the product of two factors. The first of these is a one-dimensional special case of the general relationship (4.5) which describes the image (4.3) of the angular spectrum at the processor output. The second is analogous to to the corresponding factor (3.4) for an electrooptical array with an AOM and describes the output reaction of an optical spectrum analyzer.

The implementation of the operator (4.42) is complicated by the fact that it must be reproduced not on the fixed frequency Ω' (similar to the operator [4.33]), but any frequency Ω in the operating band $\Delta\Omega = \Omega_{max} - \Omega_{min}$. However, the azimuthal operator (4.42) may be restructured with respect to frequency Ω' if we reproduce it along one of the processor dimensions [$\omega_x = \Omega'/v$ see (3.6)]. In doing this the processing algorithm will be equivalent to an algorithm "adjusted" for the current frequency Ω of the signal, since in the width $\delta\Omega = 2\pi v/\Delta x = 2\pi/\Delta\tau$ ($\Delta\tau \approx 10\ \mu$s is the AOM line filling time) of the imaging kernel F_t, the mask $\dot{T}$ in (4.42) is practically unchanged. Actually, since the argument of the mask $\dot{T}$ within the limits described above varies insignificantly [see, for example, (4.38)]

$$|KR_\circ - K'R_\circ| = \frac{R_\circ}{c}|\Omega - \Omega'| = \frac{R_\circ}{c}\delta\Omega =$$

$$= 2\pi\frac{\tau_\circ}{\Delta_\tau} \ll 2\pi$$

($\tau_\circ = R_\circ/c \approx 10$ ns), then the noted circumstance is fulfilled.

Fig. 4.7 shows an optical processor that produces algorithm (4.41) where the addressing law of the type (4.35) $\varphi = m_y y$ is placed between points on the circular array and the AOM in the object plane, Π. In correspondence with (4.41), the processor performs a one-dimensional Fourier transform of the input signal along the y-axis. To implement the one-dimensional spectrum analyzer mode of the signal expanded in the dimension x by the AOM, the processor must perform a Fourier-transform in x. Both of these transforms are done simultaneously by one spherical lens. The result of the two-dimensional Fourier-transform obtained at the spectral plane, π, is multiplied by the transmission function $\dot{T}$ with the aid of the corresponding complex mask, recorded in the scale [see (3.6) and (4.36)]

$$\omega_x = \Omega'/v, \qquad \omega_y = m_y \Omega_\varphi \tag{4.43}$$

Furthermore an astigmatic optical system may be used to perform a one-dimensional Fourier-transform on the y-axis (the cylindrical lens L_y) and to transfer the image of the frequency spectrum to the processor output plane on the x-axis (the cylindrical lenses L_{x_1}, L_{x_2}). The transformation of the planar light beam $E_\circ$ by this processor results in the formation of an image of the frequency-angular spectrum $S(\varphi, \Omega)$ of the space–time signal with the following intensity [see (3.5)]

$$\begin{aligned} I^{(1)}(\varphi, \Omega') &= \overline{\frac{1}{2\pi}\int_{-\infty}^{\infty} \dot{e}_\omega^{(1)}(\varphi', \Omega')\, \overset{*}{e_\omega^{(1)}}(\varphi', \Omega') d\omega} \\ &= I_\circ |\mathbf{F}(\varphi', \theta_\circ)|^2 \otimes |\dot{S}(\varphi', \Omega')|^2 \otimes |F_t(\Omega'/v)|^2 \end{aligned} \tag{4.44}$$

where

$$I_\circ = \frac{v}{(2\pi)^2} \left(\frac{T^{(1)} E_\circ}{\lambda f m_y} \right)^2$$

$F(\varphi', \theta_\circ)$ is the radiation pattern of a circular antenna at the section $\theta = \theta_\circ$. Note that equation (4.44) is presented as a convolution on the azimuthal coordinate in assuming the isotropy of the circular array radiators [in the more general case the processor reactive output is described by the integral of the superposition type (4.13)].

In keeping with (4.37) and (3.13), the azimuthal coordinate and frequency at the processor output is mapped according to the following scale:

$$\varphi' = m_y y, \qquad \Omega' = \omega_x v = kxv/f \tag{4.45}$$

Let us investigate the structure of the mask $\dot{T}(\Omega, \frac{\Omega'}{c} cos\theta_\circ)$ in (4.41), which agrees with (4.32), (4.34) has the form

$$\dot{T}\left(\Omega_\varphi, \frac{\Omega'}{c}\cos\theta_\circ\right) = \hat{\mathcal{F}}_\varphi\left\{J_\varphi(\varphi)\exp\left[-i\frac{R_\circ}{c}\Omega'\sin\theta_\circ\cos\varphi\right]\right\} \tag{4.46}$$

We will examine (4.46) for the special case $J_\varphi(\varphi) = \mathrm{rect}(\varphi/2\pi)$ which corresponds to the absence of radiator "shading." Then

$$\dot{T}\left(\Omega_\varphi, \frac{\Omega'}{c}\cos\theta_\circ\right)$$

$$= \exp(i\Omega_\varphi\pi)\int_\circ^{2\pi}\exp[-i(\Omega_\varphi\varphi - \Omega'\tau_\theta\cos\varphi)]d\varphi \tag{4.47}$$

where $\tau_\circ = R_\circ/c\sin\theta_\circ$ is the aperture "filling" time. When $\Omega_\varphi = n$ mask (4.7) assumes the value

$$\dot{T}\left(n, \frac{\Omega'}{c}\cos\theta_\circ\right) = 2\pi(i)^{-n}J_n(\Omega'\tau_\theta) \tag{4.48}$$

(where J_n is an nth order Bessel function of the first kind), that coincides with the corresponding values obtained by Lewis [80]. To the left of the crosshatching is a second mask which can perform the defocusing of the conjugate image, if necessary. In doing this, we must take into account the heterodyne transformation of the frequency $\Omega = \Omega_\gamma - \Omega_\Pi$ (Ω_γ is the heterodyne frequency, Ω_Π is the intermediate frequency) by the electrooptical receiver.

Like the case of the cylindrical electrooptical array (4.34), the mask of the circular electrooptical array (4.46) [or (4.48)] may be reproduced by holographic means or by computer design and subsequent photo reduction. Appendix A gives examples of circular array masks synthesized on the computer for use in sonar.

Algorithm (4.42) provides for the formation of a focused image of the type (4.5) only for signals, received from elevation $\theta \neq \theta_\circ$ with respect to which the processor mask (4.46) is implemented. Let us evaluate the deterioration of amplification when $\theta \neq \theta_\circ$ along the relative drop off of the

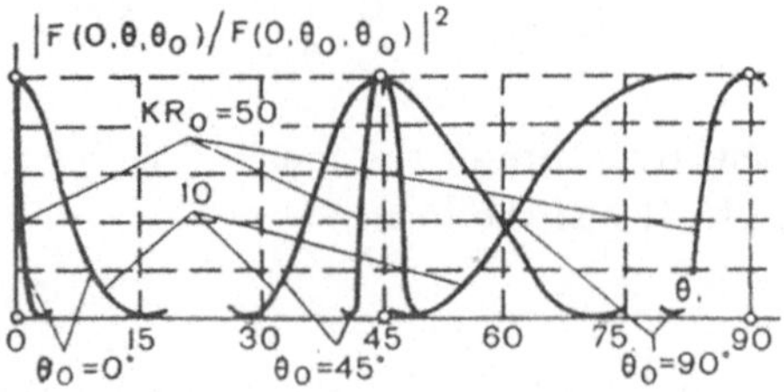

Figure 4.8: Estimating the decrease in efficiency in a circular array.

reaction maximum at its output. The reaction to the planar wave coming from point source situated in the far field coincides in the form of the array pattern

$$
\begin{aligned}
&\mathbf{F}(\varphi-\varphi_\circ,\theta,\theta_\circ)\\
&=\hat{L}_\varphi\left\{\mathcal{F}_e(\varphi-\varphi_\circ,\theta)\exp\left[i\Omega\frac{R_\circ}{c}\sin\theta\cos(\varphi-\varphi_\circ)\right]\right\}\\
&=\mathcal{F}_e(\varphi-\varphi_\circ,\theta)\exp\left[i\Omega\frac{R_\circ}{c}\sin\theta\cos(\cos(\varphi-\varphi_\circ)\right]\\
&\otimes_\varphi J_\varphi(\varphi-\varphi_\circ)\exp\left[-i\Omega\frac{R_\circ}{c}\sin\theta_\circ\cos(\varphi-\varphi_\circ)\right]
\end{aligned}
$$

where ($\theta_\circ$, $\varphi_\circ$ is in the direction of the array's maximum gain; $\mathcal{F}_e$ is the radiation pattern of the array elements (it is assumed the elements are identical and have the same polarization). Then when $\varphi=\varphi_\circ$

$$
\begin{aligned}
\left|\frac{F(0,\theta,\theta_\circ)}{F(0,\theta_\circ,\theta_\circ)}\right|^2 &= \left|\frac{\int_{-\pi}^{\pi}\mathcal{F}_e(\varphi',\theta)J_\varphi(\varphi')\exp\left[i\Omega\frac{R_\circ}{c}(\sin\theta-\sin\theta_\circ)\cos\varphi'\right]d\varphi'}{\int_{-\pi}^{\pi}\mathcal{F}_e(\varphi',\theta_\circ)J_\varphi(\varphi')d\varphi'}\right|^2\\
&\approx \left|\frac{1}{2\pi}\int_{-\pi}^{\pi}\exp[iKR_\circ(\sin\theta-\sin\theta_\circ)\cos\varphi']d\varphi'\right|^2\\
&= J_\circ^2[KR_\circ(\sin\theta-\sin\theta_\circ)] \qquad (4.49)
\end{aligned}
$$

where $J_\circ$ is a Bessel function.

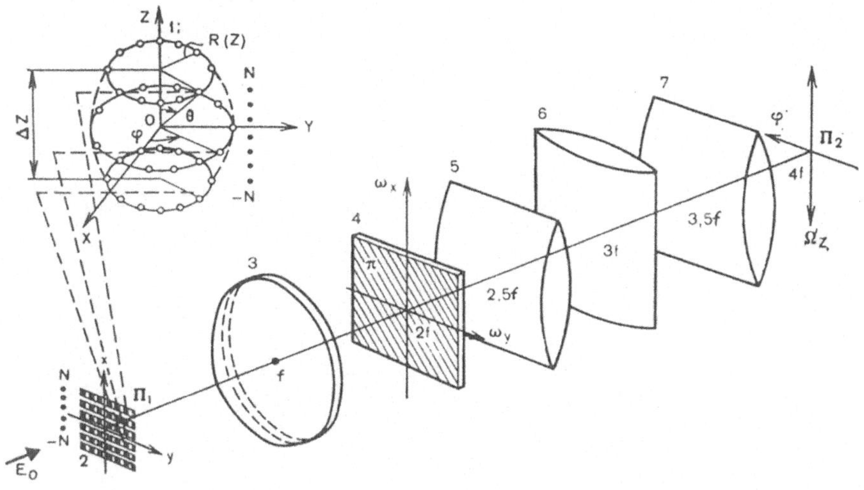

Figure 4.9: A CO processor for an axially symmetric array.

The function (4.46) is shown in Fig. 4.8 for three values of θ_o. As can be seen, it is symmetrical relative to the plane $\theta = 90°$ which points to the focusing properties of algorithm (4.42) with respect to objects with an elevation θ_o and $180° - \theta_o$ causing an ambiguity in the elevation. To avoid this problem it is necessary to use highly directional receiving elements with respect to the angle of elevation.

4.3 Random Axially-Symmetric Array Antennas

This section describes the algorithm and structure of a CO processor for processing signals from arrays whose receiving elements are arranged on a nonplanar axially-symmetric surface (a surface formed by the rotation of a flat curve around an axis) with the use of the discriminating properties of a volume hologram [146–148].

4.3.1 Description of the algorithm

Let the receiving elements be arranged continuously on a nonplanar axially-symmetric surface, described by the radius **R** (Fig. 4.9)

$$\mathbf{R} = \mathbf{n}_X R(Z) \cos \varphi + \mathbf{n}_Y R(Z) \sin \varphi + \mathbf{n}_Z Z \tag{4.50}$$

where $R(Z)$ is the radius of the circumference at section $Z = \text{const}$. It is analogous to (4.31) with due regard to the azimuthal symmetry

$$\begin{aligned} \dot{J}(\mathbf{R},\mathbf{K}') &= J_\varphi(\varphi' - \varphi) J_Z(Z) \\ &\times \exp\left\{-i\left[R(Z)\sqrt{K^2 - {\Omega'_Z}^2}\cos(\varphi' - \varphi) + \Omega'_Z Z\right]\right\} \end{aligned}$$

and

$$d^2\mathbf{R} = R(Z)\sqrt{1 + \left[\frac{\partial R(Z)}{\partial Z}\right]^2}\, d\varphi dZ$$

for (4.6) we obtain

$$\hat{L}\{\mathcal{E}(\mathbf{R})\} = \int\int_\Sigma \mathcal{E}(\mathbf{R}) R(Z)\sqrt{1 + \left[\frac{\partial R(Z)}{\partial Z}\right]^2} J_\varphi(\varphi' - \varphi) J_Z(Z)$$

$$\times \exp\left\{-i\left[R(Z)\sqrt{K^2 - {\Omega'_Z}^2}\cos(\varphi' - \varphi) + \Omega'_Z Z\right]\right\} d\varphi dZ \qquad (4.51)$$

Algorithm (4.51) simultaneously forms the spatial directional characteristics of nonplanar axially-symmetric arrays with random (equivalent and nonequivalent) arrangement of the receiving elements. For a circular cylinder $R(Z) = \text{const}$ the expression (4.51) is transformed into (4.33). In the general case, however, it is difficult to implement algorithm (4.51) by optical techniques. Calculating the discreteness of the element arrangement overcomes these difficulties.

Assume that the elements are arranged in a circle $Z = Z_n$ ($n = 0$, $\pm 1, \ldots, \pm N$; $2N + 1$ is the number of circles in Fig. 4.9; the elements in the circles are arranged randomly. Then (4.51) may be written as

$$\hat{L}\{\mathcal{E}(\mathbf{R})\} = \sum_{-N}^{+N} \hat{L}_n\{\mathcal{E}(\mathbf{R})\} \qquad (4.52)$$

Here

$$\hat{L}\{\mathcal{E}(\mathbf{R})\} = \int_{-\infty}^{\infty} \mathcal{E}(\mathbf{R}) \Phi(Z_n) J_\varphi(\varphi' - \varphi)$$

$$\times \exp\left\{-i\left[R(Z)\sqrt{K^2 - \Omega_Z'^2}\cos(\varphi' - \varphi) + \Omega_Z' Z\right]\right\} \tag{4.53}$$

where

$$\Phi(Z_n) = J_Z(Z_n)R(Z_n)\left.\sqrt{1 + \left[\frac{\partial R(Z)}{\partial Z}\right]^2}\right|_{Z=Z_n}$$

and considering that $J_Z(Z_n) = J_Z(Z)\delta(Z - Z_n)$, $\delta(Z)$ is the delta or impulse function. Transforming (4.53) analogously to (4.33) we find that

$$\hat{L}_n\{\mathcal{E}(\mathbf{R})\} = \hat{\mathcal{F}}_\varphi^{-1}\{\hat{\mathcal{F}}\{\mathcal{E}(\mathbf{R})\delta(Z - Z_n)\}\dot{T}(\Omega_Z' - \Omega_\varphi)\} \tag{4.54}$$

where

$$\dot{T}(\Omega_Z', \Omega_\varphi) = \Phi(Z_n)\hat{\mathcal{F}}_\varphi\left\{J_\varphi(\varphi)\exp\left[-iR(Z_n)\sqrt{K^2 - \Omega_Z'^2}\cos\varphi\right]\right\} \tag{4.55}$$

The implementation of operators of the (4.54)-type by optical means is treated in Section 4.2.2. Consequently, to implement algorithm (4.52) it is necessary to construct $2N + 1$ optical processors analogous to Fig. 4.4, each of which is realized by operator $\hat{L}_n$ with the weighting $\Phi(Z_n)$ corresponding to the nth circle. The optical signal from the output of each processor is summed in accordance with (4.52) at the general output plane. The complexity of realizing this processor lies in the fact that the mask-filter $\dot{T}_n$ (4.55) depends on the number of the circle n which results in the awkwardness of the processor, the need for a summing device, and the complexity of alignment. A natural alternative is to develop a single mask-filter which uses the selectivity properties of a volume hologram.

4.3.2 Coherent optical processor employing a volume filter

Fig. 4.9 shows a CO processor that implements algorithm (4.52), where $\hat{L}_n\{\mathcal{E}(\mathbf{R})\}$ is defined in (4.54). In doing this between the array receiving elements 1 and the space–time light modulator channels 2 which is at plane Π_1 and which inputs the signals to the processor, we find the addressing law:

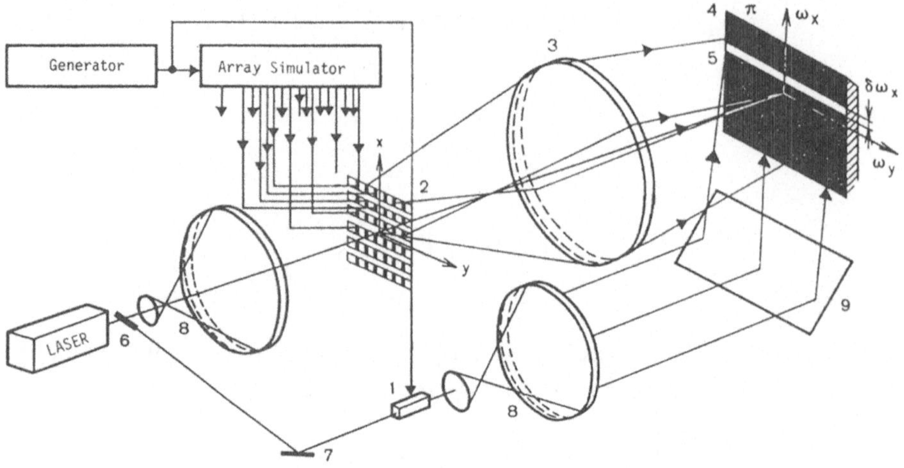

Figure 4.10: Recording a volume filter-mask.

$$Z_n = m_x x_n, \qquad n = 0 \pm 1, \ldots, \pm N; \qquad \varphi = m_y y^1 \tag{4.56}$$

In keeping with Equation (4.52), (4.54) processor performs a Fourier transform from each circle $\mathcal{E}(\mathbf{R}) = \mathcal{E}(\varphi, Z_n)$ using spherical lens 3. In the spectral plane π there is a volume hologram-mask 4 a recording of function $\dot{T}_n$ so that the Fourier response of each nth circle is multiplied by "its own" mask $\dot{T}_n$ (as we showed earlier, this can be done by using the angular discrimination factor of a volume hologram), as a result of which the sum $2N + 1$ responses are reproduced behind hologram 4. Masks $\dot{T}_n$ are written on a scale of (4.36). An astigmatic system of lenses 5, 6, 7 perform an inverse Fourier transform $\hat{\mathcal{F}}_\varphi^{-1}$ and transfers the image along ω_x. This results in the formation of light responses as an array radiation pattern at the processor output plane, Π_2, that corresponds to the emission sources in the array sweep zone.

The volume mask 4 may be made by several methods. One of the possible recording systems, the Vander Lugt system, is shown in Fig. 4.10. The amplitude-phase distribution at the space–time light modulator 2 input which corresponds to the distribution at the array output is synthesized with the help of an array simulator; the weighting function $\Phi(Z_n)$ is realized by

[1]The addressing law at every n^{th} circle on φ is shown as a continuous line for the sake of being more explicit. This should not decrease its general value.

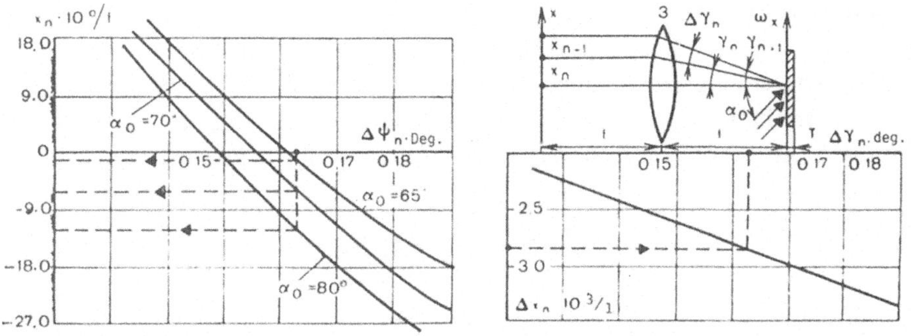

Figure 4.11: Estimating the angular discrimination factor.

modulating the input light beam $E_\circ$ or by an array simulator. The reference beam (obtained by splitting the main one with the help of mirrors 6, 7) is introduced with the help of the single-band single-channel light modulator 1 signal which comes from the array simulator element (a single band modulator provides for the formation of a stable interference pattern at plane π [73]). The volume mask is produced as follows. A diaphragm (not shown in Fig. 4.10) brings out the ruler of the space–time modulator channels 2 with coordinate x_n (4.56) corresponding to the nth circle of the array with an amplitude phase distribution

$$J_\varphi(\varphi)\left\{\exp\left[-iR(Z)\sqrt{K^2-{\Omega'_Z}^2}\cos\varphi\right]\right\}$$

where $\Omega'_Z = K cos\theta =$ const. The wavefront corresponding to a beam of light at the output of a single-channel modulator 1 is transformed by a telescopic lens system 8 into a planar one and with the help of mirror 9 is directed at an angle $\alpha_\circ$ to plane π (the angle $\alpha_\circ$ is read at the vertical plane from the processor axis, see Fig. 4.11). A thick-layered, light-sensitive material is placed at the spectral plane π to record the volume mask 4. Immediately adjacent to the light sensitive layer there is a one-dimensional diaphragm 5 with the coordinate $\omega_x = m + x\Omega'_Z = m_x K cos\theta'$ (θ' is a fixed angle) of width $\Delta \ll \delta\omega_x$ ($\delta\omega_x$ is the width of the radiation pattern at plane π along ω_x).

In exposing the hologram diaphragm 5 travels along ω_x in correspondence with changes in angle θ' in the interval $0\ldots\pi$ at a space of Δ. This cycle is repeated $(2N+1)$ times depending on the number of circles. Between cycles the circle number is changed (the rows of the space–time light modulator, the amplitude frequency modulation which is supplied by the simulator) and the position of the reference beam is constant.

The transmission function of the holographic mask obtained in this way (after the appropriate chemical treatment) in the first approximation in fulfilling the Bragg condition in the restoration stage is proportional [6, 147]:

$$\dot{T}_B \sim \sum_{n=-N}^{N} \dot{T}_n \overset{*}{\dot{R}}_p \, \delta_{np} \tag{4.57}$$

where $p = -N, \ldots, N, \dot{T}_n$ is defined by (4.55); $\dot{R}_p = A\exp[-i(f\omega_x \sin\alpha_\circ)]$ is the read beam; $*$ is the integration sign; δ is the Kronecker delta ($\delta_{np} = 1$ when $n = p$ and $\delta_{np} = 0$ when $n \neq p$).

When the array is operating, the optical signal behind the volume mask $\dot{T}_n$ (4.57), positioned in the optical processor plane π (see Fig. 4.9) has the form

$$\dot{E}\pi \sim \sum_{p=-N}^{N} \hat{\mathcal{F}}\{\mathcal{E}(\varphi, Z)\delta(Z - Z_p)\}\dot{T}_B \tag{4.58}$$

and at the processor output plane Π_2

$$\dot{E}_{\Pi_2} \sim \exp(if\omega_x \sin\alpha_\circ) \sum_{n=-N}^{N} \hat{\mathcal{F}}_\varphi^{-1}\{\hat{\mathcal{F}}\{\mathcal{E}(\varphi, Z)\delta(Z - Z_p)\}\dot{T}_n(\Omega'_Z, \Omega)\} \tag{4.59}$$

Comparison of Equations (4.59), (4.52) and (4.54) shows that they coincide with an accuracy of the phase factor. In recording the intensity of light at the plane Π_2 the phase factor may be neglected. This can be done, for example, by placing an appropriate phase mask behind the volume hologram.

4.3.3 Implementing the processor

The special characteristics of the processor that are implemented using the volume hologram are the angular selectivity or the volume hologram, its diffraction efficiency, the dynamic range, and the resolution factor of the material used for recording the hologram. The latter three parameters may be estimated by using the data in [6, 149, 150]

The angular selectivity is related to the CO processor parameters (the distance between the rows in the space–time light modulator and the focusing distance of the Fourier lens). To obtain nonoverlapping responses from adjacent circles of the array it is necessary to meet the condition (Fig. 4.11)

$$\min[\Delta\gamma_n] \geq \max[\Delta\Psi_n]$$

where $n = 0, \pm 1, \pm 2, \ldots, \pm N$; Δx_n is the distance between adjacent rows of the space–time light modulator; f is the focal length of lens 3; $\Delta\Psi_n$ is the angular deviation of the regenerationg wave corresponding to the nth row of the space–time light modulator (the nth circle), from the Bragg angle, when the diffraction efficiency drops to zero. Using the results in [151], we may write

$$\Delta\Psi_n = \frac{\lambda}{T}\left[\frac{n_c^2 - \sin^2\gamma_n}{(n_c^2 - \sin^2\gamma_n)^{1/2}\sin\alpha_\circ + (n_c^2 - \sin^2\alpha_\circ)^{1/2}\sin\gamma_n}\right]$$

where λ is the length of the light wave (in free space); T is the thickness of the recording medium; n_c is the refractive index of the medium; $\alpha_\circ, \gamma_n = \arctan(x_n/f)$ is the angle of incidence of the reference and signal (regeneration) planar waves, respectively (they are read from the optical axis). Fig. 4.11 shows the dependence $\Delta\Psi$ calculated for $\lambda = 0.5145 \cdot 10^{-6}$ m, $T = 3 \cdot 10^{4}$ m, $n_c = 1.5$, and $\alpha_\circ = 65^\circ\ 70^\circ, 80^\circ$, along which the parameters of the CO processor are determined with respect to the requirements for angular selectivity. Thus, for example, when the distance between the rows of the space–time light modulator is $\Delta x_n = 1.0 \cdot 10^{-3}$ m, $f = 0.35$ m, the processor may contain $2N + 1 = 20$ ($\alpha_\circ = 65^\circ$), 80 ($\alpha = 75^\circ$), and 150 ($\alpha_\circ = 80^\circ$) rows in the space–time light modulator (of the array circles); in Fig. 4.11 the algorithm for determining the number $2N + 1$ rows is shown as dashed lines.

4.3.4 The axially symmetric electrooptical array

Let us examine some of the features of these arrays. The mask (4.34) [see also (4.55)] as a Fourier transform of a finite partial amplitude phase distribution (4.32) is an analytic function [2]. This significantly simplifies the mask. We will reproduce, for example, only its real part $\Re e\{\dot{T} = 0.5(\dot{T} + T)$, which, according to (4.34), corresponds to the partial amplitude frequency distribution $\mathcal{F}_\varphi^{-1}\{\Re e\{\dot{T}\} = 0.5[\dot{\mathcal{J}}_\varphi(\varphi, \Omega_Z) + \overset{*}{\mathcal{J}_\varphi}\ (\varphi, \Omega_Z)]$ where we consider the evenness of the amplitude phase distribution (4.32) relative to φ. By virtue of (4.6) and (4.19) such an amplitude phase distribution is realized by

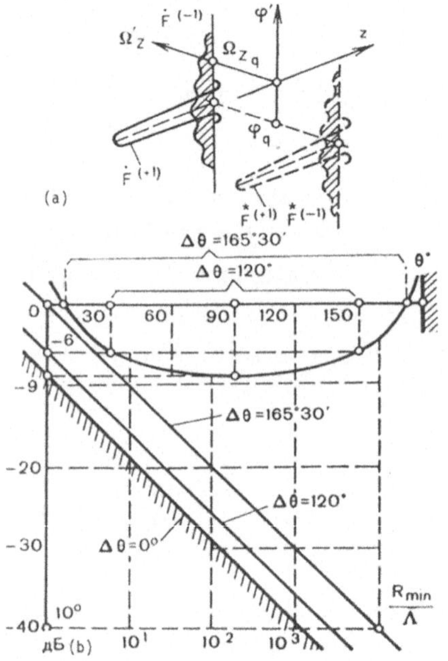

Figure 4.12: Estimating conjugate image defocusing in nonplanar arrays.

the following processing algorithm:

$$\hat{L}_{\Re e}\{\ldots\} = \frac{1}{2}\hat{L}\{\ldots\} + \frac{1}{2}\overset{*}{\dot{L}}\{\ldots\} \tag{4.60}$$

which, as opposed to (4.33), with a bipolar input, according to (4.17)–(4.21), forms a superposition of four pairs of conjugate images (pairs of focused images $\dot{F}^{+1}$, $\overset{*}{F}{}^{+1}$ and pairs of defocused images $\dot{F}^{-1}$, $\overset{*}{F}{}^{-1}$; see Fig. 4.12a). The latter results in an ambiguity of $\Omega'_Z = \pm\Omega_Z$, which is absent in a CO processor with a complex mask (4.34). Its exclusion is possible because of the methods outlined in Section 5.2.4.

Image defocusing (4.17) and (4.21) may be estimated from the relative deterioration in the gain of the imaging kernel (4.18). Using the equivalent planar aperture [133], we find

$$D_{\dot{F}(+1)}/D_{\dot{F}(-1)} \quad < \quad \left|\int_0^{\pi/2} \exp[i2KR\sin\theta\cos\varphi]\cos\varphi d\varphi\right|^2$$

$$\approx \frac{\Lambda}{8|\sin\theta|R_{min}} \tag{4.61}$$

where the stationary phase method is used ($\varphi = 0$ is the stationary point) [2], since for all practical purposes $KR_{min} \gg 2\pi$; $R_{min} = min R(Z)$ is the minimal radius of an axially symmetric array. The extent of defocusing (4.61) is shown in Fig. 4.12b depending upon the minimal effective radius of the array and the scanning sector $\Delta\theta$. This value evaluates the defocusing of a matched image with a dual band input and in the case of other nonplanar electrooptical arrays with equivalent curve radii.

Let us note one more circumstance. In addressing a space–time signal of these arrays in a space–time light modulator by the law $\varphi = m_y y$ [see (4.35)], splitting of the image of object p with the coordinate φ into parts separated from each other by a full azimuth angle 2π (Fig. 4.13a) is possible, if its azimuthal coordinate φ_p is within the limits $\pi/2 < |\varphi_p| \leq \pi$, and the length of the space–time light modulator does not exceed $\Delta y = 2\pi/m_y$. Image splitting occurs because the aperture response of the array in this kind of addressing is "cut apart" at the ends of the space–time light modulator ($\pm\pi \pm m_y\Delta y/2$) into two parts with the coordinates of their centers φ_p and $\varphi_p + 2\pi$. These parts during subsequent treatment in the CO processor with an algorithm of the type (4.33) are focused at the corresponding points (φ_p and $\varphi_p + 2\pi$) of the processor (φ') output plane. This circumstance may cause an indefiniteness in finding an object and a decrease in the gain (Fig. 4.13a) due to the splitting of the operating surfaces of the array into two disconnected parts.

This problem is eliminated by a multientry addressing ($m_y > 2\pi/\Delta_y)$), for example, double entry where one array element is addressed by two space–time light modulators ($m_y = 4\pi/\Delta_y)$). Since this lowers by a corresponding number of times the diffraction efficiency of the processor and complicates the space–time light modulator, it is more effective to use sesqui-entry addressing ($m_y = 3\pi/\Delta_y$), see Figs. 4.13b; 5.8c).

4.4 Phase Correction Method

This coherent optical method forms the antenna pattern of nonplanar arrays, which have configurations that are different from the ones we have discussed. In essence, this method is close to the holographic method of correcting phase errors in the focal region of mirror antennas that occur due to the deformations or defects in the manufacture of the mirror [73,152]. The method avoids the more significant phase errors without disrupting the

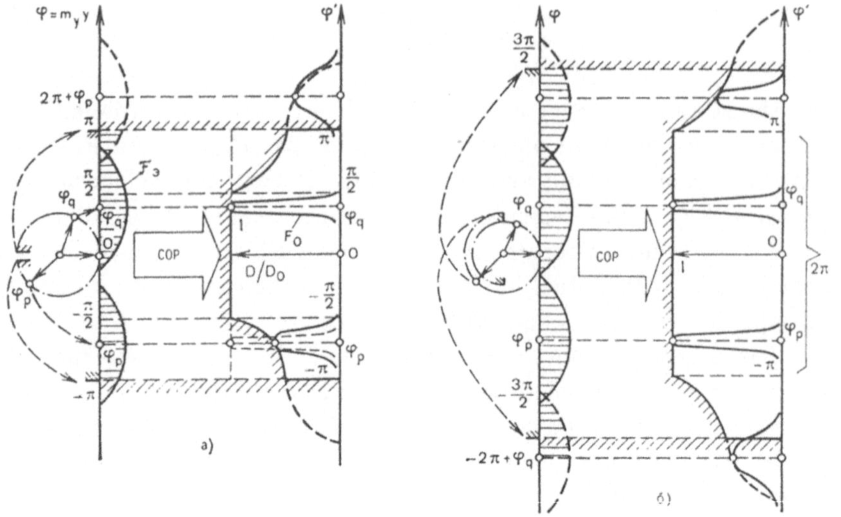

Figure 4.13: Addressing defect: (a)single entry addressing; (b) sesqui-entry addressing.

parallel spatial scanning; the method is described in detail by the authors [153–155]. The examination was conducted using the example of a planar array with random deformations of a certain size. The method, however, may also be applied to a convex planar array with an arbitrary initial surface, for example a cylindrical one, with correction for both phase and amplitude distortion. This expands the class of nonplanar arrays for the formation of a continuous beam of a highly directional radiation pattern by means of coherent optics and holography.

Chapter 5

Discreteness Effects in Planar and Nonplanar Electrooptical Antenna Arrays

In the preceding chapters we examined the pattern-forming and spectrum-forming properties of linear, planar, and nonplanar electrooptical arrays using a model of the continuous array aperture and the modeling medium of the space–time light modulator. This approach enabled us to analyze the general laws governing the formation of the spatial and spectral characteristics of electrooptical arrays, using the methods of coherent optics. The discreteness of the aperture in real arrays and space–time light modulators (see Figs. 1.3 and 1.6) gives rise to a number of new effects, which influence their pattern-forming and spectrum-forming properties. In this chapter we set forth some ways of simplifying the processors of Chapters 1–4; the basis of the approach is the invariance and redundancy of the processors with respect to the spatially discrete signal of the array. We will investigate the degradation of the pattern-forming properties of these processors due to pupil effects and the interference between channels in the space–time light modulator, and we describe ways of compensating these effects [137, 156].

5.1 Arbitrary Nonplanar Electrooptical Antenna Array

Let the receiving elements of an array (see Fig. 4.1) be located at points $\mathbf{R}_n$ of surface (4.1) and be characterized by the partial vector radiation pattern $\mathcal{F}_e(\mathbf{K})$ (which in the general case is a complex pattern) and by the maximum gain g_n; these quantities can be measured or calculated with allowance for interference (see, e.g., [51, 133, 157]). Then, as the image-forming kernel of

(4.4), (4.3) can include the vector radiation pattern of the array [52]:

$$\dot{\mathbf{F}}_{AA}(\mathbf{K},\mathbf{K}') = \langle \mathcal{F}_e(\mathbf{K},\mathbf{R}_n)\exp(-i\,\mathbf{K}\mathbf{R}_n)][\dot{\mathcal{J}}(\mathbf{R}_n,\mathbf{K}')\rangle \tag{5.1}$$

where

$$\mathcal{F}_e(\mathbf{K},\mathbf{R}_n) = i\sqrt{A_n}\mathcal{F}_n(\mathbf{K}) \qquad A_n = \Lambda^2 g_n/4\pi$$

is the effective surface of the nth element, and the change in matching of the receiving element from the reception direction $\mathbf{K}'$ is taken into account in $\mathcal{F}_n(\mathbf{K})$; $\langle\ldots]$ and $[\ldots\rangle$ are a row matrix and a column matrix, respectively.

In particular, for a linearly polarized planar array consisting of identical, evenly-spaced receiving elements, the radiation pattern of an array element in the approximation of an infinite electrodynamic array has the form [51]

$$\mathcal{F}_n(\mathbf{K}_\perp) = \mathcal{F}(\mathbf{K}_\perp) = \mathbf{p}(\mathbf{K}_\perp)[\cos\theta(1-|\Gamma(\mathbf{K}_\perp)|^2]^{1/2}\exp[i\arg(1-\Gamma)]$$

where $\Gamma(\mathbf{K}_\perp)$ is the coefficient of reflection from an element in transmission mode, which takes account of the interrelation between elements, and $\mathbf{p}(\mathbf{K}_\perp)$ is the polarization vector.

According to Section 4.1.2, the image-forming kernel (5.1) insures the acceptability of image (4.3) with respect to such parameters as resolution, directivity, and polarization efficiency. The amplitude-phase distribution $\dot{\mathcal{J}}(\mathbf{R}_n,\mathbf{K}')$ in (5.1) is generally different from the amplitude-phase distribution in (4.4) because of interference between elements in the finite planar and nonplanar arrays.

Substituting (5.1) into (4.3), we obtain by analogy with (4.5)

$$\hat{L}_{AA}\{\ldots\} = \langle\ldots]\cdot[\dot{\mathcal{J}}(\mathbf{R}_n,\mathbf{K})\rangle = \sum_{n=1}^{N}\ldots\dot{\mathcal{J}}(\mathbf{R}_n\mathbf{K}') \tag{5.2}$$

where the matrix $\langle\ldots]$ denotes the directions (4.2) to the elements at points $\mathbf{R}_n$ and N is the total number of elements in the array. Similarly, we can derive an equivalent variant of operator (5.2):

$$\overset{*}{\hat{L}}_{AA} = \langle\ldots]\cdot[\overset{*}{\mathcal{J}}(\mathbf{R}_n\mathbf{K}')\rangle \tag{5.3}$$

which, in distinction to (5.9) but similar to (4.19), has the property that, for a two-band input, it forms a focused image of the radio scene—of the type of (4.20) [not of (4.3)]—in the form of the -1 diffraction order, and a defocused image (4.21) [not (4.17)] in the form of the $+1$ order.

Let us define, on the array surface (4.1), a Dirac delta or impulse "train" $[\delta(\mathbf{R}-\mathbf{R}_n)\rangle$, which has the following filter properties [5]:

$$\hat{L}\{[\delta(\mathbf{R}-\mathbf{R}_n)\rangle\} = [\dot{J}(\mathbf{R}_n,\mathbf{K}')\rangle \tag{5.4}$$

Then, using (5.4), we can write operator (5.2) as

$$\hat{L}_{AA}\{\ldots\} = \iint_{\Sigma}\sum_{n=1}^{N}\ldots\delta(\mathbf{R}-\mathbf{R}_n)\dot{J}(\mathbf{R},\mathbf{K}')\mathrm{d}^2\mathbf{R}$$

$$= \langle\ldots]\cdot\hat{L}\{[\delta(\mathbf{R}-\mathbf{R}_n)\rangle\} = \hat{L}\{\langle\ldots]\cdot[\delta(\mathbf{R}-\mathbf{R}_n)\rangle\} \tag{5.5}$$

Similarly, we can write its alternative (5.3) as

$$\overset{*}{\hat{L}}_{AA}\{\ldots\} = \overset{*}{\hat{L}}\{\langle\ldots]\cdot[\delta(\mathbf{R}-\mathbf{R}_n)\rangle\} \tag{5.6}$$

The resulting expressions (5.5) and (5.6) link the algorithms for processing the array signal for continuous $\hat{L}(\overset{*}{\hat{L}})$ and discrete $\hat{L}_{AA}$ $(\overset{*}{\hat{L}}_{AA})$ arrangements of the receiving elements. Thus the desired coherent optical processing of the discrete signals $\langle\ldots] = \langle\mathcal{E}_n] = (\mathcal{E}_1,\ \mathcal{E}_2,\ldots,\ \mathcal{E}_N)$ reduces the processing with algorithm (4.6) [or (4.19)], using the corresponding CO processor synthesized in the continuous-aperture approximation (see Chapter 4), provided the space–time signal controlling such a processor is weighted with the impulse train defined on the array surface at each array element. The approximate impulse train may comprise a discrete input device in the form of an N-channel space–time light modulator in which the channel widths are sufficiently narrow and are placed in accordance with the addressing rule and the positioning of the elements. For example, the channels of the space–time light modulator of a planar electrooptical array should have coordinates $\mathbf{r}_{\perp m} = \widehat{m}^{-1}\mathbf{R}_{\perp n}$ [see (2.2)], while the modulator channels of a cylindrical array have coordinates $x_n = m_x^{-1}Z_n$ and $y_n = m_y^{-1}\varphi_n$ [see (4.35)].

We should note that algorithms (5.5) and (5.6) cover all special cases of element placement on the array surface (equidistant and nonequidistant); this fact shows that the starting algorithms (4.6) and (4.9) are invariant, that is, the processors corresponding to them are invariant with respect to the position of the radiators.

From (5.5) and (5.6) it also follows that all the pattern-forming properties of the electrooptical array (accuracy, energy, and range characteristics) remain the same as those studied above in the continuous-aperture approximation. Along with this generality, however, the discrete character of the element configuration gives rise to several features that affect the array's pattern-forming properties.

5.1.1 Redundancy of the starting processing algorithm

By virtue of the sifting property of the delta function (5.4), processing algorithms (5.5) and (5.6) do not change if we replace the starting operators (4.6) and (4.19) by another $\hat{L}'$ ($\hat{L} \neq \hat{L}' \neq \hat{L}$) in which

$$\dot{J}'(\mathbf{R},\mathbf{K}')\begin{cases} = \dot{J}(\mathbf{R},\mathbf{K}') & \text{for} \quad \mathbf{R}=\mathbf{R}_n \\ \neq \dot{J}(\mathbf{R},\mathbf{K}') & \text{for} \quad \mathbf{R}\neq\mathbf{R}_n \end{cases} \tag{5.7}$$

This fact shows that the starting algorithms (4.6) and (4.19) are redundant in relation to the required one (5.5) and (5.6). As a result, in order to simplify the processor corresponding to it, we can modify the algorithm $\hat{L}'$ because relation (5.7) allows some freedom in the selection of the amplitude-phase distribution outside the array elements. The processor may then lose its invariance to the configuration of the radiators, but as a rule, no such invariance is necessary. In Sections 5.2 and 5.3, this modification is illustrated in the example of the reduction (simplification) of the configuration of the starting CO processors of planar and cylindrical electrooptical arrays.

5.1.2 Effect of finite channel widths on the space–time light modulator

The impulse train in the form of an N-channel space–time light modulator is an approximate one, since each channel of the device is a distributed structure and is described not by a delta function but by a narrow pulse of finite width $\dot{J}_\delta(\mathbf{R}-\mathbf{R}_n)$ (for convenience, this function is tied to the array coordinate system). In place of (5.5) we therefore have

$$\tilde{\hat{L}}_{AA}\{\ldots\} = \hat{L}\{\langle\ldots]\cdot[J_\delta(\mathbf{R}-\mathbf{R}_n)\rangle\}$$

$$= \langle\ldots]\cdot[\tilde{J}(\mathbf{R}_n,\mathbf{K}')\rangle \neq \hat{L}_{AA}\{\ldots\} \qquad (5.8)$$

where

$$\tilde{J}(\mathbf{R}_n,\mathbf{K}') = \hat{L}\{J_\delta(\mathbf{R}-\mathbf{R}_n)\} \neq \dot{J}(\mathbf{R}_n,\mathbf{K}') = \hat{L}\{\delta(\mathbf{R}-\mathbf{R}_n)\} \qquad (5.9)$$

As can be seen from (5.9), by approximating the impulse train as a series of narrow pulses, the space–time light modulator distorts the given amplitude-phase distribution, and thus, the required CO processing algorithm. A loss of quality in the output image (4.3) inevitably results, since the directional properties of its image-forming kernel radiation pattern (5.1), are degraded. This effect can obviously be diminished by narrowing the modulator channel widths, that is, by increasing the accuracy with which the impulse train is approximated. This obvious approach is not, however, energetically appropriate, since it lowers the diffraction efficiency of the processor $[(DE)_{COP}$ is the ratio of the luminous flux at the processor output in the radio-frequency and visible ranges respectively] as found with the formula (see Appendix B):

$$(DE)_{COP} = \frac{\iint_{4\pi} |\tilde{\hat{L}}_{AA}\{\langle\dot{\mathcal{E}}_\Omega(\mathbf{R}_n)]\}|^2 d^2\mathbf{K}'}{\iint_{-\infty}^{\infty} |\tilde{\hat{L}}_{AA}\{\langle\dot{\mathcal{E}}_\Omega(\mathbf{R}_n)]\}|^2 d^2\mathbf{K}'} \approx \frac{K^\delta_{su}}{\gamma_{AA}} \qquad (5.10)$$

where K^δ_{su} is the coefficient of surface utilization of the space–time light modulator and γ_{AA} is the reactivity coefficient of the array [see, respectively, (B.3) and (B.4)].

Let us note that, to calculate the overall efficiency at which laser energy is utilized, we must take into account not only the diffraction efficiency of the processor (5.10), but also that of the modulator, the effect of masking the luminous flux that carries information from the modulator to the processor (this effect lowers the surface utilization of the modulator), and the efficiency of the entire optical system.

As (B.3) shows, $K^\delta_{su} \longrightarrow 0$ as the channels grow narrower, so that $(DE)_{COP} \longrightarrow 0$. In some cases, therefore, it is desirable to compensate this effect. To do this we introduce, for example, a special correction algorithm

$\hat{L}^I$ and require that [see (5.9)]

$$\tilde{J}(\mathbf{R}_n, \mathbf{K}') = \hat{L}^I\{J_\delta(\mathbf{R} - \mathbf{R}_n)\}$$
$$= \iint_\Sigma J_\delta(\mathbf{R} - \mathbf{R}_n) \cdot \dot{J}^I(\mathbf{R}, \mathbf{K}')\mathrm{d}^2\mathbf{R} = \dot{J}(\mathbf{R}_n, \mathbf{K}) \quad (5.11)$$

The solution of equation (5.11) for the amplitude-phase distribution $\dot{J}^I(\mathbf{R}, \mathbf{K}')$, which appears in $\hat{L}^I$, is given in Section 5.3 for the example of a cylindrical electrooptical array.

5.1.3 Interference between channels in the modulator

The space–time light modulator (whether it utilizes the acoustooptic or the electrooptical effect) is a system of closely spaced channels. In order to enhance the surface utilization factor of the modulator, it is desirable to pack the channels as close together as possible, but doing so increases the interference between channels (through either acoustic [79] or quasi-stationary electromagnetic [95] interaction) and thus leads to undesirable distortion of the required processing algorithm.

Suppose the system of interference coefficients between the channels of a two-dimensional (2-D) space–time light modulator $[C_{mn}]$ is known and is such that the actual voltages imposed on the modulator channels are

$$\langle\tilde{\dot{\mathcal{E}}}_\Omega(\mathbf{R}_n)] = \langle\dot{\mathcal{E}}_\Omega(\mathbf{R}_m)] \cdot [C_{mn}] \neq \langle\dot{\mathcal{E}}_\Omega(\mathbf{R}_n)] \quad (5.12)$$

where $[C_{mn}]$ is the square matrix of interference coefficients with $C_{mn} = C_{nm} = \mid C_{mn} \mid$ and $C_{mm} = C_{nn} = 1$ (since the interference between near channels in the modulator is quasi-stationary).

Substituting (5.12) into (5.2), we obtain

$$\hat{L}_{AA}\{\langle\tilde{\dot{\mathcal{E}}}_\Omega(\mathbf{R}_n)]\} = \langle\tilde{\dot{\mathcal{E}}}_\Omega(\mathbf{R}_m)] \cdot [C_{mn}] \cdot [\dot{J}(\mathbf{R}_n, \mathbf{K}')\rangle$$

$$= \langle\dot{\mathcal{E}}_\Omega(\mathbf{R}_m)][\tilde{\dot{J}}(\mathbf{R}_m, \mathbf{K}')\rangle = \tilde{\tilde{\hat{L}}}_{AA}\{\langle\dot{\mathcal{E}}_\Omega(\mathbf{R}_m)]\} \neq \hat{L}_{AA}\{\langle\dot{\mathcal{E}}_\Omega(\mathbf{R}_m)]\} \quad (5.13)$$

where

$$[\tilde{\dot{J}}(\mathbf{R}, \mathbf{K}')\rangle = [C_{mn}] \cdot [\dot{J}(\mathbf{R}_n, \mathbf{K}')\rangle \quad (5.14)$$

is the amplitude-phase distribution distorted by interference. The interference between modulator channels, like the pupil effect (Section 5.12), distorts the given amplitude-phase distribution and thus the required coherent optical processing algorithm. This effect is obviously weakened if the modulator channels are spaced farther apart ($C_{mn} \longrightarrow 0$ if $|\mathbf{r}_{\perp m} - \mathbf{r}_{\perp n}| \longrightarrow \infty$). This approach, however, is energetically unfavorable, since it reduces the surface utilization factor of the modulator and thus the diffraction efficiency of the processor. The distortion (5.14) can be compensated if instead of the starting amplitude-phase distribution, we use a corrected distribution

$$[\dot{J}^{II}(\mathbf{R}_m, \mathbf{K}')\rangle = [C_{mn}]^{-1} \cdot [\dot{J}(\mathbf{R}_n, \mathbf{K}')\rangle \tag{5.15}$$

where $[C_{mn}]^{-1}$ is the inverse of the matrix $[C_{mn}]$. If, along with the compensation (5.15), it is also necessary to compensate for the pupil effect (5.11), then instead of distribution (5.15) we must obviously use

$$[\dot{J}^{III}(\mathbf{R}_n, \mathbf{K}')\rangle = [C_{mn}]^{-1} \cdot [\dot{J}^{I}(\mathbf{R}_n, \mathbf{K}')\rangle \tag{5.16}$$

The implementation of the amplitude-phase distributions (5.15) and (5.16) is demonstrated in Sections 5.2 and 5.3 for examples of planar and cylindrical electrooptical antenna arrays.

Let us concretely illustrate the general principles developed in this section [137, 156].

5.2 Effects of Discreteness of Planar Electrooptical Antenna Arrays

In what follows, we systematically examine a simplification of the Fourier processor of a planar electrooptical array. Without degrading the pattern-forming properties of the array, this simplification yields a number of additional possibilities (see below). We also analyze the pupil effect and the effect of interchannel interference in the space–time light modulator.

5.2.1 Simplification of the Fourier processor

The processing algorithm for the space–time signal of the planar electrooptical array (2.7) can be written in the following way [2, 5]:

$$\hat{L}\{\ldots\} = \hat{\mathcal{F}}\{\ldots J(\mathbf{R}_\perp)\} = \exp\left[-\frac{if}{2k}(\widehat{m}\mathbf{K}_\perp)^2\right] \cdot \hat{L}_e\{\ldots\} \tag{5.17}$$

Here

$$\hat{L}_e\{\ldots\} = ||\widehat{m}|| \cdot [\ldots \dot{J}_e(\widehat{m}\hat{\mathbf{r}}_\perp)] \otimes \otimes \exp\left[\frac{ik}{2f}(\mathbf{r}_\perp)^2\right] \tag{5.18}$$

$$\dot{J}_e(\mathbf{R}_\perp) = J(\mathbf{R}_\perp) \exp\left[-\frac{ik}{2f}(\widehat{m}^{-1}\mathbf{r}_\perp)^2\right] \tag{5.19}$$

Since the starting processing algorithm (5.17) and the convolution operation (5.18) are identical up to a small phase multiplier, operator (5.18) is suitable as an equivalent algorithm. Expression (5.18) stands for the Fresnel transform, which describes the diffraction of light in a layer of space with thickness f [2]. The starting Fourier processor in Fig. 1.6 is therefore equivalent to a reduced CO processor consisting of a layer of space with half the thickness and a lens with transmittance (5.19) in the plane of the space–time light modulator. According to (5.7), it is quite possible to this lens with any mask with a transmittance equal to (5.19) just at points $\mathbf{R}_{\perp n}$. Such a mask might, for example, be a phase-only synthetic hologram [6, 143]. If the modulator channels are located on concentric circles along the antinodes of a Fresnel zone lens, even the phase-only synthetic hologram is not needed. This case, however, requires an appropriate placement of the array elements and the modulator channels.

The mask (5.19) in the equivalent processing algorithm (5.18) can also be realized at the array level [137]. This radio-frequency equivalent of a lens (consisting of phase shifters, sections of transmission line, and so forth) has a number of useful properties. Indeed, suppose that equivalent lens (5.19) has been made from sections of T-type transmission line (Fig. 5.1a) with length Z_m such that $-k/2f(\mathbf{r}_{\perp m})^2 = nKZ_m$ (where n is the refractive index of the medium filling the line). Then we can speak of an equivalent dispersing lens with an equivalent focal length of

$$f_e = \frac{k}{K}\aleph = \frac{2\pi\nu + \Omega}{\Omega} l \approx \frac{2\pi\nu}{\Omega}\aleph \tag{5.20}$$

where

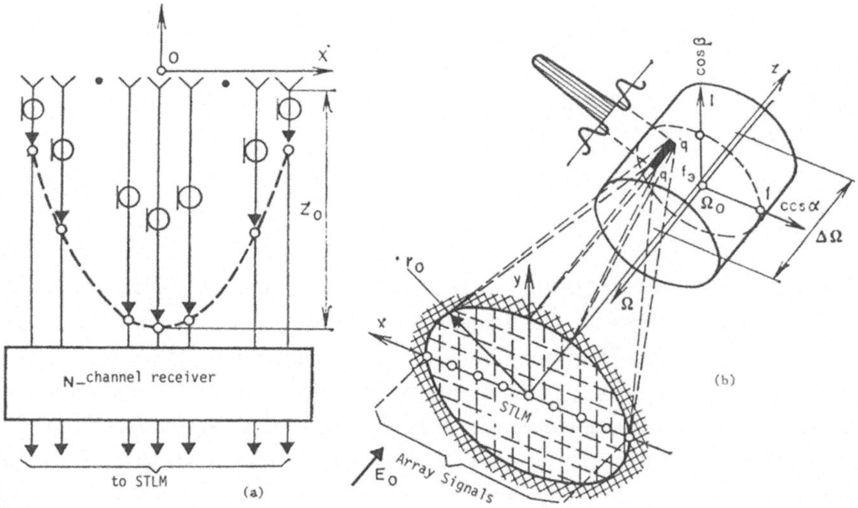

Figure 5.1: (a) Equivalent lens at the array level; (b) lensless CO processor.

$$\aleph = -\frac{(\mathbf{r}_{\perp m})^2}{2nZ_m} = -\frac{(\mathbf{r}_{\perp N})^2}{2nZ_N}$$

Using (2.15) and (5.20), we find the scale of the image in terms of the guiding cosines:

$$\left.\begin{array}{l} \cos \alpha x = lm_x l \\ \cos \beta y = lm_y l \end{array}\right\} \qquad (5.21)$$

As we see, the scale of the spectral plane of the reduced processor does not depend on the frequency, and the dispersion error (2.22) is not present. The frequency does, however, determine the position of the output plane $z = f_e$. If we imagine a 3-D screen (Fig. 5.1b), then in the x and y directions we see an image of the angular spectrum, undistorted by dispersion, while in the z direction we have the frequency spectrum of the radar sources. In other words, the geometrically undistorted 3-D image of the starting information volume shown in Fig. 2.3 is restored. To evaluate the resolution of the resulting 3-D image of the space–time spectrum in the frequency domain, we use the relation for the light distribution in the focal spot along the z-axis of the processor [1]:

$$\operatorname{sinc}\left[\frac{2r_\circ^2}{\lambda f_e}(z - f_e)\right] = \operatorname{sinc}\left(\frac{2r_\circ^2}{\lambda f_e}\frac{\delta\Omega}{\Omega}\right) = \operatorname{sinc}\left(\frac{4n}{c}Z_\circ \delta F\right) \tag{5.22}$$

where $Z_0 = r_0^2/2l$ (see Figs. 5.1a and b). From (5.24) we find that the resolving power with respect to nulls is

$$\delta F \leq c/2nZ_\circ \tag{5.23}$$

The conditions corresponding to estimate (5.23) are 100 MHz and $nZ_\circ =$ 1.5 m. This property of the lensless processor can be utilized, for example, to create coarse frequency selectivity for radar signals. Together with the circumstances noted, implementation of the amplitude-phase distribution (5.21) at the antenna-array level in the form of an equivalent lens offers yet another possibility. Indeed, in Section 2.1 it was shown that, for a two-band input of the space–time array signal to the Fourier processor, the field of view is reproduced as two equally intense ± 1 diffraction orders (2.10) differing only in frequency ($\omega = 2\pi\nu \pm \Omega$). Since the simplified processor with equivalent lens at the array level is frequency-selective, the ambiguity that arises in this situation can be eliminated. Indeed, by (5.20), the equivalent lens is convergent ($f_e > 0$) for the $+1$ diffraction order ($\Omega > 0$) and divergent ($f_e < 0$) for the -1 order ($\Omega < 0$) when $\kappa > 0$, and vice versa when $\kappa < 0$. In order to focus only the $+1$ order ($\kappa > 0$), we must lengthen the paths of all the radiators by $Z_\circ$ ($Z_0 + Z_n \geq 0$ for all n, Fig. 5.1a). Let us evaluate the defocusing of the -1 diffraction order by the "concave" equivalent lens at the array level ($\kappa > 0$). Since in this case $\Omega < 0$, in place of amplitude-phase distribution (5.19) we have

$$\begin{aligned} j_e^{(-1)}(\mathbf{R}_\perp) &= J(\mathbf{R}_\perp)\exp\left[\frac{ik}{2|f_e|}(\widehat{m}^{-1}\mathbf{R}_\perp)^2\right] \\ &= j^{(-1)}(\mathbf{R}_\perp)\exp\left[-\frac{ik}{2|f_e|}(\widehat{m}^{-1}\mathbf{R}_\perp)^2\right] \end{aligned} \tag{5.24}$$

where

$$j^{(-1)}(\mathbf{R}_\perp) = J(\mathbf{R}_\perp)\exp\left[-\frac{ik}{|f_e|}(\widehat{m}^{-1}\mathbf{R}_\perp)^2\right]$$

Representation (5.24) makes it possible to find directly the image-forming kernel of the optical image of the -1 diffraction order $F^{(-1)}$, which in the present case should figure in (2.7) instead of the kernel $F = F^{(+1)}$. By (1.1)

$$F^{(-1)}(\mathbf{K}_\perp) = \hat{\mathcal{F}}\{j^{(-1)}(\mathbf{R}_\perp)\} \neq F(\mathbf{K}_\perp) \tag{5.25}$$

If an array with a rectangular aperture is uniformly excited, then

$$F^{(-1)}(\widehat{m}^{-1}\mathbf{k}_\perp) = F_x^{(-1)}(m_x^{-1}\omega_x)F_y^{-1}(m_y^{-1}\omega_y) \tag{5.26}$$

Here

$$\begin{aligned} F_x^{(-1)}(m_x^{-1}\omega_x) &= m_x\sqrt{\frac{\lambda f_e}{4}}\exp\left(-\frac{if_e}{4k}\omega_x^2\right) \\ &\times\left\{\Phi\left[\left(\frac{f_e}{k}\omega_x - \Delta x\right)\sqrt{\frac{\pi}{2\lambda f_e}}\right] - \Phi\left[\left(\frac{f_e}{k}\omega_x + \Delta x\right)\sqrt{\frac{\pi}{2\lambda f_e}}\right]\right\} \end{aligned} \tag{5.27}$$

and similarly for $F_y^{(-1)}$, where

$$\Phi(x) = \sqrt{\frac{2}{\pi}}\int_0^x \exp(-iz^2)\mathrm{d}z$$

is the Fresnel integral and $\Delta x = \Delta X/m_x, \Delta y = \Delta Y/m_y$. Fig. 5.2a shows the focused $+1$ order and the defocused -1 order (of a point object) with envelope (2.14):

$$F_x^{(+1)}(m_x^{-1}\omega_x) = \hat{\mathcal{F}}_x\{\mathrm{rect}(X/\Delta X)\} = \Delta x\cdot\mathrm{sinc}(\omega_x\Delta x/2\pi) \tag{5.28}$$

and (5.27), respectively, where the wave dimension of the space–time light modulator is $\Delta x\sqrt{\lambda f_e} = \sqrt{2}$. Fig. 5.2b is a plot of the relation

$$|F^{(-1)}(0)/F^{(+1)}(0)|^2 = \frac{1}{4}\left(\frac{\lambda f_e}{\Delta x\Delta y}\right)^2 \tag{5.29}$$

which characterizes the extent of defocusing of the conjugate image of the 2-D electrooptical array. The same figure shows the case of a 1-D array, the evaluation of which follows in an obvious way from (5.29). We note that estimate (5.29) was actually derived in the continuous-aperture approximation, which is known to be quite accurate in the radar field of view $|\mathbf{K}_\perp| \leq K$ [52] and at the same time analytically simpler.

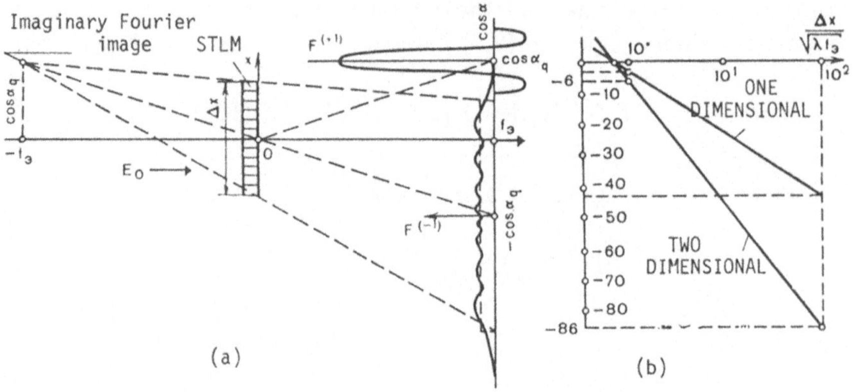

Figure 5.2: Defocusing of conjugate image in a reduced CO processor.

5.2.2 Pupil effect of modulator channels

Let the width of the channels of a space–time light modulator (Fig. 5.3) be described by the transmittance function

$$\mathcal{J}_\delta[\widehat{m}(\mathbf{r}_\perp - \mathbf{r}_{\perp n})] = \mathcal{J}_\delta(\mathbf{R}_\perp - \mathbf{R}_{\perp n})$$

which is tied to the array coordinate system [see (2.2)]. Then by (5.8) we have

$$\tilde{\hat{L}}_{AA}\{\ldots\} = \hat{\mathcal{F}}\{\langle\ldots]\cdot[\mathcal{J}_\delta(\mathbf{R}_\perp - \mathbf{R}_{\perp n})\rangle\}$$

$$= \hat{\mathcal{F}}\{\mathcal{J}_\delta(\mathbf{R}_\perp)\}\hat{\mathcal{F}}\{\langle\ldots]\cdot[\delta(\mathbf{R}_\perp - \mathbf{R}_{\perp n})\rangle\} = F_\delta(\mathbf{K}'_\perp)\hat{L}_{AA}\{\ldots\} \quad (5.30)$$

As we see, the fact that the modulator channels have a finite width leads to an envelope of the form

$$F_\delta(\mathbf{K}_\perp) = \hat{\mathcal{F}}\{\mathcal{J}_\delta(\mathbf{R}_\perp)\} \quad (5.31)$$

over the undistorted image of the field of view (2.7). If (5.31) is sufficiently uniform and does not have nulls in the radar field of view $|\mathbf{K}| \le K$, then the processing algorithm (5.30) is satisfactory as an approximation to operator

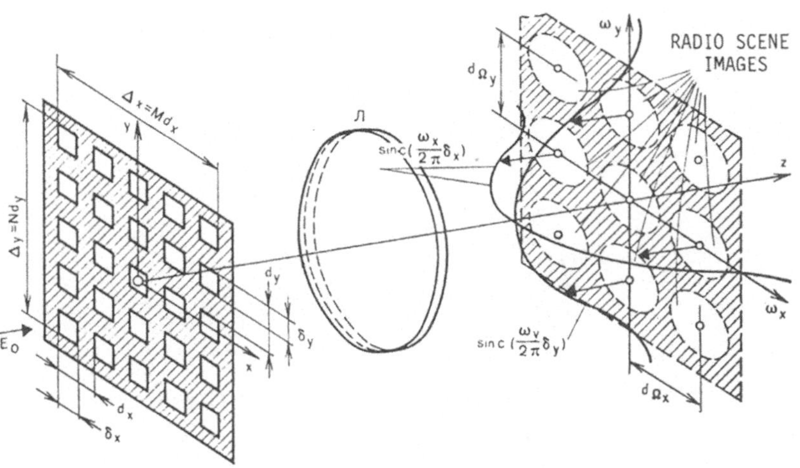

Figure 5.3: Pupil effect of modulator.

(5.5). What is more, if (5.31) falls off rapidly to zero outside the region cited, then algorithm (5.32), from an energy standpoint, is more favored than (5.5). Indeed, in Appendix B [see (B.8)], it is shown that the lower estimate for the diffraction efficiency (5.10) for a Fourier processor is

$$(DE)_{COP} \geq \left| \frac{F_{\delta min}}{F_\delta(0)} \right|^2 \frac{K^\delta_{su}}{\gamma_{AA}} \tag{5.32}$$

where $F_{\delta min}$ is the minimum value of envelope (5.32) in the region $|\mathbf{K}_\perp| \leq K$; K^δ_{su} is the surface utilization coefficient of the space–time light modulator, and γ_{AA} is the reactivity coefficient of the array.

Estimate (5.32) is plotted in Fig. 5.4a for the case of an equidistant modulator with identical and uniform channel widths

$$K^\delta_{su} = \frac{\delta x}{d_x} \frac{\delta y}{d_y}$$

for which envelope (5.31) is described by a function of the type of (5.28) (see Fig. 5.3). The resulting curve makes it possible to select the optimal modulator parameters for a given array spacing (Fig. 5.4b).

Relations (5.29)–(5.32) also hold for the reduced Fourier processor (Section 5.2.1). Indeed, by (5.9) and (5.18),

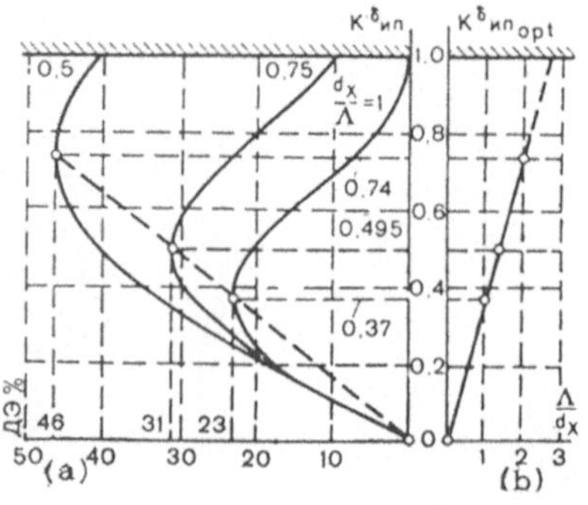

Figure 5.4: Diffraction efficiency of a CO processor.

$$\tilde{J}(\mathbf{R}_{\perp n}, \mathbf{K}') = \hat{L}_e\{J_\delta(\mathbf{R}_\perp - \mathbf{R}_{\perp n})\}$$

$$\approx ||\widehat{m}||\dot{J}_e(\mathbf{R}_{\perp n})\left\{J_\delta(\widehat{m}\mathbf{r}'_\perp - \mathbf{R}_{\perp n}) \otimes \otimes \exp\left[\frac{ik}{2f_e}(\mathbf{r}'_\perp)^2\right]\right\} \qquad (5.33)$$

Suppose, for a moderator channel width with size $\delta_x\delta_y$, the Fraunhofer diffraction condition is satisfied ($z = f_e \geq \delta_x\delta_y/\lambda \approx 0.15$ m for $\delta_x = \delta_y = 100$ μm and $\lambda = 0.63$ μm) [9]. Then the 2-D convolution in (5.33) becomes a 2-D Fourier transform:

$$\tilde{J}(\mathbf{R}_{\perp n}, \mathbf{K}'_\perp)$$

$$\approx ||\widehat{m}||\dot{J}_e(\mathbf{R}_{\perp n})\exp\left[\frac{if_e}{2k}(\mathbf{k}_\perp)^2\right]\hat{\mathcal{F}}\{J_\delta(\widehat{m}\mathbf{r}_\perp)\}\Big|_{\mathbf{k}_\perp = \widehat{m}\mathbf{K}'_\perp - k/f_e\widehat{m}^{-1}\mathbf{k}_{\perp n}}$$

$$\exp\left[\frac{if_e}{2k}(\widehat{m}K'_\perp)^2\right]\cdot F_\delta\left(\mathbf{K}'_\perp - \widehat{m}^{-1}\frac{k}{f_e}\mathbf{r}_{\perp n}\right)\cdot J(\mathbf{R}_{\perp n})\exp(-i\,\mathbf{K}'_\perp\mathbf{R}_{\perp n})$$

$$(5.34)$$

The biases of the envelope F_δ in (5.34) can be neglected provided the width of the envelope is substantially greater than the size of the light modulator (maximum bias; see Fig. 5.5); that is, we are assuming

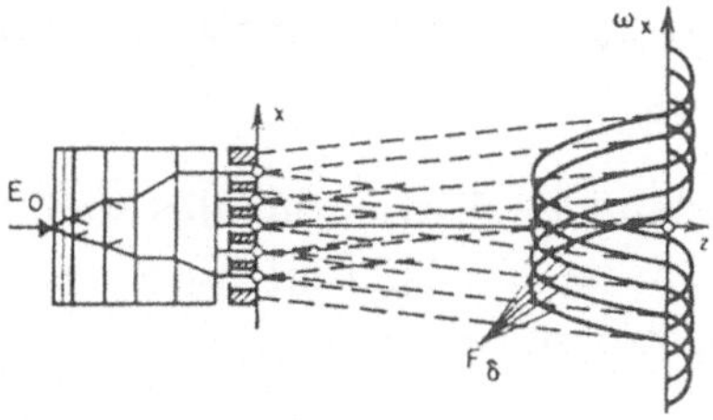

Figure 5.5: Pupil effect of modulator channels in reduced processor.

$$\Delta\Omega_{X,Y} \approx 2\pi/m_{x,y}\delta_{x,y} \geq 10\Delta x, y \cdot k/f_e m_{x,y}$$

which is equivalent to the following constraint on the width of the modulator channels:

$$\delta_{x,y} \leq 0.1\lambda f_e/\Delta x, y \tag{5.35}$$

which guarantees sufficient uniformity in the envelope F inside the radar field of view. Therefore we have the following in place of (5.34):

$$\tilde{J}(\mathbf{R}_{\perp n}, \mathbf{K}'_{\perp}) \approx \exp\left[\frac{if_e}{2k}(\widehat{m}\mathbf{K}'_{\perp})^2\right] F_\delta(\mathbf{K}'_{\perp}) J(\mathbf{R}_{\perp} n) \exp(-i\,\mathbf{K}'_{\perp}\mathbf{R}_{\perp n}) \tag{5.36}$$

Substituting (5.36) into (5.8), we obtain

$$\tilde{\hat{L}}_{AA}\{\ldots\} \approx \exp\left[\frac{if_e}{2k}(\widehat{m}\mathbf{K}'_{\perp})^2\right] F_\delta(\mathbf{K}'_{\perp}) \hat{L}_{AA}\{\ldots\} \tag{5.37}$$

As we see, processing algorithms (5.30) and (5.37) agree up to a small phase multiplier.

Since requirement (5.35) means that the simplified processor must employ a space–time light modulator with a low surface utilization factor, the reduced diffraction efficiency of the processor makes it desirable to use, instead of a collimator, a diverging system of optical radiation, such as that described in [95].

5.2.3 Effect of interference between modulator channels

By (5.14), the amplitude-phase distribution distorted by the interference between channels of the space–time light modulator has the following form in the case of a planar electrooptical array:

$$
\begin{aligned}
\tilde{\dot{J}}(\mathbf{R}_{\perp m}, \mathbf{K}'_{\perp}) &= \sum_{n=1}^{N} C_{mn} \dot{J}(\mathbf{R}_{\perp n}, \mathbf{K}'_{\perp}) \\
&= \sum_{n=1}^{N} C_{mn} \dot{J}(\mathbf{R}_{\perp n}) \exp(-i\, \mathbf{K}'_{\perp} \mathbf{R}_{\perp n}) \\
&= F_C(\mathbf{R}_{\perp m}, \mathbf{K}'_{\perp}) \dot{J}(\mathbf{R}_{\perp m}, \mathbf{K}'_{\perp}) \qquad (5.38)
\end{aligned}
$$

where

$$
\begin{aligned}
F_C(\mathbf{R}_{\perp m}, \mathbf{K}'_{\perp}) &= \sum_{n=1}^{N} C_{mn} \frac{J(\mathbf{R}_{\perp m})}{J(\mathbf{R}_{\perp n})} \exp[i\, \mathbf{K}'_{\perp} (\mathbf{R}_{\perp n} - \mathbf{R}_{\perp m})] \\
&\approx \sum_{n=1}^{N} C_{mn} \exp(i\, \mathbf{K}'_{\perp} \Delta \mathbf{R}_{\perp mn})
\end{aligned}
$$

If the number of regularly spaced modulator channels is sufficiently great, then since the coefficients C_{mn} decrease with increasing $(m-n)$ [95] we can replace the last expression by the approximation

$$
F_C(\mathbf{R}_{\perp n}, K'_{\perp}) \approx F_C(\mathbf{K}'_{\perp}) = \sum_{n=-\infty}^{\infty} C_{0n} \exp(i\, \mathbf{K}'_{\perp} \mathbf{R}_{\perp n}) \qquad (5.39)
$$

Thus, by (5.14), (5.38), and (5.39), we have

$$
\tilde{\tilde{\hat{L}}}_{AA}\{\ldots\} \approx F_C(\mathbf{K}'_{\perp}) \hat{L}_{AA}\{\ldots\} \qquad (5.40)
$$

As we see, interference at the modulator level, like the pupil effect (5.30), leads to the appearance of an envelope (5.39) over the output image. In the case of a linear equidistant modulator (Fig. 5.6), it is desirable to employ the approximation

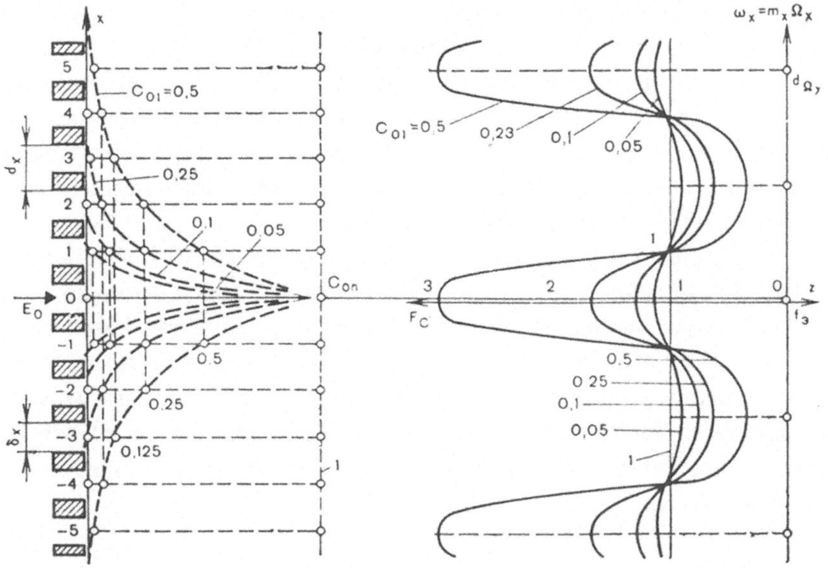

Figure 5.6: Effect of interference between modulator channels.

$$C_{0n} = \exp(-\mu|n|) \tag{5.41}$$

where μ is the rate of attenuation of the interference, which is determined experimentally or theoretically. Then from (5.41) and (5.39) it follows that

$$\begin{aligned} F_C(\mathbf{K}_\perp) &= F_C(\Omega_X) = \sum_{-\infty}^{\infty} \exp(-i\Omega_X n d_x - \mu|n|) \\ &= \left(1 - C_{01}^2\right) / [\sin^2(\Omega_X d_X) + (\cos\Omega_X d_X - C_{01})^2] \end{aligned} \tag{5.42}$$

The envelope (5.42) oscillates about unity (see Fig. 5.6), with a period equal to the period of the radiation pattern of the equidistant array [see (B.6)] and an amplitude that is proportional to the interference coefficient for adjacent channels. In (5.42) and Fig. 5.6, the envelope is represented as a function of the space frequencies (Ω_X) of the antenna array. Real space–time light modulators (such as the one described in [95]) have $C_{01} \sim 0.05 - 0.1$, so that the envelope (5.42) experiences oscillations of 30%. In contrast to (5.31), envelopes (5.39) and (5.42) do not lead to a concentration of light energy inside the radar field of view; this fact has virtually no effect on the estimation of the processor's diffraction efficiency (5.32).

If the effects discussed in Sections 5.2.2 and 5.2.3 lead to substantial distortion in the output image, then the correcting amplitude-phase distribution (5.16) must be used. This is equivalent to inserting a correcting mask in the output plane of the processor, with a transmittance function inverse to the product of envelopes (5.31) and (5.39).

In conclusion let us note that the method set forth in Sections 5.2.2 and 5.2.3 for calculating the pupil and interference effects can also be used when we analyze the same effects in an electrooptical array based on multichannel AOMs (Section 1.2.2), where "channel width" is taken to mean the transmittance function of a single channel of the acoustic waveguide (with allowance for diffractional spreading of the acoustic wave), while the interaction coefficients are determined with allowance for both electrical and acoustical "steering" [79].

5.2.4 Methods of eliminating ambiguity in the determination of coordinates

In Sections 2.1 and 3.1 it was noted that for two-band input of the array signal to the CO processor, there is an ambiguity of the form $\mathbf{K}'_\perp = \pm\mathbf{K}_\perp$. At the same time, equidistant arrays are characterized by periodic radiation patterns (B.6), which also bring in some indefiniteness. In order to insure that, for $|\mathbf{K}'_\perp| \leq K$, only a single principal maximum of the type of (B.6) will appear in the radiation pattern at the output of the CO processor, it is necessary first to satisfy the requirement $d_x/\Lambda = d_Y/\Lambda \leq 0.5$, and second to eliminate or suppress to an appropriate level one of the conjugate images (2.9). Complete elimination of the conjugate image is possible with single-band input [89] or with coherent collection of space–time information. In the first case, however, single-band modulators of complex design (an unwieldy system of a pair of two-band electrooptical modulators with orthogonal crystallographic axes and with separate control of the quadrature voltages) are necessary. The second case requires a complex superheterodyne photodetector consisting of a heterodyne laser with a frequency $\nu_h \neq \nu$ (the frequency shift $\nu_h - \nu$ is carried out by a special single-band light modulator), a semitransparent mirror and a photomatrix (see Section 8.3). The most attractive approach to an equivalent lens at the array level (Section 5.2.1), which makes it possible to suppress (defocus) the conjugate image to an extent given by (5.29), entails the "cost" of simplifying the processor, but complicating the array feeder system (see Fig. 5.1). Fig. 5.7a shows schematically the defocusing of the -1 diffraction order at the output of the planar equivalent

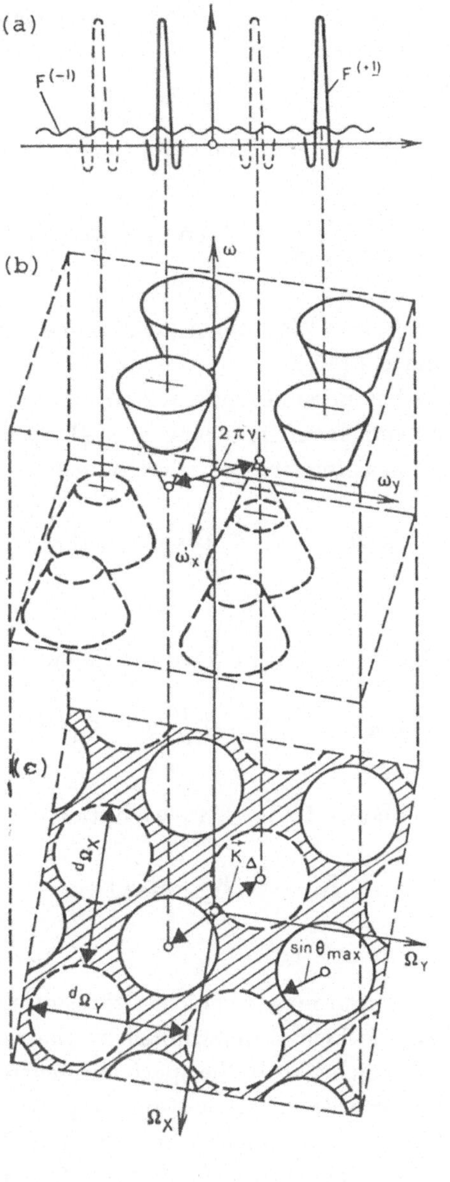

Figure 5.7: Geometric interpretation.

electrooptical array. In contrast to (5.26), the radiation pattern here is a periodic function of the type of (B.6):

$$\begin{aligned} F_{AA}^{(-1)}(\mathbf{K}_\perp) &= \hat{\mathcal{F}}\{\mathcal{J}^{(-1)}(\mathbf{R}_\perp)\mathrm{comb}(\mathbf{R}_\perp/d_X d_Y)\} \\ &= \frac{1}{(2\pi)^2 d_X d_Y} F^{(-1)}(\mathbf{K}_\perp) \otimes \otimes \mathrm{comb}\left(\mathbf{K}_\perp / d_{\Omega_X} d_{\Omega_Y}\right) \end{aligned} \quad (5.43)$$

It can be shown [52: 250] that the ratio of the maxima of the functions (5.43) and (B.6) virtually matches (5.29).

There is one other possible approach to eliminating the ambiguity: biasing with a spatial subcarrier frequency $\mathbf{K}_\Delta$ [91, 123]. For this purpose, a phase delay with a linear distribution is introduced at the array level:

$$\mathcal{J}_\Delta(\mathbf{R}_\perp) = \mathcal{J}(\mathbf{R}_\perp)\exp(i\,\mathbf{K}_\Delta \mathbf{R}_\perp) \quad (5.44)$$

Then, by the bias theorem, we have the following in place of (2.10):

$$\dot{e}^{(+1)}_{2\pi\nu+|\Omega|}[\widehat{m}(\mathbf{K}'_\perp - \mathbf{K}_\Delta)] = \overset{*}{e}^{(-1)}_{2\pi\nu-|\Omega|}\,[-\widehat{m}(\mathbf{K}'_\perp - \mathbf{K}_\Delta)] \quad (5.45)$$

and in place of the ambiguity $\mathbf{K}'_\perp = \pm\mathbf{K}_\perp$ we obtain

$$\mathbf{K}'_\perp = \pm(\mathbf{K}_\perp + \mathbf{K}_\Delta) \quad (5.46)$$

As we see, for an appropriate choice of the subcarrier $\mathbf{K}_\Delta$ and of the coverage zone $\theta \leq \theta_{max}$ (which is determined by the radiation patterns of the array elements), we can separate the space frequencies of the space–time spectrum (5.45) and prevent their intersection. In particular, if the elements of the electrooptical array are arranged in a rectangle (Fig. 1.6), the radiation pattern of the array (B.6) also has a rectangular periodicity. Thus the information volumes of the space–time spectra (5.45) have the form of periodically repeating volumes of the space–time spectrum (2.17) (see Figs. 5.7b and 2.3b). For single-band input or coherent collection, one of the halves of these volumes (e.g., the lower one, indicated by the dashed outline of the cone) is absent. But if defocusing has been done, then the complete contents of this half has been "smeared" into an information "layer" $\Delta\Omega$. From geometric arguments (Fig. 5.7c), it follows that for an azimuthally symmetric coverage zone

$$\begin{cases} \mathbf{K}_\Delta & = & (\mathbf{n}_X d_{\Omega_X} + \mathbf{n}_Y d_{\Omega_Y})/4 \\ \sin\theta_{max} & = & |\mathbf{K}_\Delta|/K \end{cases} \tag{5.47}$$

the densest packing of nonintersecting images in the zone is produced at the processor output. Given this kind of packing, the maximum dimension for an unambiguously resolved coverage zone corresponds to $d_X = d_Y$ and equals

$$\sin\theta_{max} \leq \frac{d_{\Omega X}}{4K} = \frac{\pi\sqrt{2}}{2Kd_X} \leq \frac{\Lambda_{max}}{2\sqrt{2}d_X} \tag{5.48}$$

If $d_X/\Lambda_{max} = 0.5$, then $\theta_{max} = 45$.

In the case of a non-rectangular mesh of radiator positions, we obtain a similar requirement on the coverage zone. In order to eliminate reliably the ambiguity in either case, we must use array receiving elements with radiation patterns $2\theta_{max}$ wide.

5.3 Effects of discreteness of cylindrical electrooptical arrays

The same analysis as for planar arrays is now applied to a cylindrical electrooptical array. The reduction of the CO processor (preserving the pattern-forming properties) is analyzed, the pupil effect and the effect of interference between modulator channels are investigated, and results are presented to characterize the CO processor of the cylindrical array [137].

5.3.1 Reduction of the coherent optical processor of cylindrical arrays

In the optical system of the starting CO processor (i.e., the processor synthesized in the continuous-aperture approximation; see Fig. 4.4b) we can distinguish three parts: $0 \leq z \leq 2f$, the Fourier processor; $z = 2f$, the uncontrolled mask; and $2f < z \leq z_{out}$, the astigmatic system consisting of a 1-D Fourier processor and a 1-D imaging system.

The reduction of the Fourier processor was discussed in Section 5.2.1. The equivalent spherical lens (Fig. 5.1a) can be subdivided into two cylindrical lenses (L_x, L_y), and these can be realized at different levels, for example, the L_x lens at the modulator level and the L_y lens at the array level.

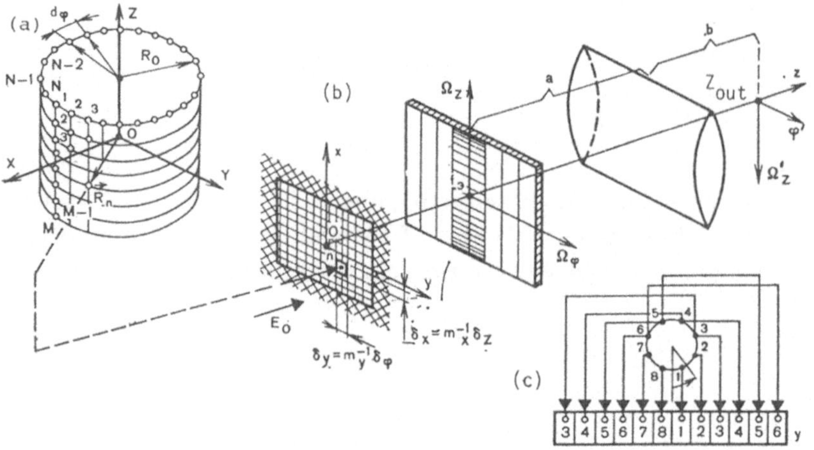

Figure 5.8: simplified processor for cylindrical array: (a) cylindrical array, (b) diagram of CO processor, (c) sesqui-entry addressing.

Because the partial amplitude-phase distributions can be varied in azimuth (4.3.2) outside the array elements, a substantial reduction is also permitted by the uncontrolled complex mask (4.34). Suppose, for example, the receiving elements of the array are equidistant in azimuth with a spacing of $d_\varphi = 2\pi/N$ (where N is the number of elements in a ring; see Fig. 5.8a).

In place of (4.32) let us form a new partial amplitude-phase distribution in azimuth, in the form of a Kotel'nikov series in functions of the readings at the element locations (Fig. 5.9a)

$$\dot{j}^I_\varphi(\varphi, \Omega_Z) = \sum_{n=-\infty}^{\infty} \dot{J}_\varphi(nd_\varphi, \Omega_Z)\mathrm{sinc}(\varphi/d_\varphi - n) \tag{5.49}$$

which matches the starting distribution at these points and is different elsewhere by virtue of a property of the function of the readings sinc $(m-n) = \delta_{mn}$ (δ_{mn} is the Kronecker delta [2, 5]). Then, by (5.7), in place of (4.34) we can employ the mask

$$\begin{aligned} \dot{T}^I(\Omega_\varphi, \Omega_Z) &= \hat{\mathcal{F}}_\varphi\{\dot{J}^I(\varphi, \Omega_Z)\} \\ &= \sum_{-\infty}^{\infty} \dot{J}_\varphi(nd_\varphi, \Omega_Z) \cdot \hat{\mathcal{F}}_q\{\mathrm{sinc}(\varphi/d_\varphi - n)\} \end{aligned}$$

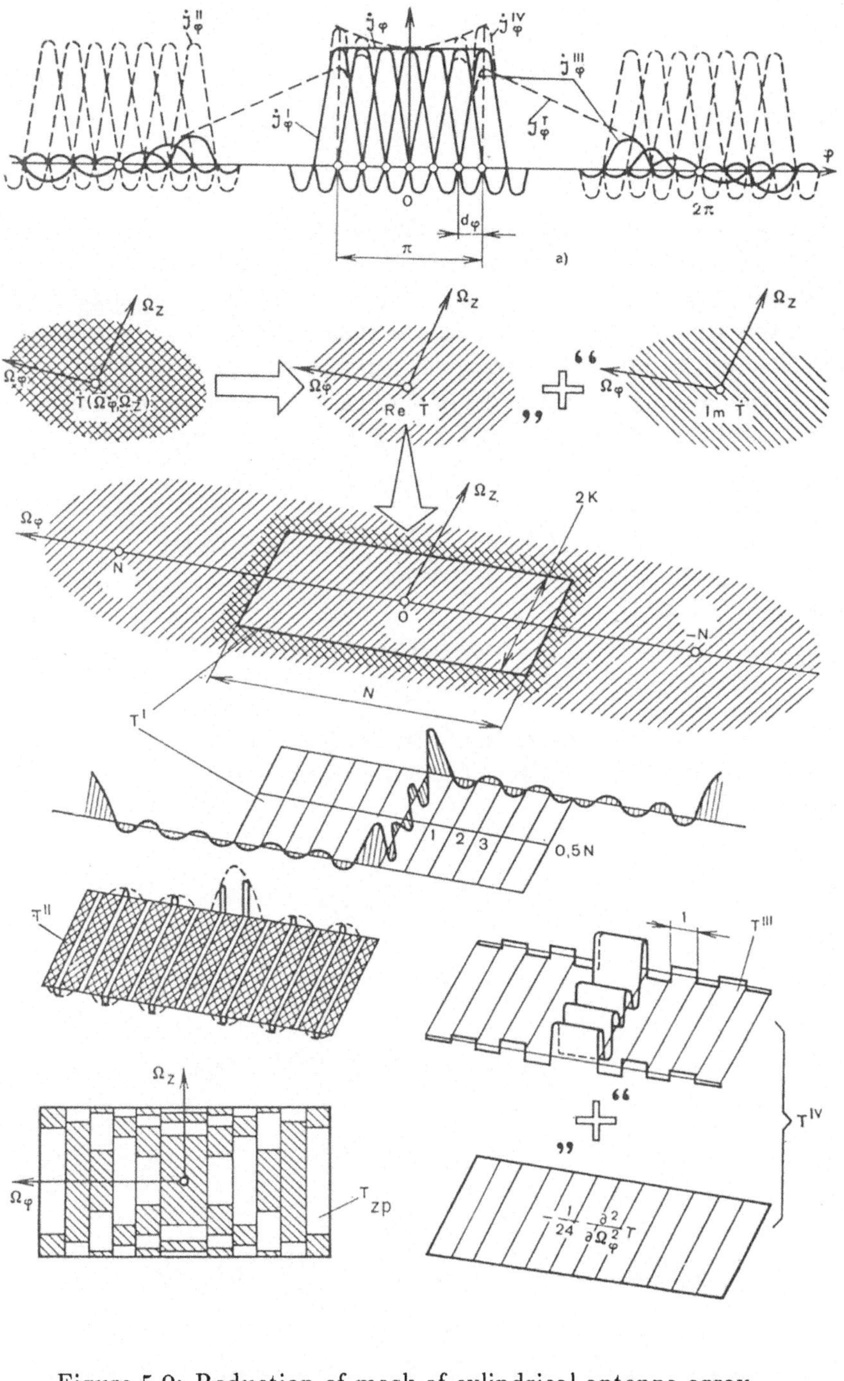

Figure 5.9: Reduction of mask of cylindrical antenna array.

$$= \frac{1}{N}\mathrm{rect}(\Omega_\varphi/N) \cdot \sum_{-\infty}^{\infty} \dot{J}_\varphi(nd_\varphi, \Omega_Z)\exp(i\Omega_\varphi nd_\varphi)$$

$$= \mathrm{rect}(\Omega_\varphi/N) \cdot \sum_{-\infty}^{\infty} \dot{T}(\Omega_\varphi + nN, \Omega_Z) \qquad (5.50)$$

where we have used the Poisson summation formula [2]. As we can see, mask (5.50) has a finite extent, in contrast to (4.43), and lies entirely within the window $-N/2 \le \Omega_\varphi \le N/2, -K \le \Omega_Z \le K$. In terms of angle measurement, this window is no wider than the radar field of view; in azimuth, it is restricted by the number of array elements in the ring (Fig. 5.9b).

To continue the reduction of the mask, let us form in place of (5.49) a new amplitude-phase distribution (Fig. 5.9a)

$$\dot{J}_\varphi^{II}(\varphi, \Omega_Z) = \sum_{m=-\infty}^{\infty} \dot{J}_\varphi^{I}(\varphi + 2\pi m, \Omega_Z)$$

$$= \dot{J}_\varphi(\varphi, \Omega_Z) \otimes \mathrm{comb}(\varphi/2\pi) \qquad (5.51)$$

which is a periodic continuation of distribution (5.49) every 2π and also satisfies condition (5.7). In place of (5.50) we can therefore employ a mask of the form

$$\dot{T}^{II}(\Omega_\varphi, \Omega_Z) = \hat{\mathcal{F}}_\varphi\{\dot{J}_\varphi^{II}(\varphi, \Omega_Z)\}$$

$$= \hat{\mathcal{F}}_\varphi\{\dot{J}_\varphi^{I}(\varphi, \Omega_Z)\} \cdot \hat{\mathcal{F}}_\varphi\{\mathrm{comb}(\varphi/2\pi)\}$$

$$= \frac{1}{2\pi}\mathrm{comb}(\Omega_\varphi) \cdot \dot{T}^{I}(\Omega_\varphi, \Omega_Z) \qquad (5.52)$$

where

$$\mathrm{comb}(\Omega_\varphi) = \hat{\mathcal{F}}_\varphi\{\mathrm{comb}(\varphi/2\pi)\} = \sum_{n=-\infty}^{\infty} \delta(\Omega_\varphi - n)$$

Thus mask (5.50) need not be reproduced in its entirety; it is sufficient to sample it along the lines $\Omega_\varphi = 0, \pm 1, \ldots, \pm 0.5N$ (a total of $N + 1$ lines). We can arrive at a similar conclusion using the Kotel'nikov theorem [5]. Mask (5.52) is weighted with a Dirac delta train (comb); in other words, it is physically unrealizable ($|\dot{T}(n, \Omega_Z)| \longrightarrow \infty$). Thus, in order to be able to realize the mask in practice, we increase the width of the samples of the mask (5.52) to a maximum value ($1/N$ of the mask size $\dot{T}^{II}$, Fig. 5.9b), and this increases its diffraction efficiency:

$$\dot{T}^{III}(\Omega_\varphi, \Omega_Z) = \sum_{-\infty}^{\infty} T_\delta(\Omega_\varphi - n)\dot{T}^I(n, \Omega_Z) \tag{5.53}$$

where

$$T_\delta(\Omega_\varphi) = \begin{cases} 1 & |\Omega_\varphi| \leq \frac{1}{2} \\ 0 & |\Omega_\varphi| > \frac{1}{2} \end{cases}$$

The approximate mask (5.53) distorts the starting amplitude-phase distribution (4.32), which in the present case takes on the form (see Fig. 5.9a)

$$\begin{aligned} \dot{J}_\varphi^{III}(\varphi, \Omega_Z) &= \hat{\mathcal{F}}_\varphi^{-1}\{\dot{T}^{III}(\Omega_\varphi, \Omega_Z)\} \\ &= \sum_{-\infty}^{\infty} \hat{\mathcal{F}}_\varphi^{-1}\{T_\delta(\Omega_\varphi - n)\} \cdot \dot{T}^I(n, \Omega_Z) \\ &= \hat{\mathcal{F}}_\varphi^{-1}\{T_\delta(\Omega_\varphi)\} \sum_{-\infty}^{\infty} \dot{T}^I(n, \Omega_Z) \exp(i\, n\, \varphi) \\ &= J_\varphi^T(\varphi) \cdot \sum_{-\infty}^{\infty} \dot{J}_\varphi^I(\varphi + 2\pi n, \Omega_Z) \\ &= J_\varphi^T(\varphi) \cdot J_\varphi^{II}(\varphi, \Omega_Z) \end{aligned} \tag{5.54}$$

where

$$J_\delta^T(\varphi) = \hat{\mathcal{F}}_\varphi^{-1}\{T_\delta(\Omega_\varphi)\}$$

is the envelope (see Fig. 5.9a) distorting the starting distribution so that condition (5.7) is not satisfied. If, however, $J_\varphi^T \approx 1$ for $|\varphi| \leq \pi/2$, then distribution (5.54) satisfies (5.7).

In particular, for $T_\delta = \text{rect}(\Omega_\varphi)$ we have $J_\delta^T(\varphi) = \text{sinc}(\varphi/2\pi) \approx 1 - \varphi^2/24 \geq 0.9$ everywhere in the interval $|\varphi| \leq \pi/2$. If this much nonuniformity (approximately 10%) in the envelope is unacceptable, for example because it reduces the directivity of the electrooptical array, then T_δ should be narrowed, a step that is energetically not favored, or else (5.54) should be replaced with the correcting distribution (see Fig. 5.9a), which on multiplication by the envelope J_φ^T gives the starting amplitude-phase distribution:

$$\dot{j}^{IV} = \dot{J}_\varphi^{III}(\varphi, \Omega_Z)/J_\delta^T(\varphi) \approx \dot{J}_\varphi^{III}(\varphi, \Omega_Z)(1 + \varphi^2/24) \qquad (5.55)$$

Corresponding to the last distribution, according to (4.34), is the mask

$$\begin{aligned} \dot{T}^{IV}(\Omega_\varphi, \Omega_Z) &= \mathcal{F}_\varphi^{-1}\left\{\dot{J}_\varphi^{IV}(\varphi, \Omega_Z)\right\} \approx \dot{T}^{III}(\Omega_\varphi, \Omega_Z) \\ &+ \sum_{-\infty}^{\infty} \text{rect}(\Omega_\varphi - n)\frac{1}{24}\frac{\partial^2}{\partial^2\Omega_\varphi}\dot{T}(\Omega_\varphi, \Omega_Z)\Big|_{\Omega_\varphi = n'} \end{aligned} \qquad (5.56)$$

where we have taken into account that the approximation

$$1/J_\delta^T(\varphi) = 1/\text{sinc}(\varphi/2\pi) \approx 1 + \varphi^2/24$$

is valid on the interval $|\varphi| < \pi/2$.

The above reduction of the complex mask (4.34) is, according to Section 4.3.4, also valid for the real (imaginary) mask $\Re e\{\dot{T}\}$ ($\Im m\{\dot{T}\}$). This fact is reflected in Fig. 5.9b.

The masks $\Re e\{\dot{T}^{III}\}$ and $\Re e\{\dot{T}^{IV}\}$ (or $\Im m\{\dot{T}\}$ are structurally simple. They represent $N+1$ parallel strips of width 1 and length $2K$, uniformly transparent along Ω_φ (see Fig. 5.9b). We recall that such masks are appropriate only for arrays with a selected radiator layout[1] (see Fig. 5.8a). They do not degrade the pattern-forming properties of the processor (provided steps have been taken to eliminate the ambiguity; see Section 5.2.4) and are

[1] If the radiators are not equidistant in the ring, the functions of the readings in (5.49) should be replaced by more complicated irregular functions of the readings, which are equal to 1 at the positions of the respective elements and 0 at the locations of the other elements [2].

easier to realize. In distinction to the photometric and holographic methods of generating mask (4.34), they can be reproduced, for example, by the method of silhouette functions [158] or, as was proposed in [159], by writing strips (5.53) individually in the form of binary silhouettes of function (5.50).

A focusing property also belongs to the zone plate $T_{zp} = \Re e\{\dot{T}\}/ \mid \Re e\{\dot{T}\} \mid$ (see Fig. 5.9b), whose binary transmittance corresponds to the sign changes of the real (imaginary) parts of the mask $\dot{T}^{III}$ (or $\dot{T}^{IV}$). This zone plate distorts the starting amplitude-phase distribution $\dot{J}_{\varphi}$ and does not satisfy (5.7), but it is useful as the simplest variant realization.

The astigmatic system of the starting processor (see Fig. 4.4) can also be simplified in the interval $2f < z \leq z_{out}$. Fig. 5.8b shows a lensless version of the 2-D Fourier processor in the interval $0 \leq z < f_e$ (by the equivalent lens method; see Section 5.2.1); a reduced real mask consisting of a system of parallel strips ($z = f_e$); an astigmatic system consisting of a 1-D Fourier processor (in the direction of Ω_{φ}); a layer of space $a+b$ and an equivalent lens in the mask plane with focal length $a + b$;[2] and the imaging system (in the direction of Ω_Z) using the cylindrical lens L_x $(1/f_x = 1/a + 1/b)$. Fig. 5.8c illustrates in detail the addressing of the signal from discrete axisymmetric arrays in the space–time light modulator (see Section 4.3.4 and Fig. 4.16), where the pattern-forming properties at all azimuthal angles are not violated.

5.3.2 Pupil effect of modulator channels

Suppose the width of a modulator channel of a cylindrical array (Fig. 5.8b) is described by the transmittance function

$$J_{\delta}(\mathbf{R} - \mathbf{R}_n) = J_{\varphi\delta}(\varphi - \varphi_n) J_{Z\delta}(Z - Z_n) \tag{5.57}$$

where $J_{\varphi\delta}(\varphi)$ and $J_{Z\delta}(Z)$ are the transmittance functions of the modulator pupil in the corresponding directions, which are tied to the array coordinate system [see (4.35)], $\varphi_n = m_y y_n$ is the azimuthal coordinate of the nth element of the array (y_n is the coordinate of the nth modulator channel), and $Z_n = m_x x_n$ is the z coordinate of the nth array element. Then, using (4.33), from (5.9) we obtain

$$\begin{aligned} &\tilde{\mathcal{J}}(\mathbf{R}_n, \mathbf{K}') \\ &= R_o \hat{\mathcal{F}}_{\varphi}^{-1}\{\hat{\mathcal{F}}\{J_{\varphi\delta}(\varphi - \varphi_n) J_{Z\delta}(Z - Z_n) J_Z(Z)\} \cdot \dot{T}(\Omega_{\varphi}, \Omega'_Z)\} \end{aligned}$$

[2] The lens is realized, for example, by appropriately correcting the phase transmittance function of the starting complex mask before reduction.

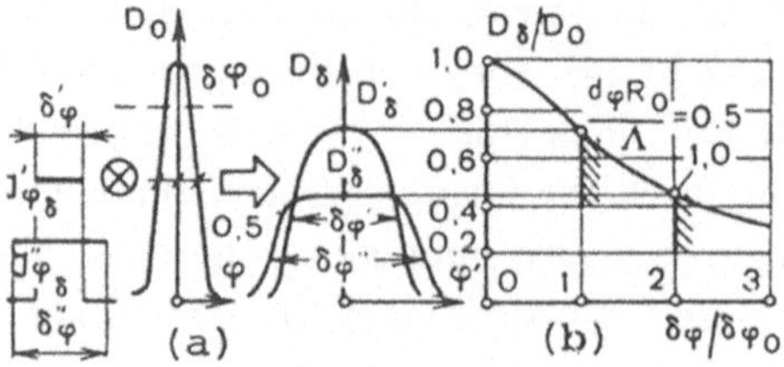

Figure 5.10: Pupil effect of modulator in CO processor of cylindrical array.

$$= F_{Z\delta}(\Omega'_Z)\{J_{\varphi\delta}(\varphi') \otimes_\varphi \dot{J}[\mathbf{R}_n, \mathbf{K}(\varphi', \Omega'_Z)]\} \tag{5.58}$$

where

$$F_{Z\delta}(\Omega_Z) = \hat{\mathcal{F}}\{\hat{J}_{Z\delta}(Z)\}$$

and we have used the commutative property of the displacement (translation) of the convolution components [5]. Substituting (5.58) into (5.8), we have

$$\tilde{\hat{L}}_{AA}\{\ldots\} = F_{Z\delta}(\Omega'_Z) J_{\varphi\delta}(\varphi') \otimes_\varphi \hat{L}_{AA}\{\ldots\} \tag{5.59}$$

Thus the pupil effect in the present case is characterized by the fact that the envelope $F_{Z\delta}$ is superimposed on the image of the field of view (4.3), which is simultaneously convoluted with the "azimuthal" pupil $J_{\varphi\delta}$. The first effect hardly reduces the quality of the optical image of the field of view; on the contrary, it leads to a gain in the diffraction efficiency of the process (see Section 5.1.2). The second effect, however, obviously [2] degrades the array's resolving power (broadens the radiation pattern) in the azimuthal direction ($\delta\varphi_\circ$). The wider the pupil (δ_φ), the greater the degradation (Fig. 5.10a). Let us evaluate the degradation with respect to the relative loss of array directivity, using the processing algorithm (5.59):

$$D_\delta / D_\circ \approx \frac{\delta\varphi_\circ}{\delta\varphi} \approx \frac{\sqrt{\sigma_\circ^2}}{\sqrt{\sigma^2}} = \frac{\sqrt{\sigma_\circ^2}}{\sqrt{\sigma_\circ^2 + \sigma_\delta^2}} \approx 1/\sqrt{1 + (\delta_\varphi/\delta\varphi_\circ)^2} \tag{5.60}$$

where we have used a relation of the type of (2.25) [52] ($\delta\varphi$ is the azimuthal width of the distorted radiation pattern) and taken into account the fact that the variance of the convolution σ^2 equals the sum of the variations of its components (variance of the undistorted radiation pattern σ_o^2 and variance of the pupil function σ_δ^2) [2]. As Fig. 5.10b shows, the loss of directivity of a cylindrical electrooptical array can be neglected only if $\delta_\varphi \ll \delta\varphi_o$, which not only limits the surface utilization factor of the space–time light modulator and the diffraction efficiency of the processor (5.10), but is also extremely difficult to insure for a narrowly directional electrooptical array ($\delta\varphi_o \longrightarrow 0$). This difficulty can be overcome only if we use the correcting algorithm (5.11). The amplitude-phase distribution $\dot{J}^I$ appearing in the algorithm can be found with an integral equation of the form (5.58), where instead of the distorted distribution $\tilde{J}$ we must use the required $\dot{J}$, and in place of the starting $\dot{J}$ we must use the desired $\dot{J}^I$. The formal solution of such an equation is known and, by the convolution theorem, is

$$J^I(\mathbf{R}_n, \mathbf{K}') = \dot{J}(\mathbf{R}_n, \mathbf{K}') \otimes_\varphi \hat{\mathcal{F}}_\varphi^{-1}\{F_\delta(\Omega_\varphi)\}/F_{Z\delta}(\Omega'_Z) \tag{5.61}$$

where

$$F_{\varphi\delta}(\Omega_\varphi) = \hat{\mathcal{F}}_\varphi\{J_{\varphi\delta}(\varphi)\}$$

According to (4.34), such an amplitude-phase distribution and thus the correcting algorithm (5.11) are realized with the help of an auxiliary mask

$$\tilde{T}(\Omega_\varphi, \Omega_Z) = \dot{T}(\Omega_\varphi, \Omega_Z) F_{\varphi\delta\mathrm{min}} F_{Z\delta\mathrm{min}} / F_{\varphi\delta}(\Omega_\varphi) F_{Z\delta}(\Omega Z) \tag{5.62}$$

where $F_{\varphi\delta min}$ and $F_{Z\delta min}$ are the minimum values of the respective envelopes inside the boundaries of the mask.

The relations and recommendations derived in this section remain in force for the simplified processor of Fig. 5.8b provided constraint (5.35) on the pupil widths of the space–time light modulator is observed.

5.3.3 Effect of interference between modulator channels

Since real two-dimensional modulators are fabricated from electrically insulated linear arrays (rulers), it is desirable to evaluate this effect separately in each of the independent directions (4.35): either in the direction $x = m_x^{-1} Z$

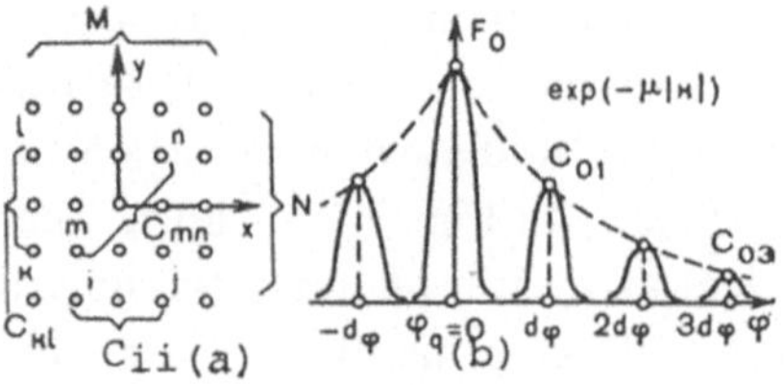

Figure 5.11: Effect of interference between modulator channels in CO processor of cylindrical array.

or in the direction $y = m_y^{-1}\varphi$, depending on the orientation of the indicated rulers.

In the first case, the amplitude-phase distribution distorted by the interaction between modulator channels has, by (5.14), the form (for equidistant channels, see Fig. 5.11a)

$$\begin{aligned}\tilde{\mathcal{J}}(\mathbf{R}_i,\mathbf{K}) &= \sum_{j=1}^{M} C_{ij}\dot{\mathcal{J}}(\mathbf{R}_j,\mathbf{K}') \\ &= \sum_{j=1}^{M} C_{ij}J_Z(Z_j)\exp(-i\Omega_Z Z_j)\dot{J}_\varphi(\varphi_i-\varphi',\Omega'_Z) \\ &= F_C(Z_i,\Omega'_Z)\cdot\dot{\mathcal{J}}(\mathbf{R}_i,\mathbf{K}') \qquad (5.63)\end{aligned}$$

where similar to (5.39)

$$F_C(Z_i,\Omega'_Z)\approx F_{ZC}(\Omega'_Z)=\sum_{j=-\infty}^{\infty} C_{0j}\exp(i\Omega'_Z Z_M) \qquad (5.64)$$

Thus, by (5.13), (5.63), and (5.64), we have

$$\hat{\tilde{\tilde{L}}}_{AA}\{\ldots\}\approx F_{ZC}(\Omega'_Z)\hat{L}_{AA}\{\ldots\} \qquad (5.65)$$

A distortion of the form (5.65) was discussed in Section 5.1.3. For equidistant linear space–time light modulators with interference law (5.41), envelope (5.64) is described by a relation of the type of (5.42), as shown in Fig. 5.6.

In the second case (linear space–time light modulators oriented along the y axis, see Fig. 5.11a), we have

$$\begin{aligned}\tilde{\tilde{J}}(\mathbf{R}_k,\mathbf{K}') &= \sum_{l=1}^{N} C_{kl}\dot{J}(\mathbf{R}_l,\mathbf{K}') \\ &= J_Z(Z)\exp(-i\,\Omega'_Z Z_k)\cdot\sum_{l=1}^{N} C_{kl}\dot{J}_\varphi(\varphi_k-\varphi',\Omega'_Z) \qquad (5.66)\end{aligned}$$

In the case of equidistant linear modulators ($\varphi_k = k\cdot d_\varphi$), by (5.13) and (5.66)

$$\hat{\tilde{\tilde{L}}}_{AA}\{\ldots\} = \hat{L}_{AA}\{\ldots\}\otimes_\varphi \sum_{l=1}^{N} C_{kl}\delta(ld_\varphi-\varphi') \qquad (5.67)$$

where δ is the delta function.

As we see, the interference between modulator channels in the azimuthal direction ($\varphi = m_y y$) leads to a periodic multiplication of the image of the field of view every $d_\varphi = 2\pi/N$ radians, the intensity declining in accordance with the law of interference [see, e.g., (5.41) and Fig. 5.11b].

This effect has an obvious physical explanation. The interference between channels in the space–time light modulator leads to multiplication of the signal from each element of the array, a few channels at a time, with an attenuation proportional to the decrease in interference. If the interference coefficients are commensurate with the side-lobe amplitude of the cylindrical array, this situation may give rise to an ambiguity of the type of $\varphi'_q = \varphi_q \pm ld_\varphi$ in the determination of the azimuth of a radiation source. Diminishing this effect by separating the modulator channels (decreasing the surface utilization factor) is energetically unfavorable, since it lowers the diffraction efficiency of the processor (5.10), which is fairly low because of losses in the mask. For example, the diffraction efficiency in the case of mask (4.38) is a factor of π^2 (an order of magnitude) lower than the diffraction efficiency of Fourier processor (5.32). In order to eliminate this unpleasant condition, we must use correcting amplitude-phase distribution (5.15). We can establish a method of realizing this distribution through the following arguments.

We multiply (5.14) by the transpose (5.15):

$$\langle \dot{J}^{II}(\mathbf{R}_m,\mathbf{K}')]\cdot[\tilde{\tilde{J}}^{II}(\mathbf{R}_m,\mathbf{K}')\rangle$$

$$= \langle \dot{J}(\mathbf{R}_n, \mathbf{K}')] \cdot [C_{mn}]^{-1} \cdot [C_{mn}] \cdot [\dot{J}(\mathbf{R}_n, \mathbf{K}')\rangle$$

where

$$\sum_{n=1}^{N} \dot{J}^{II}(\mathbf{R}_n, \mathbf{K}') \cdot \tilde{\tilde{J}}(\mathbf{R}_n, \mathbf{K}') = \sum_{n=1}^{N} \dot{J}^2(\mathbf{R}_n, \mathbf{K}') \tag{5.68}$$

If the linear space–time light modulators that make up the 2-D modulators are oriented parallel to the x-axis (see Fig. 5.11a), then by (5.63) and (5.64) we can turn Eq. (5.68) into an identity provided that

$$\begin{aligned} \dot{J}^{II}(\mathbf{R}_n, \mathbf{K}') &= \dot{J}^2(\mathbf{R}_n, \mathbf{K}')/F_{ZC}(\Omega'_Z)\dot{J}(\mathbf{R}_n, \mathbf{K}') \\ &= \dot{J}(\mathbf{R}_n, \mathbf{K}')/F_{ZC}(\Omega'_Z) \end{aligned} \tag{5.69}$$

If, however, the linear modulators are oriented parallel to the y-axis, then instead of (5.68) we have

$$\sum_{l=1}^{n} \dot{J}_{\varphi}^{II}(\varphi_l - \varphi', \Omega'_Z) \cdot \tilde{\tilde{J}}_{\varphi}(\varphi_l - \varphi', \Omega'_Z) = \sum_{l=1}^{N} \dot{J}_{\varphi}^{2}(\varphi_l - \varphi', \Omega'_Z) \tag{5.70}$$

Starting with (4.34), we further integrate (5.70) with respect to φ' and use the well-known integral relation [2]

$$\sum_l \int_{-\infty}^{\infty} \dot{T}^{II}(\Omega_\varphi, \Omega'_Z) \exp(-i\Omega_\varphi \varphi_l) \cdot \tilde{\tilde{T}}(\Omega_\varphi, \Omega'_Z) \exp(i\Omega_\varphi \varphi_l) d\Omega_\varphi$$

$$\sum_l \int_{-\infty}^{\infty} \dot{T}^{II}(\Omega_\varphi, \Omega'_Z) \exp(-i\Omega_\varphi \varphi_l) \cdot \dot{T}(-\Omega_\varphi, \Omega'_Z) \exp(i\Omega_\varphi \varphi_l) d\Omega_\varphi$$

which becomes an identity if

$$\dot{T}^{II}(\Omega_\varphi, \Omega'_Z) = \dot{T}^2(\Omega_\varphi, \Omega'_Z)/\tilde{\tilde{T}}(\Omega_\varphi, \Omega'_Z) \tag{5.71}$$

In the case of equidistant linear modulators,

$$\tilde{\tilde{T}}(\Omega_\varphi, \Omega'_Z) = \hat{\mathcal{F}}_\varphi\{\tilde{\tilde{J}}_\varphi(\varphi, \Omega'_Z)\} = F_{\varphi C}(\Omega_\varphi) \cdot \dot{T}(\Omega_\varphi, \Omega'_Z) \tag{5.72}$$

Here

$$F_{\varphi C}(\Omega_\varphi) = \hat{\mathcal{F}}_\varphi\left\{\sum_l C_{0l}\delta(ld_\varphi - \varphi)\right\} = \sum_{l=-\infty}^{\infty} C_{0l}\exp(i\Omega_\varphi ld_\varphi) \tag{5.73}$$

Substituting (5.72) into (5.71) and taking into account the normalization of the passive mask, we find

$$\tilde{\tilde{T}}(\Omega_\varphi, \Omega'_Z) = [F_{\varphi Cmin}F_{ZCmin}/F_{\varphi C}(\Omega_\varphi)F_{ZC}(\Omega_Z)]\dot{T}(\Omega_\varphi, \Omega'_Z) \leq 1 \tag{5.74}$$

where $F_{\varphi Cmin}$ and F_{ZCmin} are the minimum values of the respective envelopes within the boundaries of the mask. A mask of the form (5.74) located in the plane π of the processor in Fig. 4.4 compensates the effect of interference between channels in the space–time light modulator, in a way similar to the compensation of the pupil effect by mask (5.62). In order to eliminate these effects simultaneously, we must weight mask (5.74) with the second factor in (5.62).

Chapter 6

Rejection of Interference by Coherent Optical Methods

An electrooptical array panoramically covers a space by means of a solid beam of highly directional radiation patterns. In this type of coverage, images are simultaneously formed of all radio (acoustical) sources located in the region of coverage, including sources of active or passive noise. At high power levels,[1] the passive sources may "blind" the location system; this effect is manifest in that a background of bright optical noise image makes it difficult to detect the useful signals. In order to enhance the interference supression property of the electrooptical array, it is useful to generate a beam of radiation patterns minimized in the directions toward the noise sources (up to nulls). In phased-array systems that sequentially scan a space, this task is carried out by generating the needed amplitude phase distribution in the excitation of array elements [24, 134, 135, 160–163]. Such (aperture) methods have a number of shortcomings: First, the amplitude phase distribution necessary for this purpose is defined by solving a difficult synthesis problem; second, beam steering requires the synchronized reconstruction of the amplitude phase distribution in order to keep the gaps in the same directions as before; third, it is necessary to have equipment for controlling the signal phase and amplitude in all the array elements. What is more, such methods cannot be used in systems executing parallel surveillance of the space.

This chapter sets forth a fundamentally novel (spectral) method of generating controlled gaps in the beam of highly directional radiation patterns of an electrooptical array [38, 164–168]. The method is free of the drawbacks stated above, does not interfere with the parallel coverage regime,

[1] As before, the investigation concerns the linear regime of an electrooptical array, in which the intensity of the optical image of an object is proportional to its radio brightness.

and can be implemented with coherent optics. The essence of the method is the spatial filtering of optical images in the spectral plane of the coherent optical (CO) processor, the images being perturbed with the help of special controlled masks. The result is a redistribution of the optical noise image energy outside the space passband of the filter, and thus the suppression of noise at the processor output. It should be noted that this approach is somewhat similar to the methods of suppressing background noise in the reference component of holograms by generating an "anti-phase" hologram or by filtering [18: 89]. The method is described for an arbitrary nonplanar array and is illustrated in detail for the example of a planar array with various apertures. The results of experimental studies are presented in Chapter 9.

6.1 Statement of Problem. Choice of Criterion

The exposition, as in Chapters 2–4, is conveniently done by continuous aperture approximation (i.e., for a model of the array and the space–time light modulator in the form of continuous detecting and modulating media). This approximation can be generalized to the case of a discrete structure in accordance with the results of Chapter 5.

Suppose that an arbitrary convex array (see Fig. 4.1a) receives, in direction $\mathbf{K}_q$, a powerful blinding radio signal $|\dot{\mathbf{S}}(\mathbf{K}_q)|^2 d^2\mathbf{K}$ which we will take to be noise if it exceeds the useful signal from directions $\mathbf{K} \neq \mathbf{K}_q$ by more than the side-lobe amplitude. Then the image of form (4.3) at the output of a CO processor using algorithm (4.6) will be "fogged" by the noise image, and the signal-to-noise ratio (SNR) will be nearly zero (below the side-lobe level). Let us synthesize a processor (interference suppression) that maximizes the SNR and satisfies some additional conditions (see below). This problem can be solved in a formal way. For instance, the desired processor should be required to form an unblinded image of the field of view in the form of a solid beam, of the type of (4.3), consisting of interference supression radiation patterns with gaps in the directions $\mathbf{K}_q$:

$$\mathbf{F}_{\Pi}(\mathbf{K},\mathbf{K}') = \begin{cases} \mathbf{F}(\mathbf{K},\mathbf{K}'), & \mathbf{K} \neq \mathbf{K}_q \\ 0, & \mathbf{K} = \mathbf{K}_q, q = 1,\dots,Q \end{cases} \tag{6.1}$$

Then we solve the independent antenna synthesis problem [134, 135]: For a given radiation pattern (6.1) we find the amplitude phase distribution of array excitation $\dot{J}_{\Pi}(\mathbf{R},\mathbf{K}')$ (for all $\mathbf{K}' \in 4\pi$). We also form the operator

$$\hat{L}_{\Pi}\{\ldots\} = \iint_{\Sigma} \ldots \dot{J}(\mathbf{R}, \mathbf{K}')\mathrm{d}^2\mathbf{R} \tag{6.2}$$

which, like algorithm (4.6), generates a directive gain of acceptable resolution and a beam of radiation patterns (6.1) having acceptable polarization efficiency. Finally, we reproduce this algorithm (6.2) by a coherent optical method, using one of the suggested approaches (see Section 4.2). The method described is not, however, very suitable for the deliberate realization of an interference supression processor by coherent optical and holographic techniques, since algorithm (6.2) does not necessarily converge to a combination of transformations characteristic of it, even if such a transformation is characteristic of algorithm (4.6). What is more, we should keep in mind that the integral operator (6.2) essentially reproduces the solution of an integral equation of the type of (4.4) with the left-hand side of (6.1). This equation, in general, belongs to the class of "ill-posed" problems [19, 169] which have unstable solutions that are difficult to realize in practice.

Therefore, in order to obtain a constructive solution of the problem stated, we impose a number of conditions. First, we require the reproducibility of the starting (interference supression) algorithm (4.6) for coherent optical processing. Second, we permit the radiation pattern realized to depart from the ideal pattern (6.1) within a range of root-mean-square (rms) deviations. Third, the starting CO processor must be matched to the polarization structure of the signal. This last point calls for some explanation. Generally, as was shown in Section 4.1, the processor output for a nonplanar array always exhibits the superposition of the images of radiation polarized in the principal and crossed directions (4.9). Thus condition (6.1) cannot be fulfilled simultaneously for these polarizations, since the image-forming kernels (radiation patterns) in (4.10) are not identical ($F_{\parallel} \neq F_{\perp}$). In order to eliminate the last, it is necessary to remove from (4.9) the parasitic image $\tilde{\dot{S}}_{\perp}$, that is, to insure matched polarization reception. For this purpose, it is necessary to construct a CO processor as the superposition of two sub-processors that independently process the responses of orthogonal polarized channels of the receiving elements, using algorithms of the type of (4.6). The partial processors must have appropriate partial amplitude phase distributions. If the above conditions are satisfied, then we can synthesize an interference supression processor, employing coherent optical techniques, and we can generate an unblinded image of the field of view that is as close as possible, in the sense of the rms deviation, to (4.3).

An ideal way to protect against point noise sources would be to block them with some hypothetical point screens situated in the far field of the array and having a transmittance function

$$t(\mathbf{K},\mathbf{K}_q) = \begin{cases} 0, & \mathbf{K} = \mathbf{K}_q \\ 1, & \mathbf{K} \neq \mathbf{K}_q \end{cases} \tag{6.3}$$

since then the panoramic-surveillance regime is not disturbed, the space-time signal from other directions is not distorted, there is no need for special excitation of the starting amplitude phase distribution (all that is necessary is exact coincidence of the screens with the noise sources), suppression is stable against changes of frequency, and so forth. Obviously, the screening of noise in the far field of the array is technically unrealistic. If, however, we consider that the processor output approximately images the far field of the array, then this consideration may prove useful.

Using (6.3), we represent the ideal radiation pattern (6.1) in the form

$$\mathbf{F}_{\Pi}(\mathbf{K},\mathbf{K}')_{ideal} = F(\mathbf{K},\mathbf{K}') \prod_{q=1}^{Q} t(\mathbf{K},\mathbf{K}_q) = \mathbf{n}_{\|}(\mathbf{K}) F_{\|}(\mathbf{K},\mathbf{K}')$$

$$\times \left[1 - \sum_{q=1}^{Q} p(\mathbf{K},\mathbf{K}_q)\right] = \mathbf{n}_{\|}(\mathbf{K}) \left[F_{\|}(\mathbf{K},\mathbf{K}') - \sum_{q=1}^{Q} (\mathbf{K},\mathbf{K}_q) F_{\|}(\mathbf{K}_q,\mathbf{K}')\right]$$

$$= \mathbf{n}_{\|}(\mathbf{K}) F_{\Pi}(\mathbf{K},\mathbf{K}')_{ideal} \tag{6.4}$$

where we have taken into account that the vector radiation pattern of the array is matched with the principal polarization $\mathbf{n}_{\|}$ of the received radiation; in what follows, we will omit the $\|$ sign in $F_{\|}$.

The representation (6.4) makes it possible to consider "negative" screens

$$p(\mathbf{K},\mathbf{K}_q) = 1 - t(\mathbf{K},\mathbf{K}_q) \leq 1 \tag{6.5}$$

as some partial normalized radiation patterns. The realization of such (infinitely narrow) radiation patterns with the help of a given (finite) array is technically impossible. For this reason we use the approximation

$$p(\mathbf{K},\mathbf{K}_q) \approx F_{\circ}(\mathbf{K},\mathbf{K}_q) \tag{6.6}$$

where $F_o \leq 1$ is the normalized radiation pattern. Indeed, approximation

(6.6) implicitly has the meaning of regularizing operator (6.2). Substituting (6.6) into (6.4), we have

$$\begin{aligned}\mathbf{F}_{\Pi}(\mathbf{K},\mathbf{K}')_{real} &= \mathbf{n}(\mathbf{K})F_{\Pi}(\mathbf{K},\mathbf{K}')_{real} \\ &= \mathbf{n}_{\|}(\mathbf{K})\left[F(\mathbf{K},\mathbf{K}') - \sum_{q=1}^{Q} F_{\circ}(\mathbf{K},\mathbf{K}_q)F(\mathbf{K}_q,\mathbf{K}')\right] \quad (6.7)\end{aligned}$$

We establish the form of the partial radiation patterns (6.6) for which the realized radiation pattern (6.7) is closest to the ideal pattern (6.4) in the sense of the rms deviation:

$$\iint_{4\pi} |F_{\Pi}(\mathbf{K},\mathbf{K}')_{ideal} - \mathbf{F}_{\Pi}(\mathbf{K},\mathbf{K}')_{real}|^2 \mathrm{d}^2\mathbf{K}$$

$$= \iint_{4\pi} \left| F(\mathbf{K}_q,\mathbf{K}') \sum_{q=1}^{Q} [p(\mathbf{K},\mathbf{K}_q) - F_{\circ}(\mathbf{K},\mathbf{K}_q) \right|^2 \mathrm{d}^2\mathbf{K}$$

$$\leq \sum_{q=1}^{Q} |F(\mathbf{K}_q,\mathbf{K}')|^2 \iint_{4\pi} |F_{\circ}(\mathbf{K},\mathbf{K}_q)|^2 \mathrm{d}\mathbf{K}$$

$$= 4\pi \sum_{q=1}^{Q} |F_{\circ}(\mathbf{K}_q,\mathbf{K}')|^2 / D(\mathbf{K}_q) \quad (6.8)$$

As we see, the rms error of (6.8) is a minimum when the directive gain is a maximum [52]:

$$D(\mathbf{K}_q) = 4\pi \Big/ \iint_{4\pi} |F(\mathbf{K},\mathbf{K}_q)|^2 \mathrm{d}^2\mathbf{K} \quad (6.9)$$

but the radiation pattern (6.7) has a null in the direction toward the noise ($\mathbf{K}_q$) only for $Q = 1$. When $Q \geq 2$ (the number of noise sources is not less than two), the radiation pattern (6.7) that has partial radiation patterns $F_{\circ}$ optimal with respect to the directive gain does not have nulls in the directions toward the noise sources. Indeed,

$$|\mathbf{F}_{\Pi}(\mathbf{K}_q,\mathbf{K}')_{ideal} - \mathbf{F}_{n}(\mathbf{K}_q,\mathbf{K}')_{real}|^2$$

$$= \sum_{p \neq q}^{Q} F_{\circ}(\mathbf{K}_q,\mathbf{K}_p)F(\mathbf{K}_p,\mathbf{K}') \neq 0 \quad (6.10)$$

Since the error (6.10) does not exceed the side-lobe levels (SLL) of the array, which are of the order of $F_0(\mathbf{K}_q, \mathbf{K}_p)$ (it is assumed that the point sources are resolved by the array), the condition SNR $>$ side-lobe level insures that the radiation pattern (6.7) is satisfactory with respect to the depth of the nulls. In the contrary case (SNR $\leq$ side-lobe level), in place of the radiation pattern (6.7) we should use a pattern of the form

$$\mathbf{F}_\Pi(\mathbf{K}, \mathbf{K}')^{(N)}_{ideal} = \mathbf{n}_{\|}(\mathbf{K})\mathbf{F}_\Pi(\mathbf{K}, \mathbf{K}')^{(N-1)}_{real}$$

$$= \mathbf{F}_n(\mathbf{K}, \mathbf{K}')^{(N-1)}_{real} \sum_{p \neq q}^{Q} F_\circ(\mathbf{K}, \mathbf{K}_q) F(\mathbf{K}_q, \mathbf{K}')^{(N-1)}_{real} \tag{6.11}$$

which we call the N-approximation to pattern (6.7);

$$\mathbf{F}_\Pi(\mathbf{K}_q, \mathbf{K}')^{(1)}_{real} \equiv \mathbf{F}_\Pi(\mathbf{K}_q, \mathbf{K}')_{real}$$

The radiation pattern (6.11) differs from (6.1) by not more than

$$|\mathbf{F}_\Pi(\mathbf{K}_q, \mathbf{K}')_{ideal} - \mathbf{F}(\mathbf{K}_q, \mathbf{K}')^{(n)}_{real}|^2 \approx (SLL)^N \tag{6.12}$$

It can be shown that, for $N \geq Q$, the error (6.12) vanishes; that is, (6.11) contains ideal nulls in the directions toward Q noise sources.

6.2 Interference Suppression Processing Algorithm and Options for Its Implementation

We now establish the amplitude phase distribution corresponding to the radiation patterns (6.7) and (6.11). We begin by assuming that the lossless radiation pattern, which is optimal with respect to the directive gain [see (6.8)], coincides with

$$F(\mathbf{K}, \mathbf{K}') = F(\mathbf{K}', \mathbf{K}') F_\circ(\mathbf{K}, \mathbf{K}') \tag{6.13}$$

(in what follows we will generalize this to an arbitrary radiation pattern).

Then integral equation (4.4) with the left-hand side of (6.7) has the solution

$$\dot{\jmath}_{\Pi}(\mathbf{R},\mathbf{K}')^{(I)} = \dot{\jmath}(\mathbf{R},\mathbf{K}') - \sum_{q=1}^{Q} \dot{\jmath}(\mathbf{R},\mathbf{K}_q)F_{\circ}(\mathbf{K}_q,\mathbf{K}') \tag{6.14}$$

It is easy to show that equation (4.4) with the left-hand side of (6.11) has the solution

$$\dot{\jmath}_{\Pi}(\mathbf{R},\mathbf{K}')^{(N)} = \dot{\jmath}_{\Pi}(\mathbf{R},\mathbf{K}')^{(N-1)} - \sum_{q=1}^{Q} \dot{\jmath}(\mathbf{R},\mathbf{K}_q)^{(N-1)}F_{\circ}(\mathbf{K},\mathbf{K}') \tag{6.15}$$

As we see, interference suppression amplitude phase distributions (6.14) and (6.15) are expressed in terms of the starting amplitude phase distribution.

6.2.1 Processing algorithm

By direct substitution of (6.14) and (6.15) into (6.2) we can obtain

$$\hat{L}_{\Pi}\{\ldots\}^{(I)} = \hat{L}\{\ldots\} - \hat{I}\left\{\hat{L}\{\ldots\}\sum_{q=1}^{Q}\delta(\mathbf{K}_S,\mathbf{K}_q)\right\} \tag{6.16}$$

$$\hat{L}_{\Pi}\{\ldots\}^{(N)} = \hat{L}_{\Pi}\{\ldots\}^{(N-1)} - \hat{I}\left\{\hat{L}_{\Pi}\{\ldots\}^{(N-1)}\sum_{q=1}^{Q}\delta(\mathbf{K}_S,\mathbf{K}_q)\right\} \tag{6.17}$$

Here

$$\hat{I} = \int\!\!\int_S \ldots F_{\circ}(\mathbf{K}_S,\mathbf{K}')\mathrm{d}^2\mathbf{K}_S \tag{6.18}$$

is the "bounded reproduction" operator, which is defined on an auxiliary locus $\mathbf{K_S} \in \mathbf{S}$. We call this locus the "spectral" plane (using the same name as for the method). It is understood as the intermediate surface of the processor where the starting image (4.3) is formed for subsequent processing in order to enhance the interference suppression property. In the above formulas, $\delta(\mathbf{K}_S,\mathbf{K}_q)$ is the Dirac delta function, which is defined on plane S in the sense of a filtering property of the type of (5.4). Operators (6.17) and (6.18) are optimal in the sense of the rms deviation. They can be realized if their first approximation operator (6.16), can be realized. Let us explore the possibility of realizing this operator, noting that it contains the starting algorithm (4.6), which is a priori realizable by virtue of the conditions assumed in Section 6.1. Operators (4.6) and (6.18) can also be subtracted optically (for example, with a semitransparent mirror). As for the operator (6.18), we can use (4.4) and write it as

$$
\begin{aligned}
\hat{I}\{\ldots\} &= \iint_S \ldots [F(\mathbf{K}_S,\mathbf{K}')/F(\mathbf{K}',\mathbf{K}')]\mathrm{d}^2\mathbf{K}_S \\
&= F^{-1}(\mathbf{K}',\mathbf{K}') \iint_\Sigma \left\{ \iint_S \ldots \hat{\mathcal{F}}_e(\mathbf{K}_S,\mathbf{R}) \parallel \mathrm{d}^2\mathbf{K}_S \right\} \mathcal{J}(\mathbf{R},\mathbf{K}')\mathrm{d}^2\mathbf{R} \\
&= F^{-1}(\mathbf{K},\mathbf{K}') \cdot \hat{L}\{\hat{A}\{\ldots\}\} \qquad (6.19)
\end{aligned}
$$

where

$$
\hat{A}\{\ldots\} = \iint_S \ldots \dot{\mathcal{F}}_e(\mathbf{K}_S,\mathbf{R})_\parallel \mathrm{d}^2\mathbf{K}_S \qquad (6.20)
$$

is the "aperture" operator on the principal polarization and $\dot{\mathcal{F}}_e(\mathbf{K}_S,\mathbf{R})_\parallel = \mathcal{F}_e(\mathbf{K}_S,\mathbf{R})_\parallel \exp(-i\ \mathbf{K}_S\mathbf{R})$ is the generalized radiation pattern of an array element for the principal polarization (we recall that $\dot{\mathcal{F}}_e(\mathbf{K}_S,\mathbf{R})_\perp = 0$ by the condition assumed in Section 6.1).

Since, in accordance with (6.13), we have assumed that the starting radiation pattern of the array is optimal with respect to the directive gain, we have [133]

$$
\dot{\mathcal{J}}_e(\mathbf{K}_S,\mathbf{R}) = |\text{const}|\ \overset{*}{\mathcal{F}}_e\ (\mathbf{K}_S,\mathbf{R})_\parallel \qquad (6.21)
$$

so that (6.20) takes the form

$$
\hat{A}\{\ldots\} = \iint_S \ldots \overset{*}{\mathcal{J}}\ (\mathbf{R},\mathbf{K}_S)\mathrm{d}^2\mathbf{K}_S/|\text{const}| = \hat{L}^*\{\ldots\}/|\text{const}| \qquad (6.22)
$$

As we see, the aperture operator (6.20) is proportional to the operator conjugate to (4.6) [170] (it must not be confused with (4.19)). Thus, in the case under consideration,

$$
\hat{I}\{\ldots\} = \hat{L}\hat{L}^*\{\ldots\}/|\text{const}|F(\mathbf{K}',\mathbf{K}') \qquad (6.23)
$$

which shows that the bounded reproduction operator (6.18) can be realized, since it is the product of optically realizable algorithms.

We also note that, by (6.16), the auxiliary image of the field of view produced in the intermediate spectral plane S should be weighted with a Dirac "comb." The weighting function can be implemented only approximately, with the help of negative masks of finite extent σ_q, the amplitude transmittance function being

$$\rho(\mathbf{K}_S, \mathbf{K}_q) = \begin{cases} \rho_q, & \mathbf{K}_S \in \sigma_q \ni \mathbf{K}_q \\ 0, & \mathbf{K}_S \notin \sigma_q \end{cases} \tag{6.24}$$

where

$$\sigma_q = \int\int_S \rho(\mathbf{K}_S, \mathbf{K}_q) \mathrm{d}^2\mathbf{K}_S / \rho_q \tag{6.25}$$

Since a physically realizable processor reproduces operator (6.18) up to its norm $\| \hat{I} \|$ [170, 171], on weighting of the image in the plane S with an approximate "comb" made up of negative masks (6.24) we actually obtain the following in place of the second term in (6.16):[2]

$$\hat{I}\left\{\hat{L}\sum_{q=1}^{Q}\rho(\mathbf{K}_S, \mathbf{K}_q)\right\}/\|\hat{I}\| \approx \frac{\rho_\circ\sigma_\circ}{\|\hat{I}\|}\hat{I}\left\{\hat{L}\{\ldots\}\sum_{q=1}^{Q}\delta(\mathbf{K}_S, \mathbf{K}_q)\right\} \tag{6.26}$$

where

$$\rho_\circ\sigma_\circ = \rho_q\sigma_q (q = 1, \ldots, Q)$$

To keep the same weighting function for each term in (6.16), we must therefore increase (decrease) the first (second) by a factor of $(\rho_0\sigma_0 / \| \hat{I} \|)$. Thus in place of (6.16) we have the following according to (6.26):

$$\begin{aligned} \hat{L}\{\ldots\}^{(0)}_{real} &= \frac{\rho_\circ\sigma_\circ}{\|\hat{I}\|}\hat{L}\{\ldots\} - \hat{I}\left\{\hat{L}\{\ldots\}\sum_{q=1}^{Q}\rho(\mathbf{K}_S, \mathbf{K}_q)\right\}/\|\hat{I}\| \\ &= \frac{\rho_\circ\sigma_\circ}{\|\hat{I}\|}\left|\hat{L}\{\ldots\} - \hat{I}\left\{\hat{L}\{\ldots\}\sum_{q=1}^{Q}\frac{\rho(\mathbf{K}_S, \mathbf{K}_q)}{\rho_q\sigma_q}\right\}\right|/\|\hat{I}\| \end{aligned}$$

[2] This normalization of the operator (6.18) has to do with the necessity of allowing for the conservation of light energy (the signal energy at the output of the passive processor does not exceed the energy at the input).

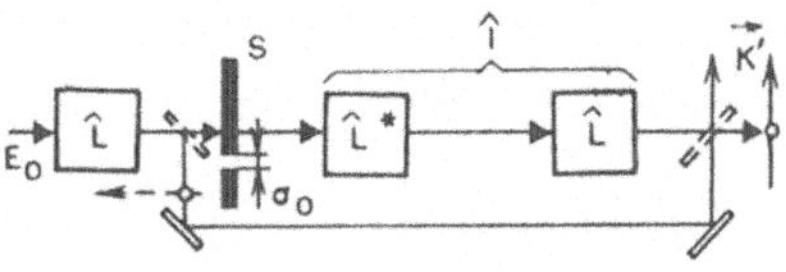

Figure 6.1: Structural diagram of an interference suppression CO processor.

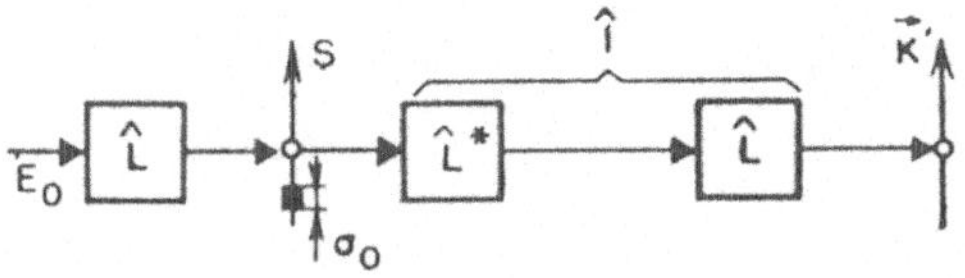

Figure 6.2: Structural diagram of a CO processor with normalized processing algorithm.

$$= \frac{\rho_o\sigma_o}{||\hat{I}||}\hat{L}_\Pi\{\ldots\}^{(0)} \tag{6.27}$$

and in place of (6.17)

$$\hat{L}_n\{\ldots\}^{(N)}_{real} \approx \frac{\rho_o\sigma_o}{||\hat{I}||}\hat{L}_\Pi\{\ldots\}^{(N)}_{real} \tag{6.28}$$

6.2.2 Options for realizing the processor

Fig. 6.1 is a structural diagram of an interference suppression processor based on algorithm (6.27). If instead of the starting CO processor, we use a processor in the circuit with the algorithm $\hat{L}_\Pi\{\ldots\}^{(N-1)}_{real}$, then algorithm (6.28) is implemented. The circuit includes one starting processor, the output of which is divided into two beams in the ratio of $\rho_o\sigma_o : 1$. The second beam is passed through a "comb" of negative masks (6.24) in the plane S. The comb has its maximum transmittance in the regions $\mathbf{K}_S \in \sigma_q$ $(\rho_q \equiv \rho_0 = 1, q = 1, \ldots, Q)$ and zero transmittance outside these regions.

The processor option under consideration is complex in realization and makes inefficient use of the light energy, since virtually the whole spectral plane is shadowed by the negative masks. The structure of the processor can be simplified with a normalized operator (4.6) [170, 171]. Then the conjugate operator coincides with the inverse:

$$\hat{L}^*\{\ldots\} = \hat{L}^{-1}\{\ldots\} \tag{6.29}$$

By (6.23), the bounded reproduction operator (6.16) then completely reproduces the result of applying the starting algorithm (4.6):

$$\hat{I}\{\hat{L}\{\ldots\}\}/||\hat{I}|| = \hat{L}\{\ldots\} \tag{6.30}$$

Therefore, instead of (6.27), we have in this case

$$\hat{L}_{\Pi}\{\ldots\}^{(1)}_{real} = \hat{I}\{\tau(\mathbf{K}_S)\hat{L}\{\ldots\}\}/||\hat{I}|| \tag{6.31}$$

where

$$\tau(\mathbf{K}_S) = \frac{\rho_\circ\sigma_\circ}{||\hat{I}||} - \sum_{q=1}^{Q}\rho(\mathbf{K}_S, \mathbf{K}_q) \tag{6.32}$$

is the transmittance function of the mask in the spectral plane. For algorithm (6.28), we obtain similarly

$$\hat{L}_{\Pi}\{\ldots\}^{(N)}_{real} = \hat{I}\left\{\tau(\mathbf{K}_S)\hat{L}_{\Pi}\{\ldots\}^{(N-1)}_{real}\right\}/||\hat{I}|| \tag{6.33}$$

Fig. 6.2 is a structural diagram of a processor based on algorithm (6.31). This scheme is obviously much simpler and, in terms of energy, better than the scheme of Fig. 6.1. Its usefulness is limited, however, by the normalized condition (6.29) of the starting algorithm (4.6); this point is characteristic of only a few optically realizable transformations (such as the Fourier, Fresnel, and Bessel transforms [5]).

If the starting operator (4.6) is not normalized, we can simplify the processor structure by using the following fact. If condition (6.30) holds, then by (6.18)

$$\hat{I}\{F_\circ(\mathbf{K}, \mathbf{K}_S)\} = \int\int_S F_\circ(\mathbf{K}, \mathbf{K}_S)F_\circ(\mathbf{K}_S, \mathbf{K}')\mathrm{d}^2\mathbf{K}_S = ||\hat{I}||F_\circ(\mathbf{K}, \mathbf{K}') \tag{6.34}$$

Here

$$||\hat{I}|| = \int\int_S |F_\circ(\mathbf{K}, \mathbf{K}_S)|^2\mathrm{d}^2\mathbf{K}_S \approx 4\pi/D(\mathbf{K}) = \sigma_{pr}(\mathbf{K}) \tag{6.35}$$

is the "mean area" of the principal maximum of the radiation pattern in the spectral plane S in direction $\mathbf{K}$ [52, p. 149]. Hence it is clear that the normalized property of the starting algorithm (4.6) has the consequence that the starting radiation pattern is reproducible, so that the whole beam of radiation patterns of the type of (4.3) is reproducible for a parallel scan. For this reason, it appears possible to make algorithm (4.6) normalized in the following way when condition (6.29) is not satisfied.

To do this, we seek some auxiliary operator $\widehat{H}\{\ldots\}$ (see below) that, acting on the starting (4.6), forms an image of the field of view of the type of (4.10) with an image-forming kernel (radiation pattern) that satisfies (6.34). An example of such a kernel is the radiation pattern of a planar equivalent aperture with a uniform amplitude phase distribution [172]. If it is desired to return to the starting radiation pattern (4.4), then operator (6.18) must be acted on by the inverse operator $\widehat{H}^{-1}$. If the operator $\widehat{H}^{-1}$ exists (if it does not, there remains only the processor option of Fig. 6.1) then, by (6.27) and the fact that $\widehat{H}\hat{L}$ satisfies (6.30), we can obtain the following operator similar to (6.31),

$$\hat{L}_{\Pi}\{\ldots\}_{real}^{(I)} = \widehat{H}^{-1}\{\hat{I}\{\tau(\mathbf{K}_S)\widehat{H}\hat{L}\{\ldots\}\}\}/||\hat{I}|| \tag{6.36}$$

which generates a beam of array radiation patterns having gaps in the directions of the noise sources, which is as close as possible (in the sense of the rms deviations) to the ideal beam of radiation patterns of the type of (6.1). Similarly, we find the following operator similar to (6.33):

$$\hat{L}_{\Pi}\{\ldots\}_{real}^{(N)} = \widehat{H}^{-1}\{\hat{I}\{\tau(\mathbf{K}_S)\widehat{H}\hat{L}_{\Pi}\{\ldots\}_{real}^{(N)}\}\} \tag{6.37}$$

In contrast to $\hat{L}$ in (6.31), the combined operator $\widehat{H}\hat{L}$ in (6.36) need not be normalized, since here we are requiring not that it be reproducible in general [see (6.30)] but only that a certain reaction of the operator [see (6.34)] on the space–time signal of the array be reproducible. This fact extends the possibilities of realizing a second processor option. We note that the operators (6.36) and (6.37) are essentially the above-mentioned generalization of the method to the case of a starting radiation pattern of arbitrary form [and not (6.13)].

6.2.3 Structure of the mask

The transmittance function (6.32) must satisfy the passivity condition

$$-1 \leq \tau(\mathbf{K}_S) \leq 1 \tag{6.38}$$

We select parameters $\rho_\circ$ and $\sigma_\circ$ for mask (6.32) in such a way as to refine its range of variation in the interval (6.38) that makes it possible to utilize the light energy most completely. Since the minimum value of the transmission function of the mask corresponds to points $\mathbf{K}_S \in \sigma_q$, from what has been said we assume

$$\tau(\mathbf{K}_S \in \sigma_q) = \frac{\rho_\circ \sigma_\circ}{||\widehat{I}||} - \rho_\circ = -1 \tag{6.39}$$

At the same time, the maximum of (6.32) corresponds to points $\mathbf{K_S} \notin \sigma_\mathbf{q}$, and therefore

$$\tau(\mathbf{K}_S \notin \sigma_q) = \frac{\rho_\circ \sigma_\circ}{||\widehat{I}||} = 1 \tag{6.40}$$

Relations (6.39) and (6.40) hold simultaneously if

$$\rho_\circ = 2, \qquad c_\circ = \sigma_\circ / ||\widehat{I}|| \approx \sigma_\circ / \sigma_{pr} = \frac{1}{2} \tag{6.41}$$

where σ_{pr} is given by expression (6.35). If $0 \leq c_o \leq 0.5$ and the minimum of (6.39) is kept, the parameter $\rho_o = 1/(1 - c_o)$ and the minimum value of (6.32) is

$$\tau(\mathbf{K}_S \notin \sigma_q) = c_\circ / (1 - c_\circ) \tag{6.42}$$

that is, the value (6.40) is not reached. If $c_0 \geq 0.5$ and the maximum of (6.40) is kept, $\rho_0 = 1/c_\circ$ and the minimum value of (6.32) is

$$\tau(\mathbf{K}_S \in \sigma_q) = 1 - 1/c_\circ \tag{6.43}$$

that is, the value (6.39) is not reached.

In Fig. 6.3, the dashed curve shows a family of masks (6.32) that satisfy condition (6.38). Here the variable is the relative size of the mask $c_\circ$. For $c_o < 0.5$, the family is bounded by envelope (6.42) (solid curve); for $c_0 \geq 0.5$, it is bounded by envelope (6.43). Obviously, the second case is preferable both energetically and from the standpoint of realizing the mask, since for $c_o < 0.5$ almost the whole plane S (Fig. 6.2) is half-shadowed by the mask with transmission (6.42), while in the second case mask (6.43) shadows only the areas σ_q. In turn, the optimal mask in the second subfamily is the one with parameter (6.41), since in this case there is no loss of light energy,

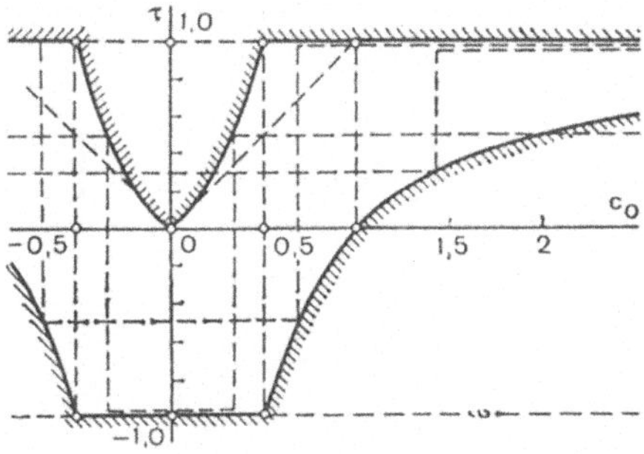

Figure 6.3: Family of masks in the spectral plane S of the CO processor shown in Fig. 6.2.

while the width of the mask is a minimum, so that the optical images of objects near the directions of the noise sources are minimally perturbed.

Because of the finite width $c_o = c_{opt} = 0.5$ of the selected mask, interference supression algorithms (6.16) and (6.17) are executed approximately [see (6.27) and (6.28)]. For this reason, the starting radiation pattern is distorted, and complete noise suppression is not insured (the nulls in the directions toward the noise sources are "filled in"). Let us investigate these distortions and point out a way of eliminating them.

6.3 Processor with Ideal Suppression of Spatial Noise Signals

To solve the problem stated, we use the apparatus of the eigenfunctions of operator (6.18). It is known [134, 172] that the self-reproducing operator in the sense of property (6.34) generates a discrete spectrum of eigenvalues $\lambda_i(c_q)$ and eigenfunctions $\Psi_i(\mathbf{K}, \mathbf{K}_q)$ corresponding to them, provided the initial action of the operator is weighted with a Π-shaped function (6.24) with a finite relative size $c_o \equiv c_q$:[3]

$$\hat{I}\left\{\frac{\rho(\mathbf{K}_S, \mathbf{K}_q)}{\rho_o}\Psi_i(\mathbf{K}_S, \mathbf{K}_q)\right\}\Big/||\hat{I}|| = \lambda_i(c_o)\Psi_i(\mathbf{K}', \mathbf{K}_q) \qquad (6.44)$$

where the eigenvalue spectrum is nondegenerate and is ordered as

$$1 > \lambda_o(c_o) > \ldots > \lambda_i(c_o) > \ldots \geq 0 \qquad (6.45)$$

[3]The eigenvalues λ_i and the eigenfunctions Ψ_i depend on c_q and $\mathbf{K_q}$ because, in general, function (6.24) depends on these parameters.

while the system of eigenfunctions is complete and orthogonal both over the entire infinite region S

$$\iint_S \Psi_i(\mathbf{K}_S,\mathbf{K}_q)\Psi_j(\mathbf{K}_S,\mathbf{K}_q)\mathrm{d}^2\mathbf{K}_S = \begin{cases} 1 & i=j \\ 0 & i\neq j \end{cases} \tag{6.46}$$

and on a finite region σ_q of S:

$$\iint_\sigma \Psi_j(\mathbf{K}_S,\mathbf{K}_q)\Psi_j(\mathbf{K}_S,\mathbf{K}_q)\mathrm{d}^2\mathbf{K}_S = \begin{cases} \lambda_i\,(c_\circ) & i=j \\ 0 & i\neq j \end{cases} \tag{6.47}$$

6.3.1 Evaluation of the depth of suppression

We expand the self-reproducing [in the sense of (6.34)] kernel $F_\circ$ with respect to the orthonormalized basis $\Psi_i(\mathbf{K},\mathbf{K}_q)$. Using (6.47) and (6.44), we obtain

$$F_\circ(\mathbf{K}_S,\mathbf{K}') = ||\hat{I}||\sum_{i=0}^{\infty}\Psi_i(\mathbf{K}_S,\mathbf{K}_q)\Psi_i(\mathbf{K}',\mathbf{K}_q) \tag{6.48}$$

Using (6.48), we establish the form of the radiation pattern (which is distorted because the parameter $c_\circ$ is finite) at the output of the processor using algorithm (6.31). This pattern is the result of the action of operator (6.18) on the starting radiation pattern (6.13) weighted with the mask (6.32)

$$\begin{aligned}\tilde{F}_n^{(I)}(\mathbf{K},\mathbf{K}') &= \hat{I}\{\tau(\mathbf{K}_S)F(\mathbf{K},\mathbf{K}_S)\}/||\hat{I}|| \\ &= \rho_\circ\sigma_\circ F(\mathbf{K},\mathbf{K})\left[F_\circ(\mathbf{K},\mathbf{K}') - \frac{||\hat{I}||^2}{\sigma_\circ}\sum_{q=1}^{Q}\sum_{i=0}^{\infty}\lambda_i(c_\circ)\Psi_i(\mathbf{K},\mathbf{K}_q)\Psi_i(\mathbf{K}',\mathbf{K}_q)\right] \\ &\approx \rho_\circ\sigma_\circ F(\mathbf{K},\mathbf{K})\left[F_\circ(\mathbf{K},\mathbf{K}') - \frac{\lambda_\circ(c_\circ)}{c_\circ}||\hat{I}||\sum_{q=1}^{Q}\Psi_\circ(\mathbf{K},\mathbf{K}')\Psi_\circ(\mathbf{K}',\mathbf{K}_q)\right]\end{aligned} \tag{6.49}$$

where $F(\mathbf{K},\mathbf{K}) = \max F(\mathbf{K},\mathbf{K}')$ and we have taken account of the asymptotic behavior [134, 172]

$$\lambda_\circ(c_\circ) \approx c_\circ \gg \lambda_1 \sim c_\circ^3 \gg \ldots \gg \lambda_i \sim c_\circ^{(2i+1)} \longrightarrow 0 \qquad (6.50)$$

as $c_o \to 0$. As we see, the limit of the distorted radiation pattern (6.49) as $c_o \to 0$ coincides with the optimal radiation pattern (6.7) up to the coefficient $\rho_o\sigma_o / \parallel \hat{I} \parallel$, since $\Psi_o \to F_\circ$. For finite $c_\circ$, however, there is no coincidence; the gaps in the radiation pattern (6.19) are finite ($\tilde{F}_n^{(1)}(\mathbf{K}_q, \mathbf{K}') \not\equiv 0$ even when $Q = 1$).

In the following case, we evaluate the depth of noise suppression for $\mathbf{K} = \mathbf{K}_q$ as the ratio of the noise image powers at the processor output with suppression (P_n) and without it ($P_\circ$):

$$\begin{aligned} \frac{P_\Pi}{P_\circ} &= \frac{\int\int_{4\pi} |\tilde{F}_\pi^{(I)}(\mathbf{K}_q, \mathbf{K}')|^2 d^2\mathbf{K}'}{\int\int_{4\pi} |\rho_\circ\sigma_\circ F(\mathbf{K}_q, \mathbf{K}')|^2 d^2\mathbf{K}'} \\ &= 1 - \frac{\lambda_\circ(c_\circ)}{c_\circ} ||\hat{I}||\, |\Psi_{omax}|^2 \left[2 - \frac{\lambda_\circ(c_\circ)}{c_\circ}\right] \geq \left[1 - \frac{\lambda_\circ(c_\circ)}{c_\circ}\right]^2 \end{aligned} \qquad (6.51)$$

where we have used (6.46) and taken into account that $||\hat{I}||\, |\Psi_{\circ max}|^2 = \sum_{i=1}^{\infty} |\Psi_i(\mathbf{K}_q, \mathbf{K}_q)|^2 \leq [\Psi_{\circ max} = \Psi_\circ(\mathbf{K}_q, \mathbf{K}_q)]$

For the cases of linear and circular arrays, relation (6.51) is described concretely in Section 6.4 and in Figs. 6.8 and 6.17, respectively. As we see, for finite $c_o \neq 0$ (in particular for $c_o = 0.5$) noise suppression is not complete, becoming worse as $c_\circ$ increases.

Thus we see the following contradiction. To generate a beam of radiation patterns with nulls in the directions of the noise sources, we must use narrow masks (6.32), but these are energetically unsuitable for $c_o \to 0$ (see Fig. 6.3), since they block nearly all the energy of the laser.

6.3.2 Generation of ideal nulls

The contradiction noted above can be resolved through one of the following approaches. For example, let us distort the Π-shaped negative masks (6.24) so as to compensate the noncoincidence of the starting ($F_\circ$) and partial ($\Psi_\circ$) radiation patterns in (6.43); then the latter will be zero for $\mathbf{K} = \mathbf{K}_q$. A different solution is also possible: replacing the optimal (with respect to directive gain) starting radiation pattern (6.13) by a partial radiation pattern $\Psi_\circ$ that is nonoptimal but close to it; this is done with the help of a modified auxiliary operator $\widehat{H} = \widehat{H}_\Psi$. There is nothing to prevent a combined approach either. We begin by examining the first possibility.

We require that the mask (6.24) satisfy the identity

$$\hat{I}\{\rho(\mathbf{K}_S,\mathbf{K}_q)F_\circ(\mathbf{K}_S,\mathbf{K}_q)\}/||\hat{I}|| = F_\circ(\mathbf{K}',\mathbf{K}_q) \tag{6.52}$$

Using expansion (6.48), we can establish that the integral equation (6.52) in ρ is satisfied by the solution

$$\rho(\mathbf{K}_S,\mathbf{K}_q) = \sum_{i=0}^{\infty}\frac{\Psi_i(\mathbf{K}_q,\mathbf{K}_q)}{\lambda_i}\Psi_i(\mathbf{K}_S,\mathbf{K}_q)/F_\circ(\mathbf{K}_S,\mathbf{K}_q) = \delta(\mathbf{K}_S,\mathbf{K}_q) \tag{6.53}$$

Thus the first approach comes back to the variant already considered, with a mask in the form of a Dirac "comb" [see (6.26)], and therefore does not contain any new solutions.

Now we examine the second option. Let the starting radiation pattern be nonoptimal with respect to the directive gain (then the interference supression radiation pattern is a nonoptimal approximation, in the sense of the rms deviations, to the ideal radiation pattern (6.4)), but proportional to the eigenfunction $\Psi_o(\mathbf{K},\mathbf{K}_S)$, that is, to a function that satisfies (6.44) for $\mathbf{K}_q = \mathbf{K}_S$:

$$F(\mathbf{K},\mathbf{K}_S) = F_\Psi \cdot \Psi_\circ(\mathbf{K},\mathbf{K}_S) \tag{6.54}$$

where F_Ψ is the proportionality constant. The change in form of the radiation pattern must obviously violate the relations (Section 6.2.3) between the mask parameters $c_\circ$ and $\rho_\circ$. We investigate this point by expanding the current eigenfunction $\Psi_0(\mathbf{K},\mathbf{K}_S)$ in series in terms of the "fixed" (relative to $\mathbf{K}_q$) eigenfunctions

$$\Psi_\circ(\mathbf{K},\mathbf{K}_q) = \sum_{i=0}^{\infty} b_i(\mathbf{K},\mathbf{K}_q)\Psi_i(\mathbf{K}_S,\mathbf{K}_q) \tag{6.55}$$

Here

$$b_i(\mathbf{K},\mathbf{K}_q) = \int\!\!\int_S \Psi_\circ(\mathbf{K},\mathbf{K}_S)\Psi_i(\mathbf{K}_S,\mathbf{K}_q)\mathrm{d}^2\mathbf{K}_S \tag{6.56}$$

It is useful to note the following property of the coefficients (6.56) [see (6.46)]:

$$b_i(\mathbf{K}_q,\mathbf{K}_q) = \begin{cases} 1, & i = 0 \\ 0, & i > 0 \end{cases} \tag{6.57}$$

By analogy with (6.49) we find

$$\Psi_{\Pi}^{(I)}(\mathbf{K},\mathbf{K}') = \hat{I}\{\tau_{\Psi}(\mathbf{K}_S)F(\mathbf{K},\mathbf{K}_S)\}/||\hat{I}||$$

$$= F_{\Psi}\left[\tau_{\circ}\Psi_{\circ}(\mathbf{K},\mathbf{K}') - \rho_{\circ}\sum_{q=1}^{Q}\sum_{i=0}^{\infty}\lambda_i(c_{\circ})b_i(\mathbf{K},\mathbf{K}_q)\Psi_i(\mathbf{K}',\mathbf{K}_q)\right]$$

$$\approx F_{\Psi}\left[\tau_{\circ}\Psi_{\circ}(\mathbf{K},\mathbf{K}') - \rho_{\circ}\lambda_{\circ}(c_{\circ})\sum_{q=1}^{Q}b_{\circ}(\mathbf{K},\mathbf{K}_q)\Psi_i(\mathbf{K}',\mathbf{K}_q)\right] \tag{6.58}$$

Here

$$\tau_{\Psi}(\mathbf{K}_s) = \tau_{\circ} - \sum_{q=1}^{Q}\rho(\mathbf{K}_S,\mathbf{K}_q) \tag{6.59}$$

is the mask in the plane S that, in the case under consideration, does not coincide with (6.32). We suppose that the sector of coverage of the electrooptical array contains one point noise source ($Q = 1$) in direction $\mathbf{K}_q$; then to suppress this noise we must require that the radiation pattern (6.58) have an ideal null in this direction, that is, $\Psi_n^{-1}(\mathbf{K}_q,\mathbf{K}') = 0$. Hence, with (6.58) and (6.57), we find that this is the case when

$$\tau_{\circ} - \rho_{\circ}\lambda_{\circ}(c_{\circ}) = 0 \tag{6.60}$$

Using (6.60) and the passivity condition (6.38), we choose parameters $\rho_{\circ}$ and $c_{\circ}$ for the mask (6.59) such that, as in Section 6.2.3, the most efficient use is made of the light energy. By analogy with relations (6.39) and (6.40), in the present case we obtain

$$\tau_{\Psi}(\mathbf{K}_S \in \sigma_q) = \tau_{\circ} - \rho_{\circ} = -1 \tag{6.61}$$

for the minimum value of the transmittance function of the mask and

$$\tau_{\Psi}(\mathbf{K}_S \notin \sigma_q) = \tau_{\circ} = 1 \tag{6.62}$$

for the maximum value. Relations (6.60)–(6.63) hold simultaneously for

$\rho_o = 2$ and

$$\lambda_o(c_o) = \lambda_o(c_{opt}) = \frac{1}{2} \tag{6.63}$$

If $0 \leq c_o \leq c_{opt}$ and the minimum value of (6.61) is kept, then by (6.60) we have $\rho_o = 1/[1 - \lambda_o(c_o)]$, and the maximum value of (6.59) is

$$\tau_\Psi(\mathbf{K}_S \notin \sigma_q) = \lambda_o(c_o)[1 - \lambda_o(c_o)] \tag{6.64}$$

that is, the value (6.62) is not reached. If $c_o \geq c_{opt}$ and the maximum value of (6.62) is kept, then by (6.60) we have $\rho_o = 1/\lambda_o(c_o)$, and the minimum value of (6.59) is

$$\tau_\Psi(\mathbf{K}_S \in \sigma_q) = 1 - \frac{1}{\lambda_o(c_o)} \tag{6.65}$$

and does not reach the value (6.61). The family of masks (6.59) has a structure analogous to that of the family (6.32), which is illustrated in Fig. 6.3. In the present case, however, there are the following differences: First, $c_{opt} > 0.5$, since $\lambda_0(c_o)/c_o > 1$ [172]; and second, $\tau_\Psi(\mathbf{K} \in \sigma_\mathbf{q}) \longrightarrow 0$ for $c_o \longrightarrow \infty$, since $\lambda_o \rightarrow 1$. The concrete form of the resulting family of masks will be discussed in Section 6.4 and illustrated in Figs. 6.7, 6.10, and 6.15 for the cases of linear, square, and circular arrays, respectively.

Thus, if our starting radiation pattern has the form (6.54), the interference supression processor with mask (6.59) in auxiliary plane S forms, at its output, an optical image of the field of view in the form of a beam of radiation patterns (6.58) with ideal nulls in the directions $\mathbf{K}_q$. To characterize the behavior of the gaps near the indicated directions (their width and the extent of their influence on reception from neighboring directions), it is appropriate to evaluate the ratio of noise powers at the processor output without suppression $P_n(\mathbf{K})$ and without it $P_o(\mathbf{K})$ [compare (6.51)]:

$$\frac{P_\Pi(\mathbf{K})}{P_o(\mathbf{K})} = \frac{\int\int_{4\pi} |\Psi_\Pi^{(1)}(\mathbf{K},\mathbf{K}')|^2 d^2\mathbf{K}'}{\int\int_{4\pi} |F_\Psi \cdot \Psi_o(\mathbf{K},\mathbf{K}')|^2 d^2\mathbf{K}'} = 1 - |b_o(\mathbf{K},\mathbf{K}_q)|^2 \tag{6.66}$$

where we have used relation (6.46) and determined b_o in (6.56) with subscript $i = 0$.

In a similar way, we can derive an analogous relation for the radiation pattern (6.7) that is optimal in the rms sense:

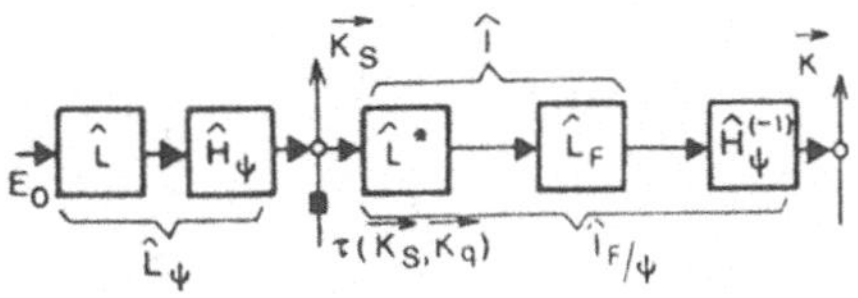

Figure 6.4: Structural diagram of a CO processor generating a null.

$$\frac{P_{\Pi}(\mathbf{K})}{P_{\circ}(\mathbf{K})} = 1 - |F_{\circ}(\mathbf{K}, \mathbf{K}_q)|^2 \tag{6.67}$$

where we have taken into account (6.34). For $c_o \longrightarrow 0$, (6.66) goes to (6.67). If $c_o \neq 0$, then the gap in (6.66) is wider than in (6.67), since by (6.56) the function $b_\circ$ is wider than $F_\circ$. In this way we have to settle for achieving ideally zero noise suppression with the use of radiation patterns that are nonoptimal with respect to the directive gain. Quantitatively, relations (6.66) and (6.67) are illustrated in Figs. 6.11, 6.12, and 6.17.

Taking account of the extreme properties of the radiation patterns (5.64), which represent the maximum energy concentration in a finite region $\sigma_\circ$ [2, 172], we can conclude that there is no third possibility based on any combination of the above two approaches that will give a better result, in the sense of ideal noise suppression and minimal perturbation of the signals near the gaps, than does the use of the auxiliary radiation pattern (6.54).

Let us sum up what has been said by means of the following representation of interference supression algorithm (6.36):

$$\hat{L}\{\ldots\}_{\Psi}^{(1)} = \hat{I}_{\Psi} \left\{ \prod_{q=1}^{Q} \tau_{\Psi}(\mathbf{K}_S, \mathbf{K}_q) \hat{L}_{\Psi}\{\ldots\} \right\} \Big/ ||\hat{I}|| \tag{6.68}$$

Here $\hat{L}_{\Psi} = \widehat{H}_{\Psi}\hat{L}$ is a modification of the starting algorithm (4.6) that generates a beam of auxiliary radiation patterns (6.54); $\hat{I}_{\Psi} = \widehat{H}_{\Psi}^{-1}\hat{I}$; and

$$\tau_{\Psi}(\mathbf{K}_S, \mathbf{K}_q) = 1 - \rho_{opt}(\mathbf{K}_S, \mathbf{K}_q) = \begin{cases} -1, & \mathbf{K}_S \in \sigma_q \\ 1, & \mathbf{K}_S \notin \sigma_q \end{cases} \tag{6.69}$$

where $\rho_{opt}(\mathbf{K}_S, \mathbf{K}_q)$ corresponds to the case (6.63). A similar form can be written for (6.37). Fig. 6.4 gives a structural scheme of a processor implementing algorithm (6.68).

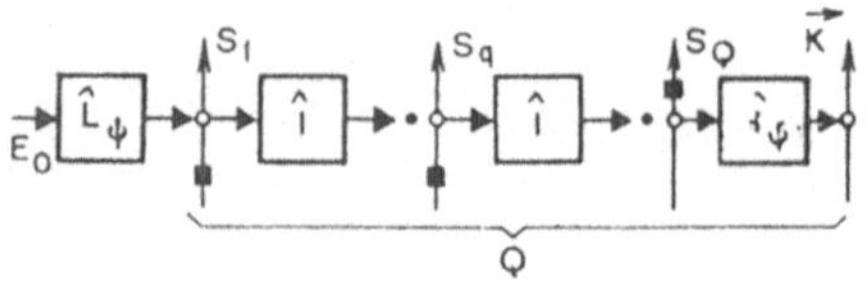

Figure 6.5: Structural diagram of a CO processor generating Q nulls.

6.3.3 Comments

Thus the most reliable suppression of signals from point noise sources is provided by point screens (6.3) situated in the far field of the array. An interference supression processor using algorithm (6.68) essentially replaces the screening of noise sources in the far field by shadowing of their optical images in the spectral plane of the processor, using extended masks (6.69) for the purpose. In the limiting case of an infinitely large array, when $F(\mathbf{K},\mathbf{K}') \longrightarrow \delta(\mathbf{K},\mathbf{K}')$ and the optical image of the field of view reproduces the original exactly ($\tilde{\dot{S}} \rightarrow \dot{S}$), we could take $\tau \equiv t$ [see (6.69) and (6.3)]. For a finite array, however, the mask (6.3) does not insure complete suppression of the noise image. Partial shadowing takes place only in the principal lobe. Here it is necessary to use more extended masks. But if the relative dimension $c_\circ$ (6.41) is increased, there is an increase in the undesirable shadowing of the images of radar objects in other directions. Therefore, in addition to "weighting" with a system of masks (6.69), the image in the spectral plane of the processor is subjected to additional processing under the algorithm $\hat{I}_\Psi$. This processing is carried out by some diffraction-limited image-forming system, whose reproducing properties are bounded in such a way that the undistorted image of the field of view is translated without hindrance to the output, while the image perturbed by a "weight" (6.69) is blocked. Essentially this optical system is a generalized spatial filter whose transfer function (non-isoplanar in the general case) is matched to the unperturbed image of the field of view and mismatched with the weighted noise image. Since the images of point noise sources in the spectral plane are extended (and even infinite), the mask (6.69) "perturbs" not only the qth image but also all the other images (though to a much lesser extent). Thus, if several noise sources are present, it may not be sufficient to use one-stage filtering with algorithm (6.68). Deeper or even complete (for $N \geq Q$) suppression is insured by successive filtering with a multi-stage processor (Fig. 6.5) using an algorithm of the type of (6.37).

6.4 Examples of Electrooptical Arrays with Noise Rejection

Let us illustrate the principles we have developed in some special cases of electrooptical array construction.

6.4.1 Planar electrooptical antenna array with arbitrary aperture shape

For this case the starting algorithm (4.6), in accordance with (2.7), takes on the form

$$\hat{L}\{\ldots\}e_{\circ}\hat{\mathcal{F}}\{\ldots \mathcal{J}(\mathbf{R}_{\perp})\} \tag{6.70}$$

Since the Fourier transform is a normalized operation [170], [4] operator (6.70) is also normalized on a set of finite space–time signals $\mathcal{E}_{\Omega}(\mathbf{R}_{\perp})$ (where $\mathbf{R}_{\perp} \in \Sigma$ is the surface of the array) provided that

$$\mathcal{J}(\mathbf{R}_{\perp}) = \mathcal{J}_{\circ}(\mathbf{R}_{\perp}) = \begin{cases} 1, & \mathbf{R}_{\perp} \in \Sigma \\ 0, & \mathbf{R}_{\perp} \notin \Sigma \end{cases} \tag{6.71}$$

is a uniform amplitude phase distribution. Thus (6.70) satisfies (6.29) and the bounded reproduction operator (6.18) is self-reproducing in the sense of (6.34). We write it in the form [see (6.23)]

$$\begin{aligned} \hat{I}\{\ldots\} &= \hat{L}\{\hat{L}^{-1}\{\ldots\}\}/|\text{const}|F(\mathbf{K},\mathbf{K}') \\ &= \hat{\mathcal{F}}\{\hat{\mathcal{F}}^{-1}\{\ldots\}\mathcal{J}_{\circ}(\mathbf{R}_{\perp})\}/|\text{const}|F(0) \\ &= \hat{\mathcal{F}}\{\hat{\mathcal{F}}^{-1}\{\ldots\}\} \otimes \otimes \hat{\mathcal{F}}\{T_{\circ}(\mathbf{R}_{\perp})\}/F(0) \\ &= \{\ldots\} \otimes \otimes F_{\circ}(\mathbf{K}_{\perp}) \end{aligned} \tag{6.72}$$

where $|$ const $| = e_{\circ}$ by the assumption that there is no light loss in the optical cascade that performs transformation (6.72). The last transformation can be realized with the help of a diffraction-limited image-forming system [172], which consists of two Fourier processors with lenses L_2 and L_3 (Fig. 6.6), with a low-frequency spatial filter in the form of a diaphragm (6.71) placed

[4] Formally, the normalized character of the operator $\hat{L}$ is determined by expression (6.29). Physically, this property results from the absence of energy losses in transformation.

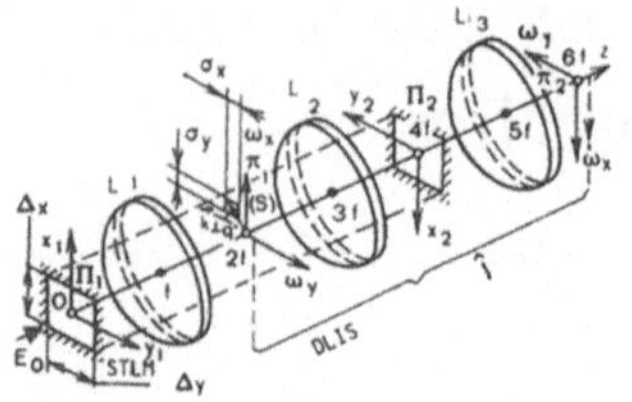

Figure 6.6: Interference supression CO processor of a planar electrooptical array with a rectangular aperture.

between the lenses in the visual plane Π_2. The diaphragm corresponds in shape to a space–time light modulator. We further specify the norm $\| \hat{I} \|$ (6.35) in expression (6.34):

$$||\hat{I}|| = \int\int_{\infty}^{\infty} |F_\circ(\mathbf{K}_\perp)|^2 d^2\mathbf{K}_\perp = 4\pi^2 \int_{\infty}^{\infty}\int \frac{J_\circ^2(\mathbf{R}_\perp)}{F^2(0)} d^2\mathbf{R}_\perp = 4\pi^2/\Sigma \qquad (6.73)$$

The auxiliary radiation pattern (6.54) is found as the eigenfunction $\Psi_\circ$ with the largest eigenvalue $\lambda_\circ$ of the following integral operator:

$$\frac{\Sigma}{\rho_\circ 4\pi^2}[\rho(\mathbf{K}')\Psi_i(\mathbf{K}', c_\circ)] \otimes \otimes F_\circ(\mathbf{K}')$$

$$= \frac{1}{4\pi^2}[\rho_\circ(\mathbf{K}'_\perp)\Psi_i(\mathbf{K}'_\perp, c_\circ)] \otimes \otimes F(\mathbf{K}'_\perp) = \lambda_i(c_\circ)\Psi_i(\mathbf{K}'_\perp, c_\circ) \qquad (6.74)$$

where $\rho_\circ(\mathbf{K}'_\perp) = \rho(\mathbf{K}'_\perp)/\rho_\circ$, $F(\mathbf{K}'_\perp) = \Sigma \cdot F_\circ(\mathbf{K}_\perp)$. Since, by (6.34) and (6.48),

$$\frac{1}{4\pi}\Psi_\circ(\mathbf{K}'_\perp, c_\circ) \otimes \otimes F(\mathbf{K}'_\perp) = \Psi_\circ(\mathbf{K}'_\perp, c_\circ) \qquad (6.75)$$

the modified operator $\hat{L}_\Psi$ in (6.68), which forms the beam of auxiliary radiation patterns (6.54), is a processing algorithm

$$\begin{aligned} \hat{L}_\Psi &= \widehat{H}_\Psi \hat{L} = \frac{1}{4\pi^2}\Psi_\circ(\mathbf{K}'_\perp, c_\circ) \otimes \otimes \hat{L}\{\ldots\} \\ &= \hat{\mathcal{F}}\{\hat{\mathcal{F}}^{-1}\{\Psi_\circ(\mathbf{K}'_\perp, c_\circ)\} \cdot \hat{\mathcal{F}}^{-1}\{e_\circ \hat{\mathcal{F}}\{\ldots J_\circ(\mathbf{R}_\perp)\}\} \\ &= e_\circ \hat{\mathcal{F}}\{\ldots J_\circ(\mathbf{R}_\perp) J_\Psi(\mathbf{R}_\perp, c_\circ)\} = e_\circ \hat{\mathcal{F}}\{\ldots(\mathbf{R}_\perp, c_\circ)\} \qquad (6.76) \end{aligned}$$

where

$$J_\Psi(\mathbf{R}_\perp, c_\circ) = \hat{\mathcal{F}}^{-1}\{\Psi_\circ(\mathbf{K}_\perp, c_\circ)\} \tag{6.77}$$

Thus, in the case of a planar electrooptical array with arbitrary aperture, algorithm (6.68), which makes it possible to form ideal nulls in the beam of highly directional array radiation patterns, takes on the following form when (6.72), (6.73), and (6.76) are taken into account:

$$\hat{L}_\Pi\{\ldots\}_\Psi^{(1)} = \frac{e_\circ \sum}{4\pi^2}\left\{\prod_{q=1}^{Q} \tau_\Psi(\mathbf{K}'_\perp - \mathbf{K}_{\perp q})\hat{\mathcal{F}}\{\ldots J_\Psi(\mathbf{R}_\perp, c_\circ)\}\right\} \otimes \otimes F_\circ(\mathbf{K}'_\perp) \tag{6.78}$$

where in place of $\hat{I}_\Psi$ in (6.68) we have used $\hat{I}$ from (6.72) and where we have determined τ_Ψ in (6.69).

Fig. 6.6 shows a CO processor that realizes the interference supression algorithm (6.78). In the algorithm, the space–time signal supplied to the processor with the help of the space–time light modulator represents an optical model of the radiation received by the array. This signal is weighted with a nonuniform amplitude phase distribution (6.77) by means of an apodization mask.[5] Then the optical signal calculated in the processor is Fourier-transformed by objective L_1 and multiplied by the transmittance function τ_Ψ of the perturbing mask in the spectral plane $\pi_1(S)$. The resulting optical image of the desired signals and noise, perturbed by the mask, is next subjected to additional processing under algorithm (6.72), which can be implemented with the help of the diffraction-limited image-forming system (objective L_2, low-frequency spatial filter in plane Π_2, objective L_3) in such a way that the image of the useful signals is translated without hindrance to the processor output (plane π_2) while the noise images perturbed by the masks are blocked. If it is necessary to get back to the starting radiation pattern $F(\mathbf{K}_\perp) = F(0) \cdot F_\circ(\mathbf{K}_\perp) = \dot{\mathcal{F}}\{J(\mathbf{R}_\perp)\}$ or to any other radiation pattern (for example, one with given side-lobe level) with corresponding amplitude phase distribution $J_F(\mathbf{R}_\perp)$, then the window of the spatial filter in plane Π_2 must contain a mask with a transmittance function of the form $J_o(\mathbf{R}_\perp)/J_\Psi(\mathbf{R}_\perp)$ or $J_F(\mathbf{R}_\perp)/J_\Psi(\mathbf{R}_\perp)$ respectively.

[5]The mask is not shown in Fig. 6.6, since it can be realized either as a photographic mask or, for example, by setting the gains of the respective receiving and amplifying modules of the array.

If the functions $\mathcal{J}_o(\mathbf{R}_\perp)$ and $\rho_o(\mathbf{K}_\perp)$ are even and geometrically suitable (the array has central symmetry), then

$$\mathcal{J}_\circ(\mathbf{R}_\perp) = \rho_\circ(\pm\sqrt{\sigma_\circ/\Sigma}\mathbf{R}_\perp) \tag{6.79}$$

where by (6.35), (6.41) and (6.73) $\sqrt{\sigma_\circ/\Sigma} = 2\pi/\Sigma\sqrt{c_\circ}$. Therefore, using the properties of the doubly orthogonal functions $\Psi_\circ$ [172], we find that

$$\mathcal{J}_\Psi(\mathbf{R}_\perp, c_\circ) = \frac{1}{\Sigma}\sqrt{\frac{c_\circ}{\lambda_\circ c_\circ}}\Psi_\circ\left(\frac{2\pi}{\Sigma}\sqrt{c_\circ}\mathbf{R}_\perp, c_\circ\right)\mathcal{J}_\circ(\mathbf{R}_\perp) \tag{6.80}$$

This last relation obviates the calculation of the auxiliary amplitude phase distribution (6.77) if the eigenfunction $\Psi_\circ$ is known.

In conclusion, let us consider the following. The analysis performed above, both in the general case and for a planar electrooptical array, does not allow for the spectral properties of the received space–time signal, since the signal considered was a monochromatic one, $\dot{\mathcal{E}}_\Omega(\mathbf{R})$. According to Sections 2.2.1 and 4.1.2, the resulting interference supression processing algorithms and the processors corresponding to them also solve the problem of noise suppression, even in the case of wideband signals (noise), provided the spectral bandwidth of the signals satisfies criteria (2.23) and (4.24). If this is not the case, the optical image of a point noise source (signal) is reproduced "smeared out" along one of the dimensions of the diffraction spot. Then the use of a perturbing mask (6.69) does not insure complete noise suppression. A partial way out of this situation is the use of an appropriately "smeared" mask (as can be seen, for example, in Fig. 2.3d); it is necessary to use a mask that is elongated along a radius at an angle φ_q and that completely covers the image of the noise source in the spectral plane S. In this case, however, it is obvious that the problem of suppressing noise and reproducing the signal in a distortion-free manner can be solved only under the condition that the latter be separated in azimuth, since the "smeared" mask covers up a wide range of position angles.

6.4.2 Linear and planar electrooptical antenna arrays with rectangular apertures

We consider this case in more detail as being of great practical interest. Since the radiation pattern of a rectangular aperture has the form of product (2.14), we separate variables in integral equation (6.44) [2, 172], expanding the operator (6.72) as

$$\hat{I}\{\ldots\}/||\hat{I}|| = \hat{I}_X\{\ldots\}\hat{I}_Y\{\ldots\} \tag{6.81}$$

where

$$\hat{I}_{X,Y} = [\ldots] \otimes \mathrm{sinc}(\Omega'_{X,Y}/\delta\Omega_{X,Y})/\delta\Omega_{X,Y} \tag{6.82}$$

The eigenfunction $\Psi_\circ$ of operator (6.81) that corresponds to the maximum eigenvalue $\lambda_\circ$ has the form

$$\Psi(\mathbf{K}_\perp, c_\circ) = \Psi_\circ^X(\Omega_X, c_X)\Psi_\circ^Y(\Omega_Y, c_Y) \tag{6.83}$$

where the $\Psi_\circ^{X,Y}$ are the eigenfunctions of the one-dimensional operators (6.82), which are solutions of the equations

$$\begin{aligned}
&[\rho_X(\Omega_X)\Psi_\circ^X(\Omega'_X, c_X)] \otimes \mathrm{sinc}(\Omega'_X/\delta\Omega_X)/\delta\Omega_X \\
&= \int_{\Delta\Omega_X/2}^{\Delta\Omega_X/2} \Psi_\circ^X(\Omega'_X, c_X) \cdot \frac{\sin[\pi(\Omega_X - \Omega'_X)/\delta\Omega_X]}{\pi(\Omega_X - \Omega'_X)} d\Omega_X \\
&= \lambda_X(c_X)\Psi_\circ^X(\Omega'_X, c_X)
\end{aligned}$$

and

$$\begin{aligned}
&[\rho_Y(\Omega_Y)\Psi_\circ^Y(\Omega'_Y, c_Y)] \otimes \mathrm{sinc}(\Omega'_Y/\delta\Omega_Y)/\delta\Omega_Y \\
&= \int_{\Delta\Omega_Y/2}^{\Delta\Omega_Y/2} \Psi_\circ^Y(\Omega'_Y, c_Y) \cdot \frac{\sin[\pi(\Omega_Y - \Omega'_Y)/\delta\Omega_Y]}{\pi(\Omega_Y - \Omega'_Y)} d\Omega_Y \\
&= \lambda_Y(c_Y)\Psi_\circ^Y(\Omega'_Y, c_Y)
\end{aligned} \tag{6.84}$$

Here

$$\lambda_\circ(c_\circ) = \lambda_X(c_X)\lambda_Y(c_Y); \qquad c_\circ = c_X c_Y \tag{6.85}$$

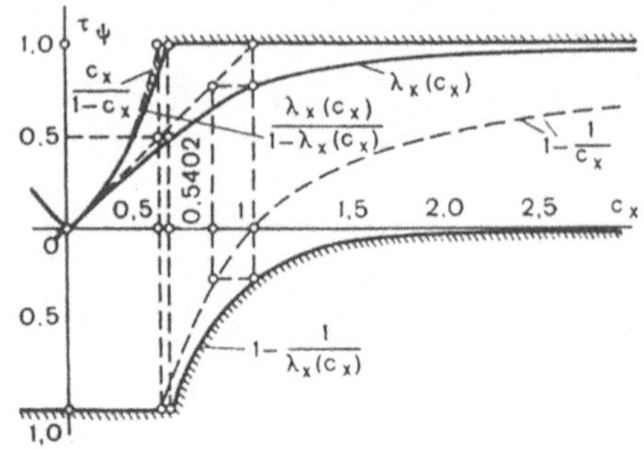

Figure 6.7: Family of masks of the interference supression CO processor of a linear array.

and, by (6.41) and (6.73)

$$c_{X,Y} = \sigma_{X,Y}\Delta X, Y/2\pi = \sigma_{X,Y}/\delta\Omega_{X,Y} \tag{6.86}$$

where $\sigma_X \cdot \sigma_Y = \sigma_o$ and $\sigma_{X,Y}$ is the width of the negative mask (6.24) in coordinates of the generalized angular variables.

It is known that the $\Psi_o^{X,Y}$, which satisfy (6.84), are extended spheroidal wave functions; a detailed description and bibliographic review of these can be found, for example, in [172] and [173].

Appendix C gives computational relations with which the expressions derived above can be numerically modified for a given case. Fig. 6.7 shows how the maximum eigenvalue $\lambda_{X,Y}$ depends on the parameter $c_{X,Y}$; the curve was calculated with (C.2). The same figure (solid curves) shows the envelopes (6.64) and (6.65) of the family of masks (6.59) that insure ideally zero noise suppression for a linear electrooptical array when $c_o = c_{X,Y}$ and accordingly $\lambda_o = \lambda_{X,Y}$. For comparison, the dashed curves in Fig. 6.7 show the envelopes (6.42) and (6.43) of the family of masks (6.32), which were derived through the use of an optimal (with respect to the directive gain) radiation pattern of the type of sinc $\{\ldots\}$, which does not insure ideal suppression (see Fig. 6.3). As the figure shows, for small values of the parameter $c_{X,Y}$ families (6.32) and (6.59) virtually coincide; as was discussed in Section 6.2.3, however, this is energetically unsuitable. The mask that is optimal with respect to efficient utilization of laser energy and perturbing action on useful signals (width of gaps) is one with the following properties: Its transmittance function has a Π shape and is equal to -1 in the vicinity of a noise image and $+1$ outside the vicinity; its relative size $c_{X,Y} = \sigma_{X,Y}/\delta\Omega_{X,Y}$ is given by transcendent

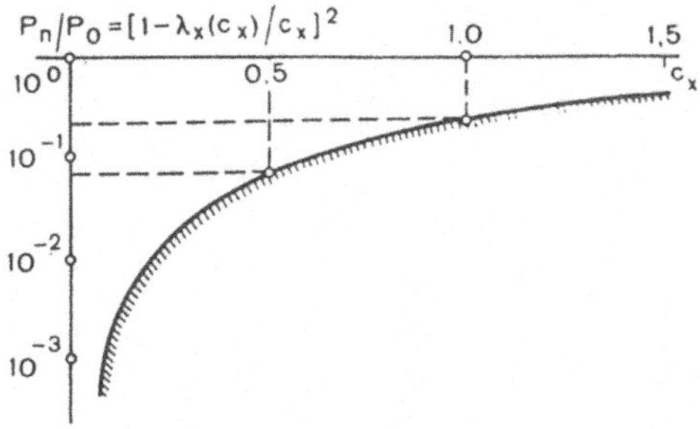

Figure 6.8: Diagram for determining the depth of noise suppression.

equation (6.63). Solution of this equation by Newton's method [171] with the use of (C.2) yields $c_{X,Y_{opt}} = 0.5402$; this is markedly different from the result (6.41), which is also plotted in Fig. 6.7. We note that, in contrast to family (6.32), the mask with zero transmittance function in the vicinity of a noise source is unsuitable, since it does not provide complete suppression of noise for a finite value of $c_{X,Y}$. Fig. 6.8 shows relation (6.51) calculated for the case of a linear electrooptical array; this function characterizes the degradation of noise suppression when the relative dimension $c_{X,Y}$ of the mask is finite in the case where the radiation pattern coincides in form with the function sinc{...} but not with the spheroidal function $\Psi_o{}^{Y,X}\{\ldots\}$.

For a planar array with a rectangular aperture the transcendental equation (6.63), by (6.86), takes the form

$$\lambda_X(c_X)\lambda_Y(c_Y) = \frac{1}{2} \tag{6.87}$$

Fig. 6.9 is a plot, in coordinates $c_{X,Y}$, of a function that solves transcendental equation (6.87). For clarity, the same figure includes the separate functions $\lambda_X(c_X)$, $\lambda_Y(c_Y)$, and $\lambda_X\lambda_Y = 0.5$. The case of greatest practical importance is $c_X = c_Y = c_{opt}$, where c_{opt} satisfies

$$\lambda_X(c_{opt}) = \frac{1}{\sqrt{2}} \tag{6.88}$$

which follows from (6.87). Solution of (6.88) by Newton's method with the use of (C.2) gives $c_{opt} = 0.847$. Here the area of the negative mask is a minimum (see the hatched square in Fig. 6.9), so that it also gives rise to a minimal perturbation of the optical images of the useful signals. Indeed, the

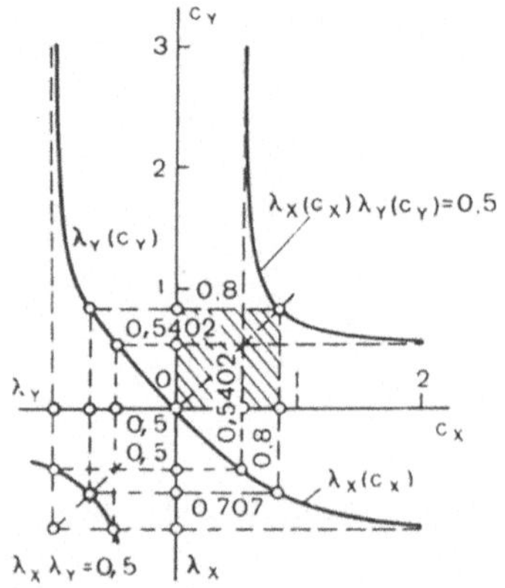

Figure 6.9: Diagram for selecting the mask of the processor of a planar array with a rectangular aperture.

relative area $c_X c_Y \geq c^2_{opt}$ under the condition (6.87), since $c_{X,Y} \geq \lambda_{X,Y}(c_{X,Y})$.

Fig. 6.10 shows a family of masks for an interference supression CO processor of a square array. This family is analogous to that in Fig. 6.7. Fig. 6.10 also contains a dashed curve like that in Fig. 6.7, representing the envelopes of the family of masks (6.32) with the use of a radiation pattern optimal with respect to the directive gain and not of functions $\Psi^X_o(\ldots), \Psi^Y_o(\ldots)$.

Fig. 6.11 shows auxiliary radiation patterns (6.54) that provide complete noise suppression and coincide in form with the function (C.1). The figure also gives plots of relation (6.66) calculated with (C.3), which characterize the depth of noise suppression at the processor output for three values of the parameter c_X: 0, 0.5402 (linear array), and 0.847 (square array). In Fig. 6.12, relation (6.66) is plotted to logarithmic coordinates. Fig. 6.13 shows an apodization amplitude phase distribution (6.77), which makes it possible to realize an auxiliary radiation pattern of the type of the fundamental extended spheroidal function $\Psi_o(\ldots)$. This distribution was calculated with formula (C.3) and is plotted in Fig. 6.13 for the same values of the parameter c_X.

Thus the structure and parameters of an interference supression CO processor (see Fig. 6.6) for a planar (linear) array are completely determined. In dynamic (real time) operation, the controlled perturbing mask τ_Ψ (spatial filter) can be suitably realized through the use of optically controlled masks. One version of such masks is considered in Chapter 9. It is necessary to know the linear dimensions (σ_x, σ_y) of the perturbing mask in plane π_1. By

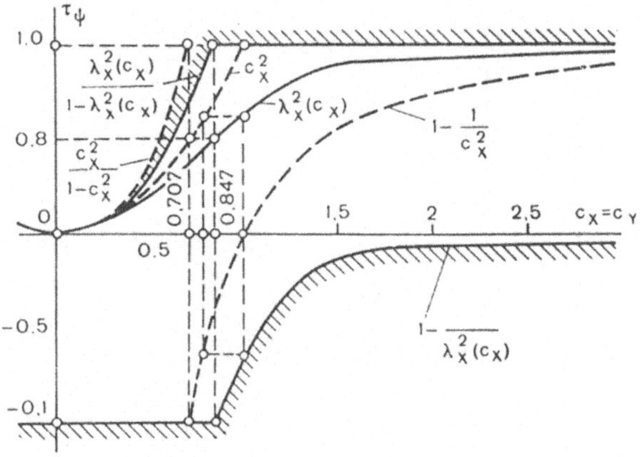

Figure 6.10: Family of masks of the interference supression CO processor of a planar array with a square aperture.

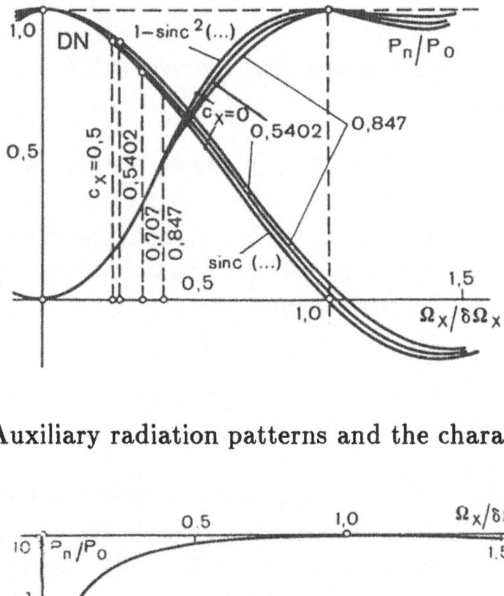

Figure 6.11: Auxiliary radiation patterns and the character of a gap.

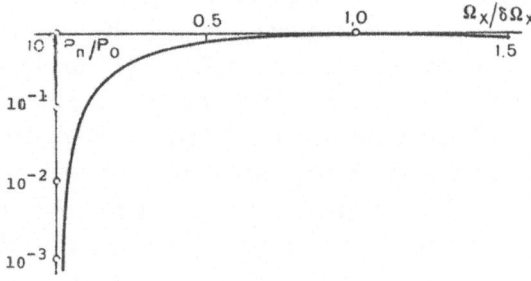

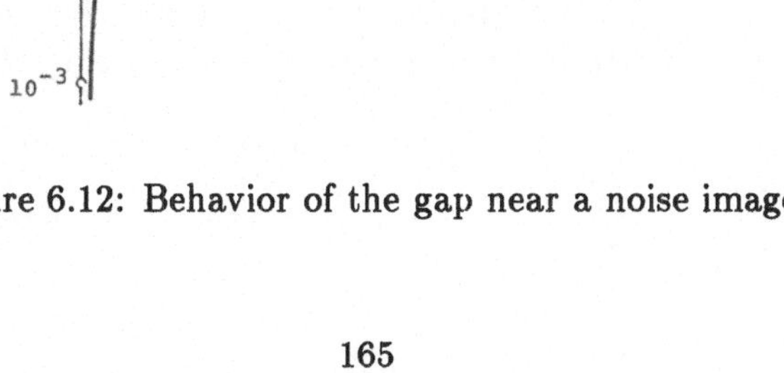

Figure 6.12: Behavior of the gap near a noise image.

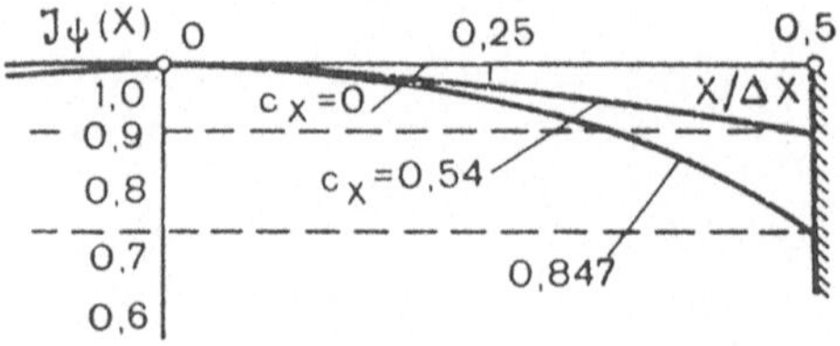

Figure 6.13: Amplitude phase distribution for an auxiliary radiation pattern.

(6.86), (2.13), and (2.2) and the discussion of (1.2)

$$\sigma_{x,y} = c_{X,Y}\lambda f/\Delta x, y \tag{6.89}$$

where $c_{X,Y} = 0.5402$ or 0.847, $\Delta x \Delta y$ is the dimension of the space–time light modulator, f is the focal length of objective L_1 in Fig. 6.6, and λ is the wavelength of the light.

6.4.3 Planar electrooptical array with circular aperture

Here the interference supression CO processor can be realized in accordance with the diagram of Fig. 6.14 (compare Fig. 6.6). By (2.2), the space–time light modulator in plane Π_1, as in the array, should have a circular (elliptical) aperture. In accordance with (6.79), the negative perturbing mask $\rho_\circ$ should also have the shape of a circle (ellipse) whose metric diameter, by analogy with (6.89), is

$$2r = c_r \lambda f / 2R_\circ \tag{6.90}$$

where $c_r = c_\circ$ is the analog of the parameter (6.86) and $R_\circ$ is the radius of the array aperture. The low-frequency spatial filter in plane Π_2 coincides in shape and dimensions with the space–time light modulator. The eigenfunction $\Psi_\circ$, which by (6.54) and (6.80) describes the form of the auxiliary radiation pattern (shape of the diffraction spot in the spectral plane π_1) and the amplitude phase distribution (transmittance function of the apodization mask in plane Π_1), can be calculated in a similar way [174, 175] with the use of earlier results [172]. In particular, the second approximation is

$$\Psi_\circ^r(\mathbf{K}_\perp, c_r) = a_\circ \left[J_\circ\left(\frac{\pi}{2} c_r \frac{|\mathbf{K}_\perp|}{K_\circ}\right) + \frac{J_2\left(\frac{\pi}{2} c_r\right) J_1\left(\frac{\pi}{2} c_r \frac{|\mathbf{K}_\perp|}{K_\circ}\right)}{B_\circ \sqrt{\frac{\pi}{2} c_r - \frac{\pi}{2} c_r J_3(c_r)}} \right] \tag{6.91}$$

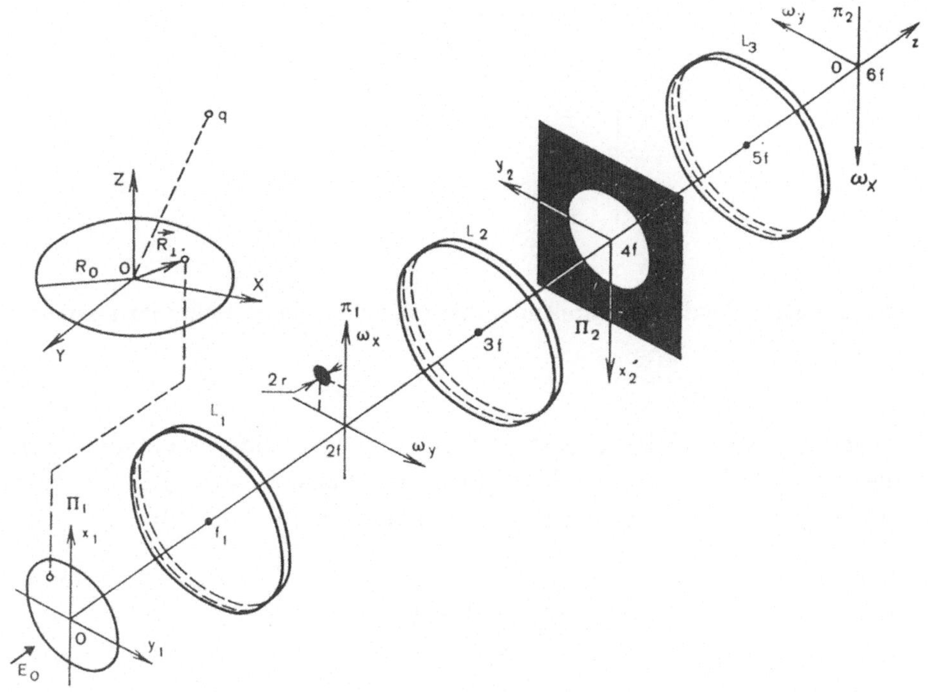

Figure 6.14: Interference supression processor of a planar electrooptical array with a circular aperture.

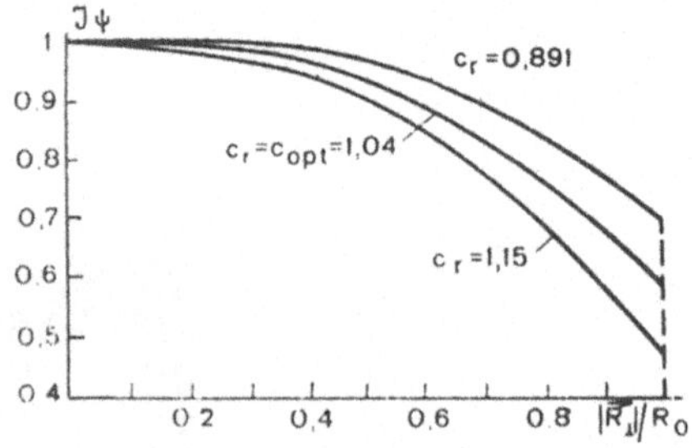

Figure 6.15: Auxiliary amplitude phase distribution of a circular aperture.

where a_o is a normalizing constant depending on c_r and selected under condition (6.46); J_0, J_1, J_2, and J_3 are Bessel functions, $K_o = rk/f$ is the radius of the circular region ρ_o in generalized angular coordinates, and

$$B_o = \frac{\sqrt{\frac{\pi}{2}c_r}\left[J_1\left(\frac{\pi}{2}c_r\right) + J_3\left(\frac{\pi}{2}c_r\right)\right]}{2} + \sqrt{\frac{\frac{\pi}{2}c_r\left[J_1\left(\frac{\pi}{2}c_r\right) - J_3\left(\frac{\pi}{2}c_r\right)\right]^2}{4} + \frac{\pi}{2}c_r J_2^2\left(\frac{\pi}{2}c_r\right)}$$

Fig. 6.15 shows an apodization amplitude phase distribution for a circular array within its aperture. The distribution was calculated with (6.91) for three values of the parameter c_r, including the optimal value (see below), with $\mid \mathbf{R}_\perp \mid / \mathbf{R}_o = \mid \mathbf{K}_\perp \mid / \mathbf{K}_o \leq 1$.

The greatest eigenvalue $\lambda_r(c_r)$, which corresponds to the eigenfunction (6.91), is given in the second approximation by

$$\lambda_r(c_r) = B_o^2 / \frac{\pi}{2}c_r \tag{6.92}$$

where B_o is the quantity expanded above.

Curve (6.92) and the envelopes (6.64) and (6.65) of the family of perturbing masks in the spectral plane π_1 are shown in Fig. 6.16. The optimal mask, from the standpoint of processor energetics and perturbing action on useful signals, is the one with the parameter $c_r = c_{opt} = 1.04$, which is obtained as the solution of transcendental equation (6.63).

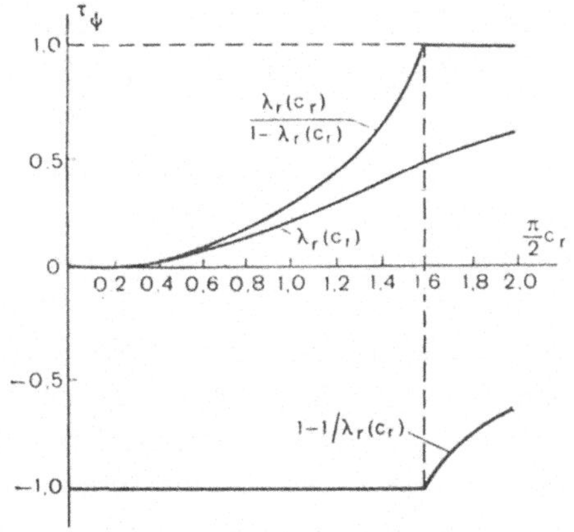

Figure 6.16: Family of masks of an interference supression CO processor of a circular array.

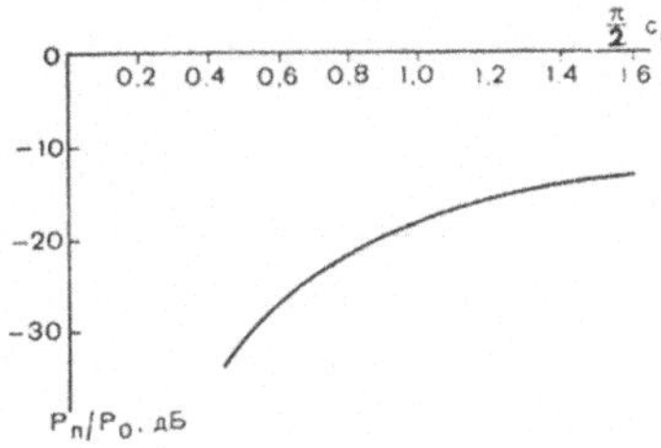

Figure 6.17: Diagram for determining the depth of noise suppression.

Fig. 6.17 shows a plot of the estimate (6.51), which characterizes the degradation in interference supression action of a CO processor with a finite parameter c_r, where in place of (6.54) we have used the radiation pattern optimal with respect to the directive gain:

$$\begin{aligned} F(K_\perp) &= \hat{\mathcal{F}}\{\mathrm{circ}(|\mathbf{R}_\perp|/R_\circ)\} = \pi R_\circ^2 J_\circ(|\mathbf{K}_\perp|R_\circ/\pi)/(|\mathbf{K}_\perp|R_\circ/\pi) \\ &= \pi^2 R_\circ J_\circ(|\mathbf{K}_\perp|R_\circ/\pi)/|\mathbf{K}_\perp| \end{aligned} \quad (6.93)$$

where

$$\mathrm{circ}(|\mathbf{R}_\perp|/R_\circ) = \begin{cases} 1, & |\mathbf{R}_\perp| \le R_\circ \\ 0, & |\mathbf{R}_\perp| > R_\circ \end{cases}$$

6.5 Generalization of the Coherent Optical Method of Noise Rejection

The results obtained above have been most fully developed in [164-168] for the following cases: rejection of noise signals from point sources (in the far field of the array) and spatial rejection of narrow-band noise signals received by an array with a planar aperture in the approximation of continuous positioning of receiving elements (space–time light modulator channels). In what follows, we present a discussion on the generalization of the results to the following cases: rejection of sectorial noise (distributed in space) and wide-band noise; linear electrooptical arrays using space–time light modulators with spatial scanning of the time signal (multi-channel AOMs, space–time light modulators with electron-beam addressing, etc.); cylindrical electrooptical arrays; and electrooptical arrays with discrete positioning of array elements (space–time light modulator channels).

6.5.1 Suppression of sectorial noise

Suppose that, in directions $\mathbf{K}_q \in \Delta_n$ (where Δ_n is the solid angle subtended by the noise sources), blinding radiation $\mid \mathbf{S}(\mathbf{K_q}) \mid^2 \gg \mid \mathbf{S}(\mathbf{K}) \mid^2$, $\mathbf{K} \notin \Delta_n$ is incident. Then, by analogy with (6.4), the ideal interference suppression radiation pattern with a null in sector Δ_n can be written as

$$F_\Pi(\mathbf{K},\mathbf{K}')_{ideal} = \mathbf{n}_\parallel(\mathbf{K})[F(\mathbf{K},\mathbf{K}') - F(\mathbf{K},\mathbf{K}')p(\mathbf{K},\Delta_\Pi)] \tag{6.94}$$

where

$$p(\mathbf{K},\Delta_\Pi) = \begin{cases} 1, & \mathbf{K} \in \Delta_\Pi \\ 0, & \mathbf{K} \notin \Delta_{pi} \end{cases} \tag{6.95}$$

is an extended negative screen covering the spectrum Δ_n in the far field of the array. As in the case of point screens (6.5), an extended screen (6.95) cannot be realized if an array of finite size is employed, since the radiation pattern of the array is an analytic function and cannot be a constant equal to zero on a finite interval [134]. Thus, to solve the problem stated, in place of the Π-shaped function (6.95) we find some other function that is realizable and is closest to (6.95) in the rms sense. To do this, we use the self-reproducibility property (6.34) of the normalized array radiation pattern that is optimal with respect to the directive gain $-F_o(\mathbf{K},\mathbf{K}')$, according to which the current radiation pattern $F(\mathbf{K},\mathbf{K}')$ can be represented as the following generalized series in sampling functions (a series of Kotel'nikov type)

[2, 172]:[6]

$$F(\mathbf{K},\mathbf{K}') = \sum_m F(\mathbf{K}_m,\mathbf{K}')F_\circ(\mathbf{K},\mathbf{K}_m) \tag{6.96}$$

where the directions $\mathbf{K}_m$ are chosen such that the sampling function (readings function)

$$F_\circ(\mathbf{K}_m,\mathbf{K}_n) = \begin{cases} 1, & m = n \\ 0, & m \neq n \end{cases} \tag{6.97}$$

Substituting (6.96) into (6.94) and summing in (6.96) over those m for which $\mathbf{K}_m \in \Delta_n$, we obtain the following realizable approximation to the radiation pattern:

$$\mathbf{F}_\Pi(\mathbf{K},\mathbf{K}')_{real} = \mathbf{n}_\parallel(\mathbf{K}[F(\mathbf{K},\mathbf{K}') - \sum_{\mathbf{K}_m\in\Delta_\Pi} F(\mathbf{K},\mathbf{K}')F_\circ(\mathbf{K},\mathbf{K}_m)] \tag{6.98}$$

which, like (6.7), is the best rms approximation to the ideal interference suppression radiation pattern (6.94), by virtue of the choice of the function $F_\circ$. In distinction to (6.7), the directions $\mathbf{K}_m$ need not coincide with the $\mathbf{K}_q$. Since interference suppression radiation patterns (6.98) and (6.7) are structurally alike, CO processors that realize them can be implemented in accordance with the diagram of Fig. 6.5.

The approach just described for suppressing sectorial spatial noise can also be recommended for combating wide-band point-source (sectorial) noise that does not satisfy criterion (4.24), since noise signals of this type are equivalent to sectorial noise because of their optical images in the processor output plane are "smeared" along the dispersion curves (2.22) or (4.23).

6.5.2 Linear electrooptical arrays with space–time light modulators using spatial scanning of a time signal

In Sections 3.1 and 3.2 it was shown that a CO processor of a linear electrooptical array with an input device that performs a spatial scan of a time signal reproduces information about the frequency and angle of radar objects in oblique-angular coordinates (3.13). The frequency spectrum $\dot{s}_q(\Omega)$ of objects with position angle $\theta_q \neq 90°$ is imaged along lines that are not parallel to the $\omega_y = \Omega/v$ axis (see Fig. 3.2). In order to achieve the spatial

[6]For the case of a linear array, where $F_\circ(\ldots) = \text{sinc}(\ldots)$, expansion (6.96) takes a well-known form [52].

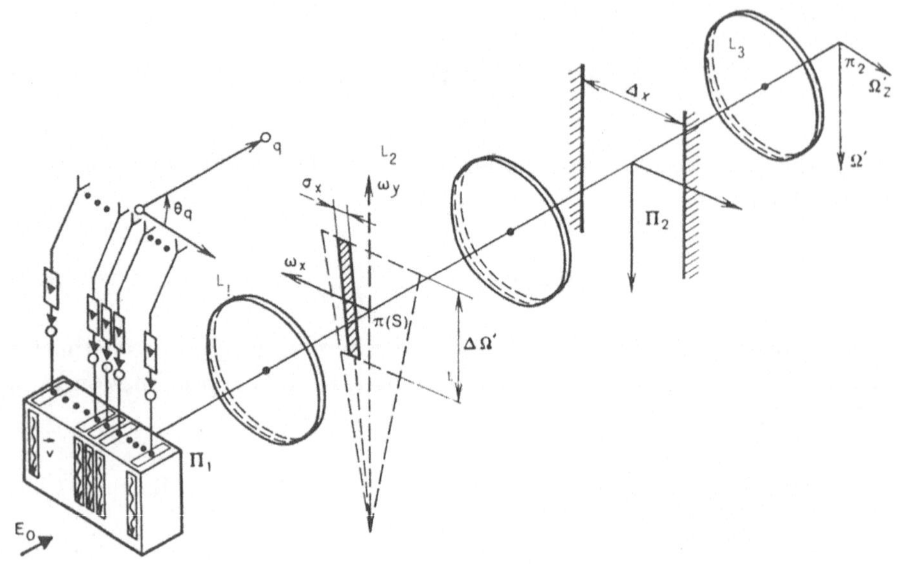

Figure 6.18: Interference suppression processor of an electrooptical array using a multi-channel AOM.

rejection of broad-band noise signals, it is necessary in the present case to place a mask (6.69) in the plane $\pi_1(S)$ along the line just mentioned. The length of the mask should ensure the complete masking of the image of the frequency spectrum $\dot{s}_q(\Omega)$. Because the mask is placed at an oblique angle, however, the rejecting spatial filter, constructed in accordance with the scheme discussed above, does not provide complete noise suppression, since the variables in the present case are not resolved with respect to the ω_x and ω_y axes.

Let us evaluate the degradation in suppression due to the fact just noted, taking as an example a point noise source with angular coordinate θ_q and frequency spectrum $\dot{s}_q(\Omega)$ [see, e.g., (3.8)]. Corresponding to this noise is a frequency-angular spectrum of the form $\dot{s}(\Omega_z, \Omega) = \dot{s}(K\cos\theta, \Omega) = \dot{s}(\Omega)\delta(\Omega_z/K - \cos\theta_q)$. For example, let the input device be a multi-channel AOM. Then, by (3.4), the radar object is imaged in the plane S of the interference suppression processor (Fig. 6.18) as the following light-field distribution:

$$\dot{e}^{(1)}_{qw}(\omega_x^{(S)}, \omega_y^{(S)}) = \frac{T^{(1)}E_\circ}{2\pi\lambda f m_x}\dot{s}_q(\Omega)F\left(\frac{\omega_x^{(S)}}{m_x} - \frac{\Omega}{c}\cos\theta_q\right)$$

$$\times \quad F_t(\omega_y^{(S)} - \Omega/v)\exp(-iy_\circ\Omega/v) \tag{6.99}$$

Assuming that the radiation pattern of the linear array $F(\Omega_z) = F_\Psi\Psi_o(\Omega_z)$ [see (6.54)], by analogy with (6.58) we obtain an expression for the noise image at the processor output:

$$\begin{aligned}
&\dot{e}_{q\omega}^{(1)}(\omega_x,\omega_y)_{\|} \\
&= \hat{I}_X\left\{\tau_\Psi\left(\frac{\omega_x^{(S)}}{m_x} - \frac{v\omega_y^{(S)}}{c}\cos\theta_q\right)\dot{e}_{q\omega}^{(1)}[\omega^{(S)},\omega(S)]\right\}\Big/\|\hat{I}\| \\
&\approx \frac{T^{(1)}E_\circ}{2\pi\lambda f m_x}F_\Psi\left\{\Psi_\circ\left(\frac{\omega_x}{m_x} - \frac{\Omega}{c}\cos\theta_q\right)\right. \\
&\left.-b_\circ\left[\frac{\cos\theta_q}{c}(\Omega - v\omega_y)\right]\cdot\Psi_\circ\left(\frac{\omega_x}{m_x} - \frac{v\omega_y}{c}\cos\theta_q\right)\right\} \\
&\times\dot{s}(\Omega)F_t(\omega_y - \Omega/v)\exp\left(-\frac{iy_\circ}{v}\Omega\right)
\end{aligned} \tag{6.100}$$

where, by (6.69), $\tau_\Psi(\Omega_z) = 1 - 2\mathrm{rect}(\Omega_z/\Delta\Omega_z)$, $\Delta\Omega_z = 2\pi c_z/\Delta Z$, and c_z is a parameter analogous to $c_{X,Y}$. The final estimate is found by analogy with (6.66):

$$\begin{aligned}
\frac{P_\Pi}{P_\circ} &= \frac{\int\int\int_{-\infty}^{\infty}|\dot{e}_{q\omega}^{(1)}(\omega_x,\omega_y)_\Pi|^2 d\omega d\omega_x d\omega_y}{\int\int\int_{-\infty}^{\infty}|\dot{e}_{q\omega}^{(1)}(\omega_x,\omega_y)|^2 d\omega d\omega_x d\omega_y} \\
&= \frac{\int_{-\infty}^{\infty}\{[1 - b_\circ^2(\cos\theta_q/cv\omega_y)]\,F_t^2(\omega_y)\}\otimes|\dot{s}(v\omega_y)|^2 d\omega_y}{\int_{-\infty}^{\infty}F_t^2(\omega_y)\otimes|\dot{s}(v\omega_y)|^2 d\omega_y} \\
&= \int_{-\infty}^{\infty}\left[1 - b_\circ^2\left(\frac{\cos\theta_q}{c}v\omega_y\right)\right]F_t^2(\omega_y)d\omega_y\Big/\int_{-\infty}^{\infty}F_t^2(\omega_y)d\omega_y \\
&= 1 - \int_{-\infty}^{\infty}J_\Psi^4(Z)J_t^2\left(Z\frac{c}{cos\theta_q v}\right)dZ\Big/\int_{-\infty}^{\infty}J_t^2\left(Z\frac{c}{cos\theta_q v}\right)dZ \\
&= 1 - \int_{-\Delta y\cos\theta_q v/2c}^{\Delta y\cos\theta_q v/2c}J_\Psi^4(Z)dZ/\Delta y\frac{\cos\theta_q v}{c}
\end{aligned}$$

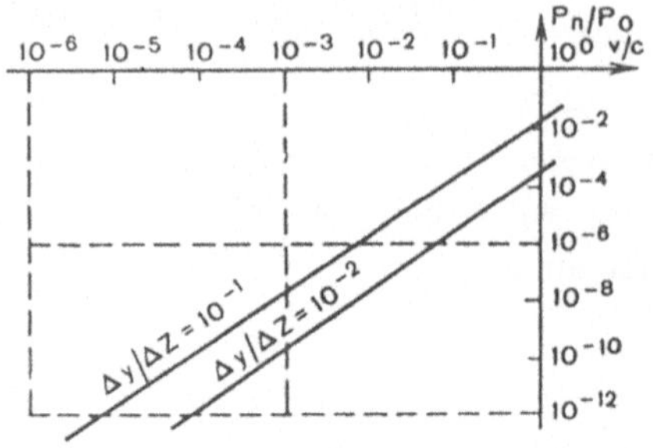

Figure 6.19: Depth of noise suppression.

$$\approx \; 1.8\left(\frac{\Delta y^2}{\Delta Z}\right)\left(\frac{v}{c}\right)^2 \cos^2\theta_q \tag{6.101}$$

where we have used relations (6.99), (6.100), (6.77), and (C.4) and the integral properties of the scanning transformation, as well as the power-series expansion of the amplitude phase distribution (C.3).

Fig. 6.19 is a plot of relation (6.101) for $\cos\theta_q = 1(\theta_q = 90°)$, that is, for normal incidence of the wave on the array. The figure contains curves for two possible values of the ratio $(\Delta y/\Delta Z)$ of the acoustical waveguide length (spatial scanning of the time signal) Δy to the array length ΔZ. The evaluation confirms that the rejection method under discussion is efficient even for wide-band noise signals.

6.5.3 Cylindrical electrooptical array

Processing algorithm (4.39) for a space–time signal received by a cylindrical array does not satisfy the normalized property (6.29), since the passive complex mask (4.34) entails uncompensated light losses ($\mid \dot{T}(\Omega_\varphi, \Omega_Z) \mid \rightarrow 0$ as $\Omega_\varphi \rightarrow \infty$). For this reason, we cannot construct an interference suppression processor with the structural scheme of Fig. 6.2. If, however, in place of the starting processor realizing algorithm (4.33) we use a CO processor with the reduced mask shown in Fig. 5.9, then such an auxiliary transformation $\widehat{H}_\Psi$ is possible in algorithm (6.68). This transformation makes it possible to realize a CO processor with the structural scheme of Fig. 6.4. Indeed, the reduced mask $\dot{T}^{(I,II,III)}$ in Fig. 5.9b is finite (its dimension with respect to Ω_φ is proportional to the number N of elements in the rings of the cylindrical array). The radiation pattern formed in the output plane of the processor in Fig. 4.4b is therefore finite in its spectrum and can be transformed to the auxil-

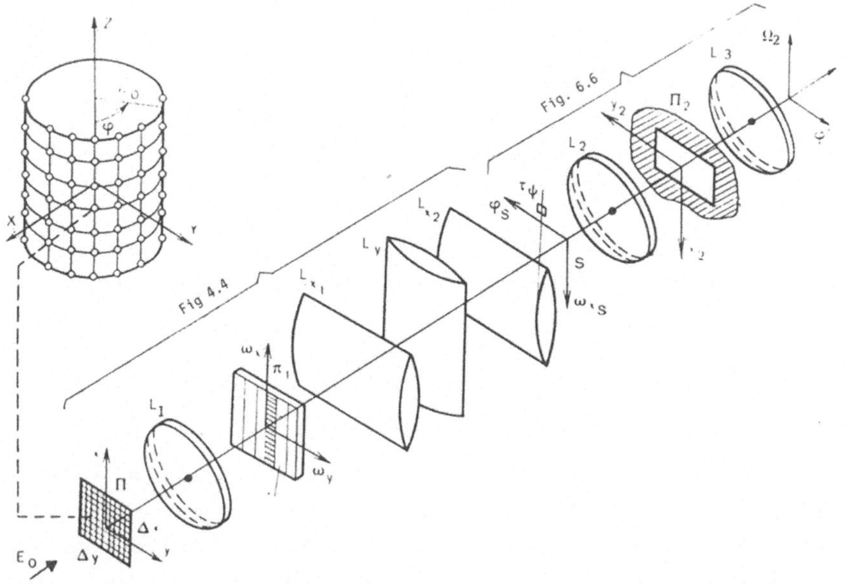

Figure 6.20: Interference suppression CO processor of a cylindrical electrooptical array.

iary radiation pattern Ψ_o (see Appendix C and Fig. 6.11).[7] Fig. 6.20 shows an interference supression processor of a cylindrical array. The processor is constructed in accordance with algorithm (6.68). To generate the auxiliary radiation pattern Ψ_o in the plane S, the transmittance function of the reduced mask in plane π_1 of the CO processor $\dot{T}$ is corrected $(\sim J_\Psi / \mid \dot{T}^{(III)} \mid^2)$ in such a way that the light flux passing through it is described by (6.80). The possibility of such a correction is guaranteed because the modulus of the transmittance function of reduced mask (5.53) [or (5.56)] has no zeroes. The diffraction-limited image-forming system in Fig. 6.20 matches that of Fig. 6.6 except that the dimensions of the low-frequency spatial filter in the plane Π_2 are equal to Δx (the dimensions of the space–time light modulator) in the x_2 direction and $m_y N f/k = N\lambda f/\Delta y$ in the y_2 direction [where we have used formulas (1.2), (4.35) and (4.36) and the fact that $\mid \Omega_{varphi} \mid \leq N/2$; here Δy is the orthogonal dimension of the space–time light modulator].

[7]In the Ω_Z direction, the structure of the interference supression processor is similar to the linear-array case analyzed in Section 6.4.2, since the variables Z and φ are separated in algorithm (4.33).

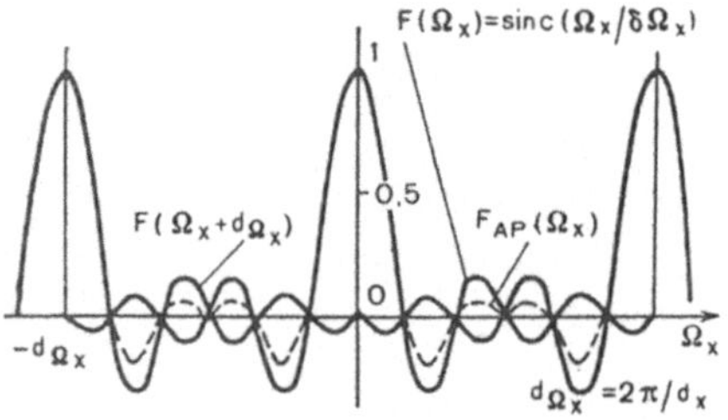

Figure 6.21: Effect of array discreteness on depth of gap.

6.5.4 Effect of discreteness of the array and the space–time light modulator

As we showed in Chapter 5, the discrete structure of the array and of the space–time light modulator does not alter the optical scheme of the starting CO processor (interference suppression), but it does offer further possibilities for the deliberate modification of the processor (for example, the modification employed in Section 6.5.3). But the radiation pattern of the electrooptical array with discrete positioning of the array elements and the channels of the space–time light modulator differs from a radiation pattern with a continuous structure [compare (5.1) and (4.4)]. The greater the number of elements, however, the less difference there is in the region of radio visibility [52, p. 240]. The interference suppression processors synthesized above in the continuous-aperture approximation, although they do not afford theoretically ideal suppression, they do make it possible to obtain results that can be applied in practice. We illustrate what has been said in the example of a planar equidistant array whose radiation pattern, by (C.6), is the superposition of continuous-aperture radiation patterns shifted by $2\pi/d_{X,Y}$, where $d_{X,Y}$ is the spacing between elements of the array (see Fig. 6.21). If a single-cascade processor based on Fig. 6.6 is used, then from the superposition just mentioned there will be "removed" a principal radiation pattern $F(\mathbf{K}_\perp)$, whose principal maximum lies in the region of radio visibility ($\mathbf{K}_\perp \leq K$). The shifted radiation patterns $F(\mathbf{K}_\perp \pm \mathbf{n}_x m 2\pi/dx \pm \mathbf{n}_Y n 2\pi/d_Y)$ are not suppressed, but their maximum level in the radio visibility zone does not exceed the amplitude of side-lobe number $N/2$, whose relative intensity [52] is of the order of $[2/(N+1)\pi]^2 \approx 1/N^2$ for a linear array of N elements and $\approx 1/N^2M^2$ for a rectangular array of MN elements. The resulting estimates characterize the depth of noise suppression by a single-cascade processor. If the degree of noise suppression is not adequate (e.g., for $N = 30$ the depth of suppression is approximately -30 dB), it is necessary to use the multi-cascade processor illustrated in Fig. 6.5.

6.6 Effect of Errors in the Realization of the Spatial Filter on the Depth of Null Formation

As we have already shown, the perturbing mask with transmittance function (6.69) in the plane π_1 of the processor (see Fig. 6.6) is implemented in real time with the help of optically controlled masks (see Chapters 1 and 9). The actual technical characteristics of the devices lead to a distortion of algorithm (6.68) for the suppression of spatial noise signals. In [168], for the example of a linear electrooptical array, the effects on the depth of formation of the controlled gaps $(P_n/P_\circ)$ due to several factors are studied in detail. The factors are the random phase variations in the optically controlled mask, the inaccuracy of placement of the filter in the direction of the noise source, the distortion of the filter edges due to the finite resolving power of the optically controlled mask, and the distortions of the amplitude phase distribution at the levels of the array, the receiving and amplifying modules, and the space–time light modulator. Let us look at the main ones of the resulting estimates.

The transmittance function $\tau_{\Psi}(\mathbf{K}_S, \mathbf{K}_q)$ (6.69) for the case of a linear array oriented, say, along the OX axis can be written in the following way with allowance for the phase variations in real optically controlled masks:

$$\tilde{\tau}_{\Psi}(\Omega_{XS} - \Omega_{X_q}) = [1 - \rho(\Omega_{XS} - \Omega_{X_q})] \exp[i\xi(\Omega_{XS})] \tag{6.102}$$

where $\xi(\Omega_{X_S})$ is a random function, stationary with respect to the spatial variable Ω_{X_S}, characterizing the phase variation of the optically controlled masks; and $|\xi(\Omega_{X_s})| < 1$; $\Omega_X = K\cos\alpha = K\sin\theta\cos\varphi = -(\mathbf{n}_X\mathbf{K})$ is the one-dimensional analog of vector $\mathbf{K}$.

A distorted mask of the form (6.102) leads to a distortion of the ideal interference suppression radiation pattern (6.58), which is defined as the reaction of a CO processor to plane waves arriving from various directions Ω_{X_q} [action of operator (6.78) on plane waves]. Therefore, writing expression (6.78) for the one-dimensional case, substituting into it (6.102), and using (6.48) and (6.55), we obtain relation (6.66), which characterizes the degree of noise suppression for $\Omega_X = \Omega_{X_q}$:

$$\Pi_\circ = \left.\frac{P_\Pi}{P_\circ}\right|_{\Omega_X = \Omega_{X_q}} \approx 4[\lambda_X(2c_X) - \lambda_X(c_x)]^2 \xi^2 \tag{6.103}$$

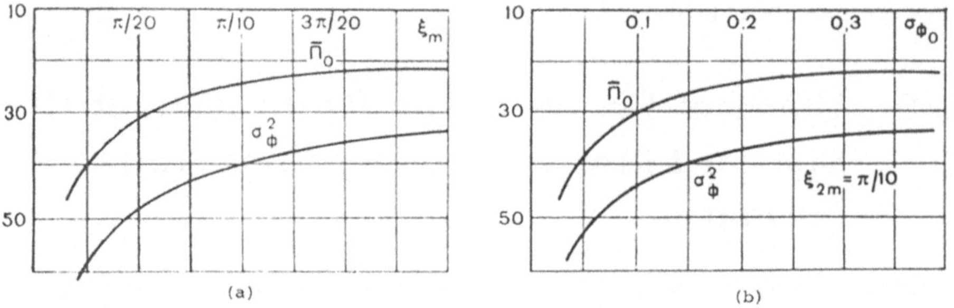

Figure 6.22: Effect of random phase variations of optically controlled mask.

where λ_X and c_X are as defined above and are found in accordance with Appendix C. Here we have taken into account that, by the bounding property of the diaphragm in plane Π_2 of Fig. 6.6, the formation of the output response at the output of the CO processor includes a contribution only from the spectral components of the random function $\xi(\Omega_{X_S})$ lying in the interval $|X| \leq \Delta X$ (where ΔX is the linear dimension of the array).

Expression (6.103) makes it possible to find the mean $\overline{\Pi}_\circ$ and the variance $\sigma_\phi^2 = \overline{\Pi^2}_\circ - (\overline{\Pi}_\circ)^2$ if we know how the random quantity $\xi(\Omega_{X_S})$ is distributed. In particular, Fig. 6.22a and b are plots of $\overline{\Pi}_\circ$ and σ_ϕ^2 for the following two distributions $p[\xi(\Omega_{X_S})]$: for the uniform distribution

$$p[\xi(\Omega_{X_s})] = \begin{cases} (1/2)\xi_m, & |\xi| \leq \xi_m, \\ 0, & |\xi| \leq \xi_m, \end{cases}$$

and for the "random grid" distribution (model of an optically controlled mask made of liquid crystals [176])

$$\begin{aligned} \xi(\Omega_{X_S}) &= \xi_\circ + \xi_1(\Omega_{X_S}) \sin[\xi_2(\Omega_{X_S})], \qquad v = 0, 1 \\ p[\xi_v(\Omega_{X_S})] &= \frac{1}{\sqrt{2\pi}\sigma_{\phi_v}} \exp[-\xi_v^2/2\sigma_{\phi_v}^2]; \\ p[\xi_2(\Omega_{X_S})] &= \begin{cases} \frac{1}{2}\xi_m & |\xi_2| \leq \xi_{2m} \\ 0 & |\xi_2| > \xi_{2m} \end{cases} \end{aligned}$$

For a uniform phase variation and $\xi_m = \pi/20$, suppression is not complete

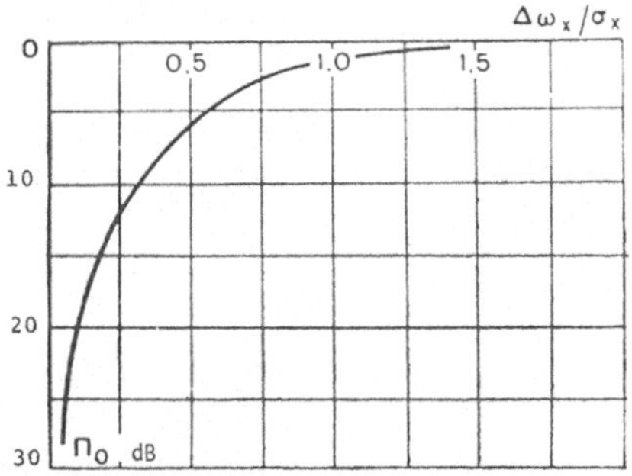

Figure 6.23: Evaluation of accuracy of filter placement in direction of noise source.

($\overline{\Pi}_\circ \approx 32$ dB); for the "random grid" distribution with $\sigma^2_{\phi_o} = \sigma_{\phi_1} = 0.1$ and $\xi_{2m} = \pi/10$, $\overline{\Pi}_\circ \approx 30$ dB.

To evaluate the effect of the accuracy of filter placement, we use expression (6.66), which we can obviously write in the following form for the linear-array case under consideration:

$$\Pi_\circ(\Omega_X - \Omega_{X_q}) = \frac{P_\Pi(\Omega_X - \Omega_{X_q})}{P_\circ(\Omega_X)} = 1 - b^2_\circ(\Omega_X - \Omega_{X_q}) \tag{6.104}$$

where $b_\circ(\Omega_X - \Omega_{X_q}) = \int_{-\infty}^{\infty} \Psi_\circ(\Omega_X - \Omega_{X_S})\Psi(\Omega_X - \Omega_{X_q}) d\Omega_{X_S}$ (the computational formula is given in Appendix C).

For accurate placement of the filter in the direction of the noise source $\Omega_X = \Omega_{X_q}$, an null is formed: $\Pi_0(\Omega_{X_q}) = 0$. Fig. 6.23 shows (6.104) as a function of the parameter $\Delta\omega_x/\sigma_x = \mid \omega_x - \omega_{x_q} \mid /\sigma_x$, which characterizes the accuracy of filter placement. Here $(\Omega_X - \Omega_{X_q})/\delta\Omega_X = \Delta\omega_x/\delta\omega_x$, where $\delta\Omega_X = 2\pi/\Delta X, \delta\omega_x = 2\pi/\Delta x, \omega_x$ is the spatial frequency read off in plane π_1 of Fig. 6.6; ω_{x_q} is the spatial frequency of the noise sources; and $\sigma_x = c_X\delta\omega_x$ is the width of the filter (6.89). In particular, for $\Delta\omega_x/\sigma_x = 0.05$ the depth of suppression is about 30 dB. The factor under consideration is decisive with regard to the effect on the depth of suppression and imposes stringent requirements on the characteristic of the adaptive feedback loop.

Because the material of the optically controlled mask has a finite resolution, the edge of the filter is not "sharp" but has a smeared-out front of decreasing phase (the dimension σ_x is "grown over"). In [168], for an exponential edge model [176] with an exponential "tail" roughly $0.1\sigma_x$ wide, the depth of suppression was found to be about 40 dB.

The effect of the distortion of the amplitude phase distribution on the depth of the gap in the direction of the noise source is analyzed with the

Table 6.1: Sources of mask distortion.

General Distortion Factor	Specific Distortion Factor	Π_o, $\tilde{\Pi}_o$, dB
Phase Variations in Optically Controlled Mask	Uniform Distribution for ξ_m: Not more than $\pi/20$ Not more than $\pi/40$	 -30 -40
	Random Array Distribution Type for $\xi_{2m} \leq \pi/10$ and $\sigma^2_{\Phi_o}$ equal to: Not more than 0.1 Not more than 0.04	 -30 -40
Accuracy of Filter Installation	Deflector Error $\delta\omega_x/\sigma_x$: Not more than 0.005 Not more than 0.01	 -30 -40
Imperfect Filter Edge	Maximum resolution of optically controlled mask with exponentially decreasing phase of not more than $0.1\sigma_x$	-40
Amplitude Phase Distribution Error	Normalized distribution for: $r_A/\Lambda \leq 0.8$ and $\sigma^2 \leq 1.0$ $r_A/\Lambda \leq 0.9$ and $\sigma^2 \leq 0.1$	 -30 -40

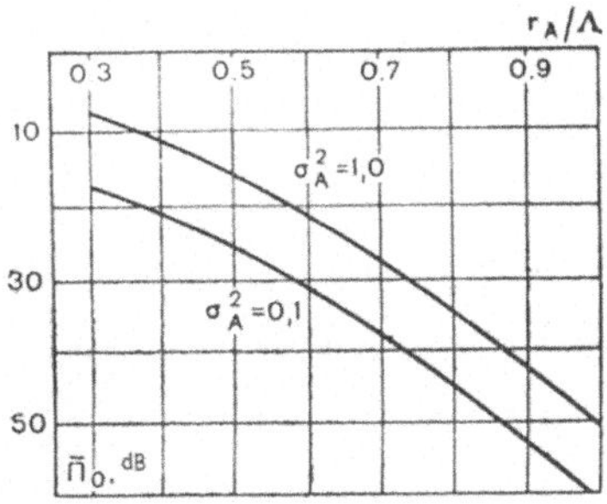

Figure 6.24: Effect of distortion of amplitude phase distribution on depth of gap.

same scheme that was used to evaluate the phase variation of the optically controlled mask. We assume that the distortion of the amplitude phase distribution is due only to phase errors, which are normally distributed (with variance $\sigma_A^2(X)$ and mean zero). For $\sigma_A^2(X) = \sigma_A^2$ and $r_A \ll \Delta X$ (where r_A is the correlation radius of the errors), the following estimate was obtained in [168] for the mean depth of suppression:

$$\bar{\Pi}_\circ = \sigma_A^2[1 - \Phi(\pi r_A/\Lambda)] \tag{6.105}$$

where $\Phi(z) = \frac{2}{\sqrt{\pi}} \int_0^z e^{z^2} dz$ and Λ is the radio wavelength. Fig. 6.24 plots the depth of suppression of the noise signal versus the relative correlation radius r_A/Λ for $\sigma_A^2 = 1.0$ and 0.1. In particular, if the rms phase error is $\sigma_A = \pi/3$ and $r_A/\Lambda = 0.9$, then $\overline{\Pi}_\circ \approx 40$ dB.

Table 6.1 makes it possible to assess the possibility of suppressing noise signals by the method set forth in this chapter.

Chapter 7

The Influence of Uncertainties on Electrooptical Arrays

Previously, it was supposed that the electrooptical array is an ideal transforming system: distortionless and noiseless. However, the coherent-optical processor acting as a precision analog type computing device is quite sensitive to the influence of various distorting factors, particularly due to the shortness of the optical carrier wavelength (about 1 μm) and to the inhomogeneities of the same order [1, 4, 8, 177]. In this case, along with the output radio scene distortion by the non-ideal elements in the processor, the signal-to-noise ratio worsens due to the emmergence of local (space and time) various natural noises at the output of the electrooptical array antenna. Apparently, these circumstances degrade the pattern- and spectrum-forming properties of real electrooptical arrays. Such abilities as the resolving power, gain, the side lobe level, dynamic range, sensitivity, etc. are affected.

This chapter successively discusses the following coherent optical processor distortion factor influences on the electrooptical array characteristics [178, 179]. These factors are: (1) The effect of the pupil and space–time light modulator channel interference; (2) the optical processor systems aberrations; (3) elements' misalignment; (4) the reading light spatial incoherence; (5) time incoherence; (6) Fresnel reflections in the optical system; (7) light diffusion on the optical elements' inhomogeneities; (8) the existence of a conjugate image during dualband input; (9) the space–time light modulator non-linearities; and (10) the zero order of diffraction.

The distorting factors which emerge on the level of the array receiving elements and the reception amplification modules (errors due to element excitation, channel nonidentity, etc.) are not addressed in this chapter because these are unavoidable in other systems as well (phased array antennas, multibeam array antennas) and could be considered through known methods [180]. Considering factors 1 through 10, this chapter develops a set of requirements for the coherent optical processor parameters which guarantee the best accuracy (the radiation pattern reproduction accuracy) of about 1% and the best dynamic operating range of the device, approximately 60 dB. The obtained results can be used during the first-order electrooptical array coherent optical processor design stage.

7.1 The Electrooptical Array Antenna Gain

The quality of the optical image is judged by its contrast [16]. This contrast, however, is not convenient for the use with the electrooptical arrays, because it is not directly tied with the main array power parameters. During linear reproduction of the source radio brightness it is wise to introduce a concept of optical source "average contrast" which will stand for the relation of the diffractional maximum to the average side-lobe level. In fact, this magnitude coincides with the electrooptical array gain, G_{EOAA}. The electrooptical array gain is determined not only by the array directivity, D_{AA}, but also by the engineering design of its coherent optical processor. The latter is the main limitation and, in fact, defines the dynamic range of the electrooptical array, DR, which is impossible to improve through subsequent processing.

The antenna system gain is defined by the product of its directivity and its overall efficiency [52]. Since the electrooptical array does not have phase rotators or a branched feed system, the receiving elements efficiency is close to being equal to one. However, the efficiency is lower than the array's due to radiation pattern distortion (defocusing) and also the parasitic noise of the relevant optical image at the real processor output. Therefore, the actual electrooptical array gain practically coincides with its actual directivity through the relation of the maximal electrooptical array reaction $\tilde{I}_q = (DF)_{COP} \cdot I_q$ (where $(DF)_{COP} = |\tilde{F}(\mathbf{K}_q, \mathbf{K}_q)/F(\mathbf{K}_q, \mathbf{K}_q)|^2$) to the planar wave from the source q to the sum of the source intensities which is equal to the uniform radio noise planar wave (in the array limits).

$$\begin{aligned} \tilde{I}_\phi &= \tilde{I}_q \int\!\!\int_{4\pi} |\tilde{F}(\mathbf{K}, \mathbf{K}_q)|^2 d^2\mathbf{K}/4\pi |\tilde{F}(\mathbf{K}_q, \mathbf{K}_q)|^2 \\ &\approx DF_{COP} I_q \int\!\!\int_{4\pi} |F(\mathbf{K}, \mathbf{K}_q)|^2 d^2\mathbf{K}/4\pi DF_{COP} |F(\mathbf{K}_q, \mathbf{K}_q)|^2 = I_\phi \end{aligned}$$

and statistically averaged exposure I_{ex} caused by the inidealities:

$$
\begin{aligned}
(G)_{EOAA} &\approx D_{EOAA} = \frac{\tilde{I}_q}{\tilde{I}_\phi + I_z} = \frac{DF_{COP} \cdot I_q}{I_\phi + I_z} \\
&= DF_{COP} \left(\frac{I_\phi}{I_q} + \frac{I_z}{I_q} \right)^{-1} = DF_{COP} \left[D_{AA}^{-1} + DR_{COP}^{-1} \right]^{-1} \\
&= e_{EOAA} \cdot D_{AA} \qquad (7.1)
\end{aligned}
$$

Here $D_{AA} = I_q/I_\phi = D_\circ$; $(DR)_{COP} = I_q/I_z < \infty$ is the processor dynamic range; DF_{COP} is the radiation pattern defocusing by the CO processor;

$$
e_{EOAA} = \frac{DF_{COP}}{1 + D_{AA}/DR_{COP}} \qquad (7.2)
$$

is the equivalent electrooptical array efficiency.

The dynamic range is defined as the relation of the signal intensity maximum to the averaged intensity of the background given by this signal. The dynamic range with such a definition coincides with the observation of the single signal of maximal intensity on the background of its own coherent optical processor noise [4]. Curves (7.1) and (7.2) are given in Fig. 7.1. As can be seen, as the array directivity approaches infinity (in case of a large array), the electrooptical array directivity approaches the dynamic range of the coherent optical processor. Therefore the processor's dynamic range defines the limiting electrooptical array gain. If the array directivity does not exceed the coherent optical processor dynamic range, then the electrooptical array is more energy effective since its efficiency is, in this case higher than the phased array's (the electrooptical array efficiency $\geq 0.5 \sim$ the phased array efficiency). Since the actual coherent optical processor dynamic range is about 30 to 50 dB, the use of arrays whose directivity is higher then the values given in the electrooptical array structure is not wise from an energy standpoint.

Improving the coherent optical processor optical system makes it possible to in principle increase its dynamic range to values on the order of 50 to 60 dB [178, 179]. However, the operation precision of the processor becomes worse due to increasing complexity (see below).

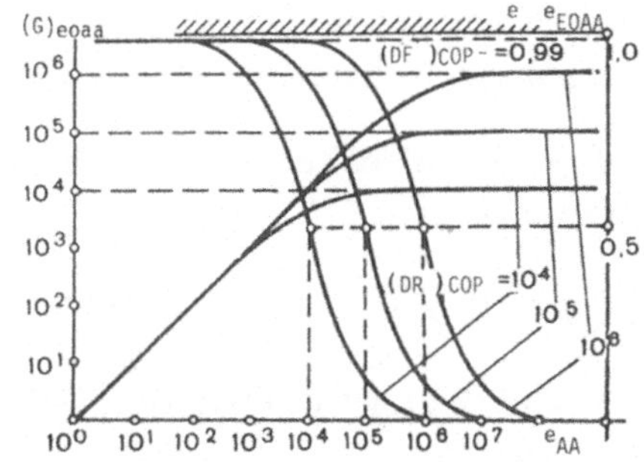

Figure 7.1: Gain and equivalent electrooptical array efficiency evaluation.

7.2 Coherent Optical Processor Errors

Supposing that the distortion of the receiving array (the amplitude-phase distribution errors) are taken into account in its directivity, let us turn to evaluating the influences on the electrooptical array pattern-forming characteristics enumerated by the ten coherent optical processor distortion factors listed above. Since these factors are statistically independent and do not have a single probabilistic model they cannot be eliminated or attenuated by some common empirical processing such as a coordinated spatial filtration [2, 19]. Therefore, let us look at attenuating each factor separately.

Since each is statistically independent, they contribute the following to the overall coherent optical processor defocusing, which was previously defined as the ratio of the intensities in the maximum of the diffraction spot at the outputs of a real and an ideal processor

$$DF_{COP} = \prod_{n=1}^{10} DF_n \tag{7.3}$$

where DF_n is the radiation pattern defocusing by the nth distorting factor.

The overall processor dynamic range, also defined above, is

$$DR_{COP} = I_q/I_z = I_q/\sum_n I_n = \left[\sum_n DR_n\right]^{-1} \tag{7.4}$$

where $DR_n = I_q/I_n$ is the coherent optical processor dynamic range with only the factor n.

Let us note that each of the distortions appears in (7.3) and (7.4) simultaneously since three situations listed below are possible:

(1) The parasitic noise is not present [see factors 1–4, $I_n = 0$, $DR_n = \infty$], but the radiation pattern is distorted [$DF_n < 1$];

(2) the radiation pattern is not distorted [see factors 6–8, 10 $DF_n = 1$], but there is exposure [$I_n \neq 0$, $DR_n < \infty$]; or

(3) both together [$DF_n < 1$; $DR_n < \infty$, see factors 5, 9].

Considering (7.3) and (7.4), let us transform (7.2) into a suitable form for the following evaluations:

$$e_{EOAA} = \prod_{n=1}^{10} DF_n \Big/ \left[1 + D_{AA} \sum_{n=1}^{10} DR_n^{-1}\right] \approx \prod_{n=1}^{10} e_n \tag{7.5}$$

Here

$$\begin{aligned} e_n &= DF_n/[1 + D_{AA} DR_n^{-1}] \\ &= \begin{cases} DF_n, DR_n = \infty \\ 1/[1 + D_{AA} DR_n^{-1}], DF_n = 1 \end{cases} \end{aligned} \tag{7.6}$$

7.2.1 The effect of the pupil and space–time light modulator channel interaction

This factor is accompanied by distortion of the necessary amplitude-phase distribution (see Sections 5.1.2 and 5.1.3), thus causing radiation pattern defocusing [$DF_1 < 1, DR_1 = \infty$]. The higher the degree of space–time light modulator surface usage, the greater the degree of defocusing. However, decreasing the degree of this surface usage leads to the fall of the processor diffractional effectiveness [see (5.10)]. This in turn lowers its dynamic range.

In accordance with Sections 5.1.2 and 5.1.3 these effects may be compensated for. If the latter is accomplished, then $DF_1 = 1$, $DR_1 = \infty$. However according to (5.10) the diffraction spot intensity is at its maximum at the coherent optical processor with a discrete space–time light modulator output $I_q \approx DE_{COP} I_q$ (I_q is the intensity with a continuous space–time light modulator). With other distortion factors present, there is a background noise $I_z \neq 0$, which practically coincides with noise $\tilde{I}_z$ at the output of the processor with continuous space–time light modulator ($\tilde{I}_z \approx I_z$). This occurs due to the suggested compensating methods not influencing accidental background illumination. Therefore the following evaluation is in order:

$$DR_{COP} = I_q/I_z \geq DE_{COP}\tilde{I}_q/I_z \approx K_{su}^{\delta c}\widetilde{DR} \tag{7.7}$$

where $\widetilde{DR}$ is the dynamic range of a coherent optical processor with a continuous space–time light modulator whose $K_{su}^{\delta c} = 1$. Supposing $K_{su}^{\delta c} \sim (0.7)^2 \approx 0.5$ (see Fig. 5.4) and considering (7.4), we obtain the following:

$$DR_{COP} \geq 0.5/\sum_{n=2}^{10} \widetilde{DR}_n^{-1} \tag{7.8}$$

where $\widetilde{DR}_n$ is the dynamic range of the processor with continuous space–time light modulation with only the nth factor present.

Evaluation (7.8) allows us to, in the future, disregard the effects of the space–time light modulator digitalization and to use a more convenient model of it appearing as a continuous modulating medium.[1]

7.2.2 Potential accuracy and the coherent optical processor optical system aberrations

As known, the given factor distorts the radiation pattern image at the coherent optical processor output $[DF_2 < 1]$, without the emergence of a parasitic background $[DR_2 = \infty]$. Integral transformations, performed by optical cascades, are researched in a paraxial approximation. Not following this condition causes systematic amplitude spatial-frequency and phase errors. A sufficiently complete and strict analysis of similar errors and their effect on the parameters of coherent optical processors which perform spatial filtration and correlations and spectral analysis is presented in a number of works, for example [16].

A generalized coherent optical processor (see Fig. 1.1) for an arbitrary electrooptical array could be represented as a series of simple optical cascades—Fourier processors. A detailed analysis of optical Fourier transform errors is given in [181]. However, in practice, for evaluating the coherent optical processor quality, it is wise to have an integrated concept of the errors with regard to all types of optical cascade errors listed above as well as their number.

For evaluating the radiation pattern defocusing due to aberrations an expression [1: 502] is used that is correct for small random aberrations of

[1] Henceforth the sign ~ will be omitted.

the form

$$DF_2 \approx 1 - \alpha \tag{7.9}$$

where $\sqrt{\alpha}/k$ is the rms aberration of the wavefront at the output of the optical cascade; $k = 2\pi/\lambda$.

Aberrational optimization of long focal length ($f \approx 1$ m) Fourier processor lenses showed [8: 70], that with the size of its operational region in its input plane (frontal background) $\Delta x \Delta y < 40 \times 40$ mm the rms aberration[2] is determined by only the spatial bandwidth of the Fourier cascade passband:

$$\begin{aligned} SBW &= \frac{1}{2\pi}\Delta\omega_x \Delta x = \frac{1}{2\pi}\Delta\Omega_X \Delta X/m_x \\ &= \frac{1}{2\pi}2K\Delta X = 2\Delta X/\Lambda \end{aligned}$$

where$\Delta\omega_x$, $\Delta\Omega_X \leq 2K$ are the spatial frequency band of light and radio waves ($K = 2\pi/\Lambda$); $m_x \Delta X/\Delta x$. Taking into account that gain $G_{AA} \approx 4\pi \Delta X \Delta Y/\Lambda^2 \sim SBW^2 < DR_{COP}$ [see Section 7.1] and aiming at attaining a processor dynamic range of 60 dB, a requirement that spatial passband $< 10^3$ is derived.

Figure 7.2 shows the curve describing the dependence of the minimal attainable rms aberration $\sqrt{\alpha}/k$ on the spatial passband. The curve was constructed using the results of [8: 70]. As can be seen, $\sqrt{\alpha}/k < 0.01\lambda$ coincides with values of the spatial passband $4 < 10^3$. Similar characteristics are only found in long focal length astronomy lenses with removed aberrations on an infinitely remote point on the optical axis. Defocusing $DF_2 \geq 0.996$ coincides with such aberrations in virtue of (7.9). A real electrooptical array processor, as a rule contains two or three optical Fourier-cascades (see, for example, Figs. 4.4 and 6.6). Since small aberrations of such processor are practically uncorrelated, while the initial spatial bandwidth parameter varies [2] (similarly to the constant analogous time signal parameter during its linear transformations [56]), in this case, the resulting electrooptical array radiation pattern defocusing is

$$DF_2 \approx (1-\alpha)^{2\ldots 3} \geq (0.996)^{2\ldots 3} \approx 0.99 \tag{7.10}$$

[2]Preston [8] uses the concept of a root-mean-square (rms) difference of optical paths rather than a mean-square-aberration.

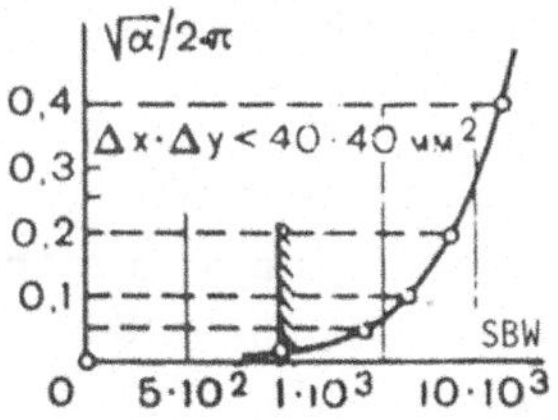

Figure 7.2: Selecting a basic Fourier-lens.

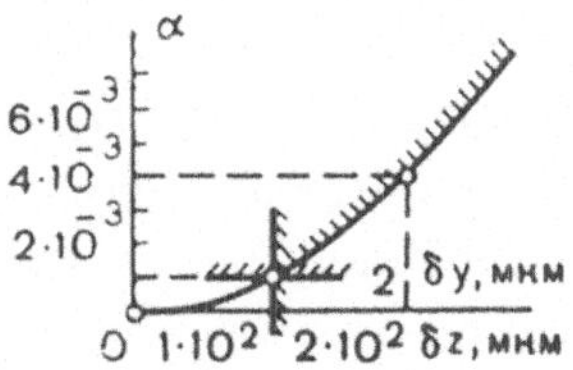

Figure 7.3: Coherent optical processor element misalignment.

This factor is a determining one and unlike others is not subject to significant attenuation. The attained evaluation, in essence, characterizes the electrooptical array radiation pattern maximum reproduction potential accuracy (about 1%) and further as the reference criterion for the coherent optical processor development.

Useful information on optimizing the optical system for holographic recording and information processing is presented in [12, 181].

7.2.3 Coherent processor element misalignment

This factor may be related to the previous one. Thus, misalignment of longitudinal element (lenses, masks etc.) leads to a spherical wavefront aberration at the optical Fourier cascades' outputs, and the transverse increase the nonsymmetrical aberrations (coma, astigmatism). In order to determine the required accuracy in the processor element mounting let us look at each misalignment element separately.

For evaluating the influences of longitudinal element misalignment δz, let us use the equation for the light intensity distribution in the focal spot along the processor axis [1: 477] [see also (5.22)]:

$$DF_3 = \text{sinc}^2(\Delta x^2 \delta z / 2\lambda f^2) \tag{7.11}$$

where $\Delta x = \Delta y$ is the size of the operational region in the optical plane of a Fourier cascade whose focal length is f; λ is the light wavelength.

Figure 7.3 shows the error in the radiation pattern reproduction $\alpha \approx 1 - DF^5{}_3$ [see equation (7.9), where the power, 5, is the average number of processor elements] with $\Delta x^2/\lambda f \approx (4 \cdot 10^{-2})^2/0.63 \cdot 10^{-6} \cdot 1 < 3 \cdot 10^3$. As can be seen $\alpha \approx 0.1\%$, that is one order of magnitude higher than the potential accuracy, if $\delta z \leq 10^{-4} f_{eff} \approx (50...100)\mu m$. This agrees with the results of similar research [5: 545].

Let us evaluate the influence of the transverse coherent optical processor element mixing. Apparently, such mixings in a single Fourier cascade do not lead to distortions, but simply cause a corresponding displacement of the output diffraction pattern or its weighting by a linear phase multiplier. The coherent optical processors which perform the spatial filtration or correlation (see Figs. 1.4, 1.8, 4.4, 4.7, and 4.9) are the ones most critical to the transversal displacements. Let, for example, the mask (4.34) in the processor plane π of Fig. 4.4 be displaced along the ω_y axis by $\delta\omega_y = m_y \delta\Omega_\phi \phi$ [see (4.37)]. Then, instead of the necessary transformation L (4.33), the coherent optical processor will carry out the transformation of type:

$$R_\circ \hat{\mathcal{F}}_\phi^{-1}\{\hat{\mathcal{F}}\{...J_Z(Z)\}\dot{T}(\Omega_\phi + \delta\Omega_\phi, \Omega_z)\} = \hat{L}\{... \exp(-i\delta\Omega_\phi \phi)\} \tag{7.12}$$

This could be proved using, for example, random functions. The phase multiplier $\exp(-i\delta\Omega_\phi\phi) = \exp(-i \times \delta\omega_y y)$ is a type of an initial Seidlitz aberration [1]. Therefore, the processor transversal mixings' δy defocusing influence can be evaluated according to (7.9) where magnitude α, which is input into it, is equal to

$$\alpha = \frac{1}{\Delta y}\int_{-0.5\Delta y}^{0.5\Delta y} |\delta\omega_y y|^2 \text{d}y$$

$$= \frac{(\delta\omega_y \Delta y)^2}{12} = \left(\frac{k\Delta y \delta y}{f}\right)^2 /12 = 10^9(\delta y)^2 \tag{7.13}$$

With $\alpha < 10^{-3}$ [$DF_3 > 0.999$], the reproduction error will not be less than 0.1% if $\delta_y \approx 1\mu$m. The evaluation is more strict than the analogous results of [5: 540]. This is given by the set processor operation accuracy.

7.2.4 Spatial incoherence of the reading light

In the common case, space–time light coherence is described by the mutual coherence function [2, 182]. It is wise, for convenience to (as before in Section 5.1.2) use it in the array antenna coordinate system

$$\gamma(\mathbf{R}_1, \mathbf{R}_2; t_1, t_2) = \gamma_{12}(\mathbf{R}_1, \mathbf{R}_2)\gamma_{11}(t_2 - t_1) \tag{7.14}$$

Here, $\mathbf{R}_1$ and $\mathbf{R}_2$ are the array surface points. They correspond to the space–time light modulator points between which the spatial coherence is set; t_1 and t_2 are the time points between which the time coherence of the reading light is set.

$$\gamma_{12}(\mathbf{R}_1, \mathbf{R}_2) = \overline{\dot{E}(\mathbf{R}_1) \overset{*}{E} (\mathbf{R}_2)}/|\dot{E}(\mathbf{R}_1) \overset{*}{E} (\mathbf{R}_2)| \tag{7.15}$$

The above equation describes the degree of the spatial coherence. $\dot{E}(\mathbf{R})$ is the reading light beam complex amplitude distribution at the space–time light modulator aperture (the overbar above signifies averaging). The distribution is tied to the array antenna coordinate system.

$$\gamma_{11}(t_2 - t_1) = \overline{\dot{E}(\mathbf{R}_1, t_1) \overset{*}{E} (\mathbf{R}_1, t_2)}/|\dot{E}(\mathbf{R}_1)|^2 \tag{7.16}$$

Equation (7.16) describes the degree of the reading light beam time coherence. Giving $\gamma(\mathbf{R}_1, \mathbf{R}_2, t_2, t_1)$ as a product of functions (7.15) and (7.16) is only possible in case of "spectrally clean" noise which is a characteristic of a laser-collimator system [2: 383; 182].

A reduction in the electrooptical arrays' directional properties due to partial spatial light coherence $\dot{E}(\mathbf{R}, t)$ can be characterized by the following relationship (see the Hopkins formula in [2, 182]):

$$DF_4 = \frac{\hat{L}_1\{\mathcal{E}_\Omega(\mathbf{R}_1)_q\ \hat{L}_2\ [\overset{*}{\mathcal{E}}_\Omega\ (\mathbf{R}_2)_q \gamma_{12}(\mathbf{R}_1, \mathbf{R}_2)]\}}{|\hat{L}\{\mathcal{E}_\Omega(\mathbf{R})\}|^2} \tag{7.17}$$

where $\hat{L}_1$, $\hat{L}_2$ are operators (4.6), given by sets $\mathbf{R}_1$ and $\mathbf{R}_2$ respectively; $\mathcal{E}_\Omega(\mathbf{R})_q$ is the response (4.2) to the planar wave q, which arrives from direction $\mathbf{K}_q$; $\mathbf{K}'$ is the direction of the radiation pattern maximum.

In a common case an analytical evaluation of relation (7.17) does not appear to be possible (below a result just for the planar electrooptical array will be given). Let us bypass this difficulty using a convenient evaluation given in [180, pp. 53] [see also (7.6)]

$$e_4 \approx DF_4 \geq \exp(-\alpha) \tag{7.18}$$

(α is the amplitude-phase distribution deviation). This equation is also true for errors with a correlation radius $R_\alpha \geq \Lambda$. In the case under consideration, this constraint is ensured by the continuous nature of the function $\gamma_{12}(\mathbf{R}_1, \mathbf{R}_2)$ at the output of the light collimator. Let us find the correspondence between the deviation and the degree of spatial coherence (7.15), using a known model of partially coherent light as a superposition of a completely coherent wave and weakly correlated spatial noise [182]

$$\begin{aligned}
&\gamma_{12}(\mathbf{R}_1, \mathbf{R}_2) \\
&= \overline{|\dot{E}(\mathbf{R}_1)| \exp[i\Phi(\mathbf{R}_1)] \cdot | \overset{*}{\dot{E}} (\mathbf{R}_2)| \exp[-i\Phi(\mathbf{R}_2)] / |\dot{E}(\mathbf{R}_1) \overset{*}{\dot{E}} (\mathbf{R}_2)|} \\
&= \overline{\exp i[\Phi(\mathbf{R}_1) - \Phi(\mathbf{R}_2)]} \\
&\approx 1 + i\overline{[\Phi(\mathbf{R}_1) - \Phi(\mathbf{R}_2)]} - \frac{1}{2}\overline{[\Phi(\mathbf{R}_1) - \Phi(\mathbf{R}_2)]^2} = 1 - \frac{\alpha}{2}
\end{aligned}$$

where $\Phi(\mathbf{R}) = \arg[\dot{E}(\mathbf{R})]$. Therefore,

$$\alpha \leq 2 - 2|\gamma_{12}|_{min} \tag{7.19}$$

where $|\gamma_{12}|_{min}$ is the minimum value for the degree of spatial coherence within limits of the space–time light modulator aperture. Using (7.18), (7.19) let us find

$$DF_4 \geq \exp(2|\gamma_{12}|_{min} - 2) \geq 2|\gamma_{12}|_{min} - 1 \tag{7.20}$$

The degree of single mode laser spatial coherence (7.15) within the limits of the beam width (approximately 1 mm) at a level of 0.5 does not fall bellow 0.9991 ± 10^{-4} [182, 183]. Suppose that the collimator in Fig. 1.2 widens the laser beam by a factor of 50 (that is, the ratio of the large and small lens focal lengths is 50:1). Then $|\gamma_{12}|_{min} \geq 0.9991$ within limits of surface $\Delta x \Delta y \leq 40 \times 40$ mm. This gives a value of DF_4 one order of magnitude better then the potential accuracy (approximately 1%).

It is wise to compare the common evaluation (7.20) with magnitude (7.17), calculated for example for planar electrooptical array. For this purpose let us specify the values input into (7.17) ($\hat{L}_{12} = \hat{\mathcal{F}}_{12}$, $\mathbf{R}_{12} = \mathbf{R}_{\perp 1,2}$) and take into consideration constancy of the degrees of spatial coherence $\gamma_{12}(\mathbf{R}_1, \mathbf{R}_2) = \gamma_{12}(\mathbf{R}_{\perp 1} - \mathbf{R}_{\perp 2})$. Then after the corresponding calculations, we get

$$DF_4 = \left. \frac{\hat{\mathcal{F}}\{\gamma_{12}(\mathbf{R}_\perp)\} \otimes \otimes F^2(\mathbf{K}_\perp)}{4\pi^2 F^2\left(\mathbf{K}'_\perp - \mathbf{K}_{\perp q}\right)} \right|_{\mathbf{K}'_\perp = \mathbf{K}_{\perp q}} = |\gamma_{12}|_{cp} \tag{7.21}$$

where $\mathbf{K}_\perp$ is the vector spatial frequency,

$$F(\mathbf{K}_\perp) = \hat{\mathcal{F}}\{J(\mathbf{R}_\perp)\} \text{ – radiation pattern;}$$

$$|\gamma_{12}|_{cp} = \iint\limits_{\Sigma} \frac{\gamma_{12}(\mathbf{R}_\perp) d^2\mathbf{R}_\perp}{\Sigma}$$

is the average value of the spatial coherence within the limits of the space–time light modulator; Σ is the array area. As can be seen, the coherent evaluation (7.21) is less rigid and does not contradict (7.20), since $|\gamma_{12}|_{cp} > 2|\gamma_{12}|_{min} - 1$ (in virtue of the fact that $|\gamma_{12}|_{min} < |\gamma_{12}|_{cp} < 1$).

In order to improve the spatial coherence of light being output from the collimator (see Fig. 1.2), while using multimode lasers (or thermal sources) a round diaphragm is installed in the common focal point of its co-focal lenses. Let the radius of this diaphragm be r_k, and the focal length of the large collimator lens, f_k. Then in virtue of Van Zitter-Zernik [1, 2, 5]

$$|\gamma_{12}| \geq \left| \frac{2J_1(kr_k r_\circ / f_k)}{kr_k r_\circ / f_k} \right| \geq 1 - \frac{1}{8}\left(\frac{kr_k r_\circ}{f_k}\right)^2 \tag{7.22}$$

(J_1 is the first order Bessel function; $r_\circ$ is the space–time light modulator radius) with any degree of spatial light coherence, at the collimator output, as well as inside its diaphragm. In this case, according to (7.20), we have (Fig. 7.4)

$$DF_4 \geq 1 - \frac{1}{4}\left(\frac{kr_k r_\circ}{f_k}\right)^2 \tag{7.23}$$

Using the concept of "coherence area" [182]

$$\sigma_{ca} \approx \frac{\lambda^2 f_k^2}{\pi r_k^2} = 4\pi(f_k/kr_k)^2 \tag{7.24}$$

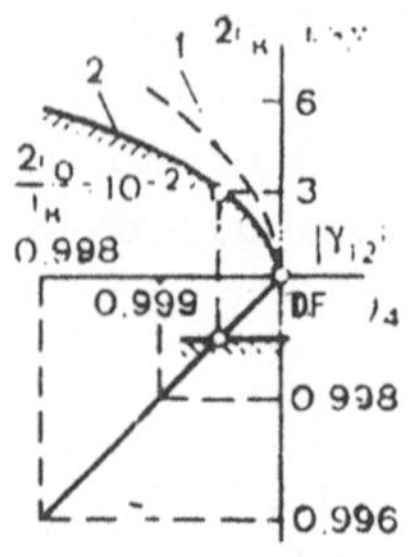

Figure 7.4: The effect of partial read light spatial coherence: (1) Planar electrooptical array; (2) arbitrary form EOAA.

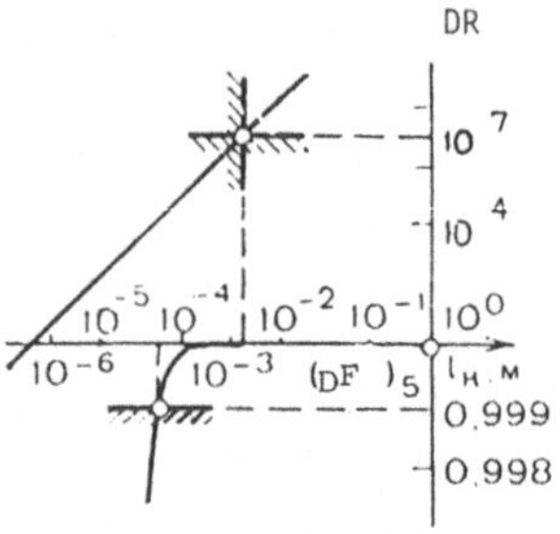

Figure 7.5: The effect of the reading light time coherence.

with the help of which equation (7.23) can be written as:

$$DF_4 \geq 1 - \sigma_{STLM}/\sigma_{ca} \tag{7.25}$$

where $\sigma_{STLM} = \pi r_\circ^2$ is the area of the space–time light modulator.

7.2.5 Time incoherence of the light source

This factor distorts the electrooptical array radiation pattern for two reasons: due to optical signal spatial frequency deviation [see for example (1.2)] and also due to chromatic processor aberration [1]. The first effect can be evaluated by converging it to the preceding (spatial coherence). Indeed, the coherent optical processor, acting as a pattern-forming system, compensates for the phase delays in signals received by the array by delaying the corresponding light wave

$$\mathbf{K}'(\mathbf{R}_2 - \mathbf{R}_1) = (2\pi\nu + \Omega) \cdot (t_2 - t_1) = \omega(t_2 - t_1)$$

where $2\pi\nu$ and Ω are the radian frequencies of the light and the radiowave, respectively. Taking into consideration the quasimonochromatic nature of light, let us present the degree of time coherence (7.16) as [3, 182]

$$\gamma_{11}(t_2 - t_1) = \gamma_\circ(t_2 - t_1) \exp[i2\pi\nu(t_2 - t_1)]$$
$$= \gamma_\circ[\mathbf{K}'(\mathbf{R}_2 - \mathbf{R}_1)/\omega] \exp[i2\pi\nu(t_2 - t_1)] \quad (7.26)$$

where $\gamma_\circ(t_2 - t_1) = \overline{\exp i[\phi(t_2) - \phi(t_1)]}$.

The form of the obtained representation coincides with the form of the "spectrally clear" noise mutual coherence function (7.14) . The first multiplier $\gamma_\circ[...]$ acts as the degree of spatial coherence (7.16), and the second, exp[...], the degree of ideally monochromatic wave [3] time coherence (7.16). Therefore, the radiation pattern defocusing caused by time incoherence can be evaluated according to (7.17) substituting $\gamma_\circ[....]$ for $\gamma_{12}(...)$. However, due to the noted difficulties in evaluating such an expression in its general form, it is advisable to use a close approximation of type (7.20) with a corresponding substitution, that is

$$DF_5 \geq 2|\gamma_\circ|_{min} - 1 \quad (7.27)$$

where $|\gamma_\circ|_{min} = \min\{\gamma_\circ[\mathbf{K}'(\mathbf{R}_2 - \mathbf{R}_1)/\omega]\} = \gamma_\circ(K\Delta D/\omega)$, and ΔD is the array diameter. Since according to (7.26) and [3, 182], $\gamma_\circ(t_2 - t_1) = |\gamma_{11}(t_2 - t_1)| \approx \exp[-(t_2 - t_1)^2 \tau^2 k]$ (τ_k is the coherence time), then

$$|\gamma_\circ|_{min} \approx \exp[-(K\Delta D/\omega\tau_k)^2 \geq 1 - \left(\frac{l}{\Lambda} \cdot \frac{\Delta D}{l_k}\right)^2$$

$l_k = c\tau_k$ is the coherence length. Substituting $|\gamma_\circ|_{min}$ into (7.27), we get:

$$DF_5 = \geq 1 - 2\left(\frac{l}{\Lambda} \cdot \frac{\Delta D}{l_k}\right)^2 \quad (7.28)$$

The obtained evaluation depends on the coherent optical processor parameters (λ, l_k) as well as on the array's $(\Lambda, \Delta D)$. However, the dependency of

DF_5 on array parameters turns out to be illogical to a known extent. This is so because here the distortions caused directly by the coherent optical processor are observed. Therefore, it is advisable to proceed from a hypothesis that the light source time incoherence leads not to the output image defocusing but to its noise, that is, to the limiting of the dynamic range. Because evaluation (7.28) is the direct result of (7.18) and thereby characterizes the gain attenuation, which in this case is defined by the electrooptical array efficiency (7.2) and (7.5), and then in accordance with (7.6), it is possible to obtain an expression for DR_5, considering that defocusing and noise should lead to a similar drop in e_5:

$$DR_5 = \frac{e_5 D_5}{1 - D_5} \geq \frac{e_5}{2(\lambda \Delta D / \Lambda l_k)^2} \approx 2\pi (l_k/\lambda)^2 \tag{7.29}$$

where it is taken into account that the $D_5 \approx 4\pi(\Delta D/\Lambda)^2$ [52].

In this way, the light source time incoherence is equivalent to the additive coherent optical processor output parasitic noise. This corresponds with the well-known model of quasi-coherent light exposure as a superposition of an ideal monochromatic with a noisy one [182]. With coherence length of $l_K > 10^{-3}$ m; $DR_5 \geq 10^7$ (Fig. 7.5). The magnitude of l_K is over 10^2 for He-Ne lasers, about 1 m for ruby lasers, about 10 mm and higher for semiconductor lasers [183] and about 1 mm for fluorescent lamps [4]. From the latter, it follows that the light source time incoherence effect on the coherent optical processor dynamic range is not a determining one.

Let us, as before, test evaluation (7.28) and, therefore (7.29), on a planar electrooptical array example. Since concept (7.26) is of stable form (depending on the difference $t_2 - t_1$), then calculating the numerator in expressions of type (7.17) is analogous to (7.21) and yields

$$\begin{aligned} DF_5 &= \left. \frac{\Gamma_{11}(\Delta\Omega) \otimes (\Omega_q \rho_q \Delta\Omega/\omega c)}{2\pi F^2(0)} \right|_{\Delta\Omega=0} \\ &\approx \frac{1}{\sqrt{1 + (\Omega_q \cos\theta_q \Delta/\omega c\tau_k)^2}} \approx 1 - \frac{1}{2}\left(\frac{\lambda}{\Lambda}\frac{\Delta D}{l_k}\cos\theta_q\right)^2 \end{aligned} \tag{7.30}$$

where $\Gamma_{11}(\Delta\Omega) = \hat{\mathcal{F}}_t\{\gamma_\circ(\Delta t)\}$, and an evaluation of (5.60) -type is also used (the remaining definitions are given in Section 2.2). Analogously, for a planar electrooptical array

$$DR_5 \approx \frac{\pi}{2}(l_k/\lambda)^2 \tag{7.31}$$

As can be seen, the coherent evaluation (7.31) is less rigid and does not contradict the overall evaluation (7.29).

For an exact chromatic aberration evaluation, it is necessary to consider the coherent structure of the coherent optical processor and the refraction indicator deviation of each of the composing lenses. However, since the chromatic influence on the image quality is relatively low, let us consider just the latter evaluation. Let us use formula (7.9), which is true for small random type aberrations. The relative focal length, $\Delta f/f$, error due to the change in the wavelength of light $\Delta\lambda$, m, is $\Delta f/f \approx \Delta_l \cdot \Delta\lambda/0.2 \cdot 10^{-6}$, where $\Delta_l$is the relative lens dispersion (1/30 with heavy flint glass [1]). Therefore, at the coherent optical processor lens, whose focal length is f and the diameter is d_l, output, the light wave phase error dispersion α is determined as an rms light wave $\left[\frac{k}{2f}(\mathbf{r}_\perp)^2\right]$ phase error in the lens output plane

$$\alpha \approx \overline{\left[\Delta f \cdot d\left(\frac{k}{2f}\mathbf{r}_\perp^2\right)/\mathrm{d}f\right]^2} < (kd_l^2\Delta f/8f^2)^2 \qquad (7.32)$$

where the overbar signifies averaging in the lens output plane; d(...)/df is the derivative with respect to f; $|\mathbf{r}_\perp| = r_\perp \leq d_L/2$; $r_\perp$ is the vector radius of points in the lens output plane. Supposing that $d_L \approx 50^{-3}$ m, and that the number of lenses in the processor ranges from three to five, in accordance with (7.9) and (7.32), we obtain (see Fig. 7.5)

$$DF_5 \approx (1-\alpha)^5 \geq 1 - 10^{-12}/l_k^2 \qquad (7.33)$$

where l_k m is the coherence length.

Let us summarize the requirement for the light source coherence volume, which would allow us to choose it with the maximum light output [5,82]. According to (7.25) and (7.29), $DF_4 \geq 0.999$ and $DR_5 \geq 10^7$, if the reading light beam space–time coherence volume is not lower than

$$V_k = \sigma_k \cdot l_k \geq 10^{-2}\,\mathrm{m}^3 \qquad (7.34)$$

Research of the light source coherence effect on the characteristics of the optical processors which perform the spectral or correlation analysis and on the holographic storing devices is done in a number of papers [e.g., 184, 12].

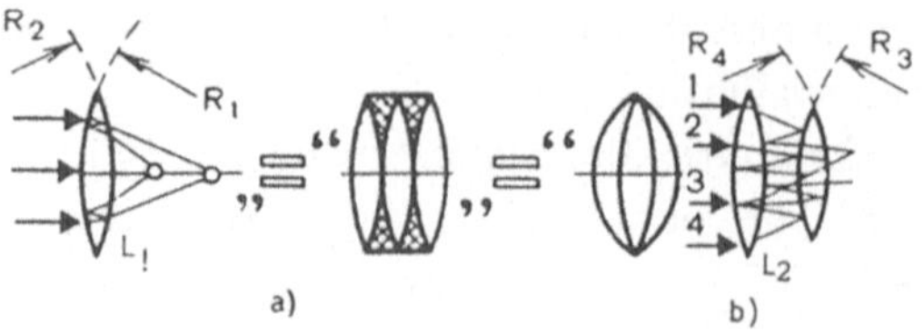

Figure 7.6: (a) Intra-lens and (b) inter-lens reflections.

7.2.6 Fresnel light reflection from coherent optical processor elements

This factor lowers coherent optical processor's dynamic range $[DR_6 < \infty, DF_6 = 1]$ for two reasons: because of the weakening of the processor's luminosity and because of the emergence of a false image ("ghost") background. The first reason can be checked using the light transmission coefficient [1, 177]

$$t_{COP} = (1 - r_{CT})^{2N_{CT}}(1 - r_L)^{2N_L}(1 - r_\tau)^{2N_T} \approx 0.15...0.80 \qquad (7.35)$$

where $N_{CT} = 1...3$ is the number of controlled masks (CT), [3] $N_{CT} = 2$ (e.g., Fig. 8.15); $N_L = 3...5$ is the number of lenses in the coherent optical processor; N_t is the number of the uncontrolled masks ($N_t = 1$ in the electrooptical array processors which look like rotational bodies); $r_{CT} = 0.0278$ to 0.184; $r_L = 0.0400$ are the reflection coefficients of the front and the rear surfaces of the respective coherent optical processor elements. The processor "ghosts" emerge due to multiple internal and between-element re-reflections (Fig. 7.6b). The two-time re-reflections have the prevalent meaning, since the one-, three-, five- time etc. ones fall on the processor input, but the four-, six-time etc. ones are two or three times weaker than the two-time ones. Indeed, the relative level of the latter is $r^2 = r_L^2...r_L r_t = 0.00688...0.05 < r_{CT}^2 = 0.034$, and of the quadruple, the second degree of these values. The most felt are the re-reflections inside the space–time light modulator (approximately 3%). They do not, however, cause "ghosts" because the multiple reflections between the planar-parallel space–time light modulator faces do not distort the reading light phase front, but simply amplify (attenuate) the light modulation effect, provided that the rereflected waves are in phase or have opposite phase. Therefore, for consideration of the summed level of "ghosts" it is important to know the internal lens (approximately 0.5%) and the between lens (approximately 1.5%) re-reflections. The twofold rereflection inside lens

[3] The controlled mask may be replaced by an optically controlled mask with a semiconductive layer—liquid crystal Prom.

L_1 (Fig. 7.6a) having focal length f could be analyzed using an equivalent lens model with focal length $f' = f(N_L - 1)/(3N_L - 1)$ ($N_L = 1.4...1.8$ is the refraction index of the lens). The relative intensity of the "ghost," in this case, can be evaluated as a square of the normalized phase error integral coefficient using the Gaussian comparison sphere [1]:

$$\left| \int\!\!\int_{L_1} \exp\left[\frac{ikr_\perp^2}{2}\left(\frac{1}{f'} - \frac{1}{f}\right)\right] \mathrm{d}^2\mathbf{r}_\perp \Big/ \int\!\!\int_{L_1} \mathrm{d}^2\mathbf{r}_\perp \right|^2$$

$$= \left| \int_0^{d_L/2}\int_0^{2\pi} \exp\left[\frac{ikr_\perp^2}{2}\left(\frac{1}{f'} - \frac{1}{f}\right)\right] r_\perp \mathrm{d}r_\perp \mathrm{d}\phi \Big/ \frac{\pi \mathrm{d}_L^2}{4} \right|^2$$

$$= 8f(N_L - 1)/k\mathrm{d}_L^2 \qquad (7.36)$$

To consider the between element re-reflections it is advisable to limit ourselves to the review of the most adverse situation. The situation when the re-reflections are not attenuated by vignetting (the re-reflections between closely housed elements, for example the collimator and the space–time light modulator, lenses [8]). As can be seen from Fig 7.6b, depending upon the rereflection type 1, 2, 3, or 4, the optical path in such a case lengthens or shortens by:

1. [$(kr_\perp^2/2)2\,(1/R_2 + 1/R_3) \sim kr_\perp^2/[f(N_L - 1)]$]
2. [$(kr_\perp^2/2)2\,(-1/f_1 - 1/R_1 - 1/f_2 - 1/R_4) \sim (kr_\perp^2/f)\cdot[2N_L - 1/(N_L - 1)]$]
3. [$(kr_\perp^2/2)2\,(1/R_2 - 1/f_2 - 1/R_4) \sim -kr_\perp^2/f$]
4. [$(kr_\perp^2/2)2\,(-1/f_1 - 1/R_1 - 1/R_3) \sim -kr_\perp^2/f$]

where $1/f_1, 2 = (N_L - 1)(1/R_{1,3} + 1/R_{2,4})$; f_{12} is the lens L_{12} focal length whose front surface curvature radius is $R_1, 2$ and the back face's $R_2, 4$. The following normalized "ghost" intensities correspond to re-reflectionsy 1, 2, 3, 4 [see (7.36)]:

1. [$[8f(N_L - 1)/kd_L^2]^2 \sim 10^{-5}$]
2. [$\{[8f(N_L - 1)/[kd_L^2(2N_L - 1)]\}^2 \sim 10^{-6}$]
3. [$[8f/kd_L^2]^2 \sim 5 \cdot 10^{-5}$]

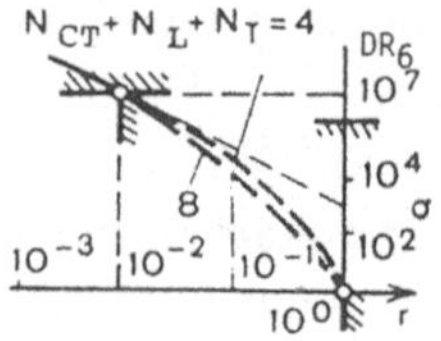

Figure 7.7: The dynamic range lowering due to reflections

4. $[[8f/kd_L^2]^2 \sim 5 \cdot 10^{-5}]$

whose sum is approximately 10^{-4}.

In this way we finally have:

$$DR_6 > t_{COP}/r^2[N_L \cdot 10^{-5} + (N_{CT} + N_L + N_T) \cdot 10^{-4}] \geq 10^3 t_{COP}/r^2 \quad (7.37)$$

Since according to (7.35), $t_{COP}/r^2 \approx 1/r^2 - 2(N_{CT} + N_L + N_t) \approx 10...100$, then $DR_6 \approx 10^4...10^5$ (Fig. 7.7), which corresponds with the evaluations given in [8, 18]. Using special antireflection coatings [8, 177], the DR_6 value could be improved up to the value of about 10^6.

7.2.7 Limiting the dynamic range

The light refraction on the coherent optical processor inhomogeneities, for example dust particles, scratches, bubbles or grains, together with the preceding factor determine the limiting processor dynamic range. Precision qualitative optical devices are characterized by inhomogeneities, the sizes of which do not exceed $r_\alpha \leq 10\ \mu$m. The amplitude-phase distribution errors caused by such inhomogeneities, have a correlation radius on the order of $R_\alpha \approx mr_\alpha \approx 1.0\ \mu\text{m} \ll \Lambda \approx 3\ cm$. This allows the use of an evaluation more accurate than (7.18) [180]:

$$e_7 \geq 1 - 3\pi^2 \alpha R_\alpha^2 / 4\Lambda^2 \quad (7.38)$$

According to (7.6), Fig. 7.8 corresponds to evaluation (7.38)

$$DR_7 \approx \frac{4\Lambda^2}{3\pi^2 \alpha m^2 r_\alpha^2} \cdot \frac{4\pi \sum}{\Lambda^2} = \frac{1.7\sigma_{STLM}}{\alpha r_\alpha^2} \quad (7.39)$$

where $\sum = m^2 \sigma_{STLM}$ is the array area ($\sigma_{STLM} = 5.5 \cdot 10^{-4} m^2$ is the area of the space–time light modulator). Evaluation (7.39) coincides with the

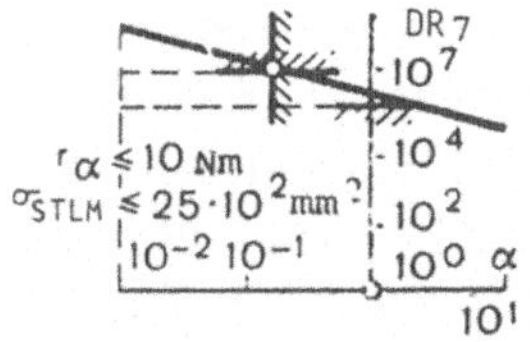

Figure 7.8: The dynamic range lowering due to light diffusion

respective result for the spectro analyzer [4] with accuracy coefficient of 1.7. If a $DR_7 \geq 10^7$ were to be set, then, according to evaluation (7.39) $\alpha \leq 1.7 \cdot 25 \cdot 10^{-4}/10^7 10^{-19} \approx 4$. Taking into consideration that the inhomogeneities of the various coherent optical processor elements are not correlated, let us find their allowable rms size

$$\bar{r}_\alpha \leq \frac{1}{k}\sqrt{\frac{\alpha}{N_{CT}+N_L+N_T}} \approx \frac{\lambda}{2\pi}\sqrt{\frac{4}{8}} \approx 0.2\lambda \tag{7.40}$$

Let us note that simultaneous rereflection, attenuation, and light diffusion on the processor optical systems' inhomogeneities is difficult because increasing the number of the antireflection coatings increases light diffusion. Theoretical evaluations show that the $\max\{[DR_6^{-1} + DR_7^{-1}]\} \approx 60$ dB. This value of the dynamic range should be viewed as the limiting one since other factors are subject to significant attenuation.

7.3 Space–time Light Modulator Distortion Factors

The factors reviewed above are described only by the coherent optical processor distortion and do not completely reflect the effects exerted by the space–time light modulator input device on the electrooptical array characteristics in question. Some space–time light modulator distortion factors are considered below: the existence of an adjacent image with dual-band signal input, the nonlinearity of the space–time light modulator modulating characteristic and the noise by the zero order diffraction.

7.3.1 Dual-band input

The given factor leads to the emergence of parasitic noise $[DR_8 < \infty]$ in the form of adjacent order of diffraction and is characterized by undistorted radiation pattern $[DF_8 = 1]$ reproduction in form of the main order of diffraction.

Since the adjacent image is energywise equal to the main one, $DR_8 \geq e_{AA}$.[4] According to (7.6), the electrooptical array directivity here does not exceed 0.5 even in absence of other distorting factors. The noted deficiency could be eliminated in one of the ways described in Section 5.2. An example is single-band input or coherent noise (see Section 8.3), during which the adjacent image is completely eliminated $[DR_8 = \infty]$. If an equivalent lens is used at the array level (see Fig. 5.1a), then according to (5.29) the adjacent image is suppressed by $DR_8 \geq 70$ dB. For example, with $\sigma_{STLM} = \Delta x \Delta y = (5 \cdot 10^{-2})^2\,\mathrm{m}^2$ and $f_e \leq 2.5$ m or $\sigma_{STLM} = 10^{-4}\,\mathrm{m}^2$, $f_e \leq 0.1$ m.

In accordance with Section 5.2.1, the given focal length values, f_e, for the equivalent lens could turn out to be unconstructive (the longitudinal coherent optical processor is unjustifiably high: $z > f_e = 2.5$ m; or the diffraction pattern size in the processor output plane is excessively small). Therefore, in such cases, it is expedient to introduce a correction (increasing or decreasing the diffraction pattern) by using an auxiliary lens with focal length f_a, installed, for example, in a space–time light modulator plane which does not change evaluation (5.29). Indeed, in this case, some new lens with focal length $f_+ = f_e f_a/(f_a + f_e)$ corresponds to the $+1^{\mathrm{st}}$ order of diffraction, and one with $f_- = -f_e f_a/(f_a - f_e)$ to the -1st order of diffraction [9].

Let, for example, the auxilary lens be a diverging one $(f_e < 0)$. Then with $f_e < |f_a|$, the $+1^{\mathrm{st}}$ will be focused in $z = f_+ > 0$, and the -1st in the false Fourier image plane $z = f_- < 0$. Then, the latter will be defocused in the $z = f_+$ plane to the following degree

$$
\begin{aligned}
& |F^{(-1)}(0)/F^{(+1)}(0)|^2 \approx DR_8^{-1} \\
& \leq \left(-\frac{f_-}{f_+ + f_-}\right)^2 \Big/ \left(\frac{\Delta x \Delta y}{\lambda f_-}\right)^2 \\
= \; & 0.25(\lambda f_e/\Delta x \Delta y)^2 \qquad (7.41)
\end{aligned}
$$

which coincides with (5.29). An analogous result is true also for the converging $(f_e > 0)$ auxiliary lens. This case is even preferable to the last one because of a higher diffractional efficiency of the processor in virtue of a narrower envelope (5.31). Equivalent and auxiliary lenses are also used for convex electrooptical arrays. For insufficient natural defocusing of their adjacent image [see (4.61) and Fig. 4.12b], the reviewed method increases defocusing by (7.41) (Fig. 7.9).

[4] Equality is achieved with a planar EOAA with a non-reducing processor.

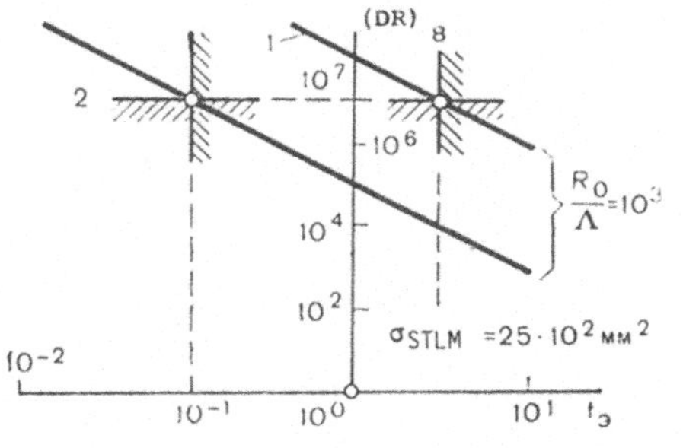

Figure 7.9: The influence of the conjugate image (1) nonplanar EOAA; (2) planar EOAA.

7.3.2 Space–time light modulator modulation characteristic nonlinearity

This factor leads to parasitic noise $[DR_9 < \infty]$ and to radiation pattern $[DF_9 < 1]$ distortion. The first is defined by the higher $\pm n$ order of diffractions ($n \geq 2$), which are brought about by the spectral components of the read into the processor space–time signals on even frequencies $n\Omega_o$. If the space–time light modulator nonlinearity is significant $[\Gamma \approx \pi$ cm see (1.6)], then the approximation of its transmission function $\dot{T}$ by a two-term exponential series (2.3) becomes invalid. Considering a higher number of terms, expansion (2.3) is not too constructive due to unavoidable difficulties in subsequent calculation of the space–time spectrum of type (2.7). Researching the nonlinear distortions becomes simpler if a supposition about the previously known narrow width of the space–time signal is made (2.1).

If limitations (2.23) and (4.24), which allow us to disregard the aperture dispersion, are in effect, the following expression is justified [138][5]

$$u(\mathbf{R},t) = \frac{\pi}{2}\mathcal{E}(\mathbf{R},t)/\mathcal{E}_{\lambda/2} = U(\mathbf{R},t)\sin[\Omega_o t + \mathbf{\Phi}(\mathbf{R},t)] \tag{7.42}$$

where $\mathcal{E}_{\lambda/2}$ is the received signal half-wave voltage, with $\Gamma = \pi$; $U(\mathbf{R},t)$ is the slow changing normalized amplitude; $\mathbf{\Phi}(\mathbf{R},t)$ is the slow changing phase of the normalized signal. Using expression (7.42), transmission functions (1.6a) and (1.6b) for amplitude $\dot{T}_a$ and phase $\dot{T}_p$ space–time light modulator can be expanded using the Jacobi-Anger formula [144]

$$\dot{T}_a(R,t) \;=\; \dot{T}_o \exp(i\Gamma_o/2)\sin[\Gamma_o/2 + \Gamma(\mathbf{R},t)/2]$$

[5]The space–time signal of (7.42) can generally be described as the superposition of plane waves of different frequencies at the receiving elements of the array.

$$= \dot{T}_\circ \exp(i\Gamma_\circ/2)\{\sin(\Gamma_\circ/2)J_\circ[U(\mathbf{R},t)]$$

$$+ \; 2\sum_{n=1}^{\infty}(-1)^{n-1}\sin(\Gamma_\circ/2)J_\circ[U(\mathbf{R},t)]$$

$$\times \; \sin[n\Omega_\circ t + n\Phi(\mathbf{R},t) + (n-1)\pi/2]\} \qquad (7.43)$$

$$\dot{T}_\phi(\mathbf{R},t) = T_\circ \exp(i\Gamma_\circ)\exp[i\Gamma(\mathbf{R},t)/2]$$

$$= \; T_\circ \exp(i\Gamma_\circ)\exp[iu(\mathbf{R},t)] = T_\circ \exp(i\Gamma_\circ)\{J_\circ[U(\mathbf{R},t)]$$

$$-\sum_{n=1}^{\infty}(-1)^n J_n[U(\mathbf{R},t)]\sin[n\Omega_\circ t + n\Phi(\mathbf{R},t) + (n-1)\pi/2]\} \; (7.44)$$

Expansions (7.43) and (7.44) provide a visual image about the spectral composition of signals read into the processor from non-linear amplitude and phase space–time light modulators, respectively. As can be seen, the case at hand differs from a linear one in that, first of all, new spectral signal components emerge (on frequencies $n\Omega_\circ$, $|n| \geq 2$) and, second, in that distortion of the signal itself is observed. This signal, unlike a linear member of expansion (2.3) is

$$T_\circ \exp(i\Gamma_\circ/2)\cos(\Gamma_\circ/2)\; 2J_1[U(\mathbf{R},t)] \times \sin[\Omega_\circ t + \Phi(\mathbf{R},t)]$$

$$= \; T^{(1)}\frac{J_1(U)}{U}\mathcal{E}(\mathbf{R},t) \qquad (7.45)$$

in the amplitude space–time light modulator case, and

$$T_\circ \exp(i\Gamma_\circ)\; 2i\; J_1[U(\mathbf{R},t)]\sin[\Omega_\circ t + \Phi(\mathbf{R},t)]$$

$$= T^{(1)}\frac{2J_1(U)}{U}\mathcal{E}(\mathbf{R},t) \qquad (7.46)$$

in a phase space–time light modulator case.

The spectral components of the read beam into the processor signal give birth to parasitic noise of the radio range image $[DR_9 < \infty]$ as higher

$\pm n^{\text{th}}$ orders of diffraction. Each of them is defocused at the output of any nonplanar electrooptical array (see Section 4.3.4). Only in the case of a planar electrooptical array with no attenuation (that is with no equivalent lens) is the focusing possible. Indeed, for this, it is necessary that

$$\begin{aligned} & J_n[U(\mathbf{R},t)]\exp\{i\,n|\Omega_\circ t + \Phi(\mathbf{R},t)|\} \\ = \; & J_n[U(\mathbf{R},t)] \times \exp\{i[\Omega_\circ nt + \Phi(n\mathbf{R},nt)]\} \\ = \; & \frac{J_n[U(\mathbf{R},t)]}{U(n\mathbf{R},nt)} u(n\mathbf{R},nt) \end{aligned} \tag{7.47}$$

i.e., it is necessary to provide a space–time analog to the $+n^{\text{th}}$ and $+1^{\text{st}}$ components of expansions (7.43) and (7.44). However (7.47) is only carried out in the case of $\Phi(\mathbf{R},t) = -i\mathbf{KR}$, i.e., in the case of a single monochromatic wave (or a number of waves, but with one dominating) falling on the electrooptical array. Only in this condition could the $+n^{\text{th}}$ component of (7.47) be focused using the previous coherent optical processor, at a frequency of $n\Omega_\circ$, since $n\Phi(\mathbf{R},t) = -n\,\mathbf{K}_q\mathbf{R} = -n\Omega_\circ\mathbf{n}_q\mathbf{R}/c$. This also comes from the principle of electrodynamic likeness. This retuning of the unreduced processor lies in recalibrating its output plane, stretching the scale by n times. Retuning of the nonplanar electrooptical array coherent optical processor, or the attenuated processor of a planar one deals with changing the parameters of its elements (the uncontrolled mask, the equivalent lens, and so on). For this reason, such processors always focus the main and defocus the higher orders of diffraction. Let us note that changing the processor operational frequency to a higher one could turn out to be useful when, for example, it is necessary to improve the electrooptical array angular clearance (by n times). Figures 7.10 and 7.11 show the point image q responses at the output of a non-attenuated planar electrooptical array Fourier processor with a non-linear space–time light modulator with phase and amplitude light modulation respectively. As can be seen, all order of diffractions have the same shape (width). The nth one, however, is measured along a scale extended n times, and it therefore reconstructs the point source, q, n times more accurately.

And so, since the higher orders diffraction at the output of any nonplanar electrooptical array are defocused, to evaluate the defocusing it is wise to use the relation of type (4.61), from which we can determine that "the washing out" of the $\pm n^{\text{th}}$ order of diffraction makes up $2/|\pm n - 1| \cdot DR_8$. Thus,

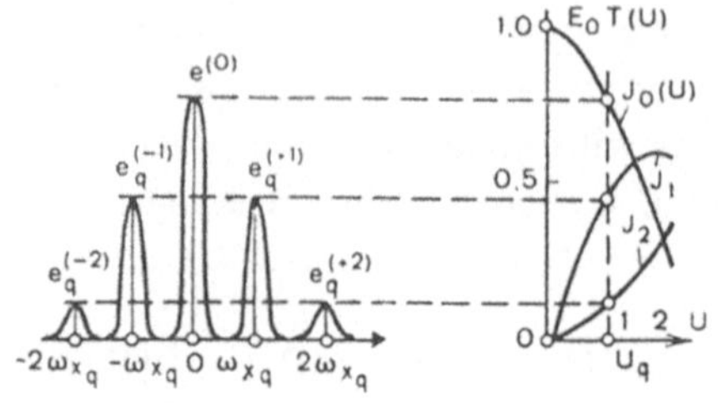

Figure 7.10: The diffraction pattern at the output of a processor with a phase STLM.

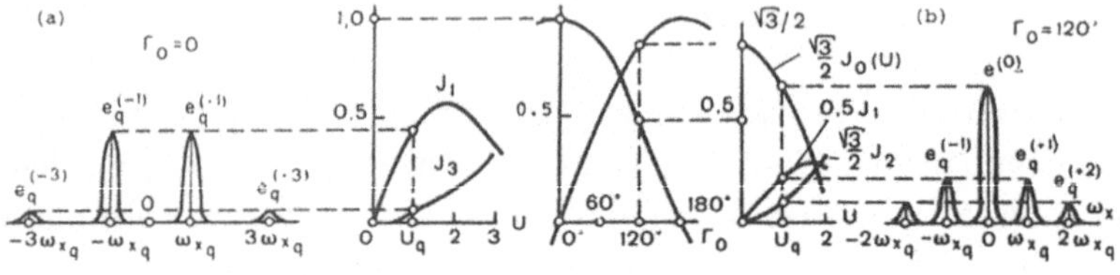

Figure 7.11: A diffraction pattern at processor output: (a) amplitude STLM; (b) balanced STLM.

$$
\begin{aligned}
DR_9 &= \frac{\cos^2(\Gamma_\circ/2)J_1^2(U)}{\sum_{n=2}^{\infty} DR_8^{-1}\sin^2(\Gamma_\circ/2+n\pi/2)} \\
&\times \; J_n^2(U)[2/(n-1)+2/(n+1)] = DR_8\cos^2(\Gamma_\circ/2) \\
&\times \; J_1^2(U)/4\sum_{n=2}^{\infty}\frac{n}{n^2-1}\sin^2(\Gamma_\circ/2+n\pi/2)J_n^2(U) \qquad (7.48)
\end{aligned}
$$

in the case of an amplitude space–time light modulator with transmission function (7.43); and

$$
DR_9 = DR_8 J_1^2(U)/4\sum_{n=2}^{\infty}\frac{n}{n^2-1}J_n^2(U) \qquad (7.49)
$$

in the case of a phase space–time light modulator with transmission function (7.44). As can be seen, the dynamic range (7.48) is maximum when $\Gamma_\circ = 2\pi m$ $(m = 0, 1, 2, \ldots)$, which corresponds to the case of balanced light modulation (see Fig. 7.11b) [89], where the first component of the expansion (2.3) is ascent and the light carrier is suppressed.

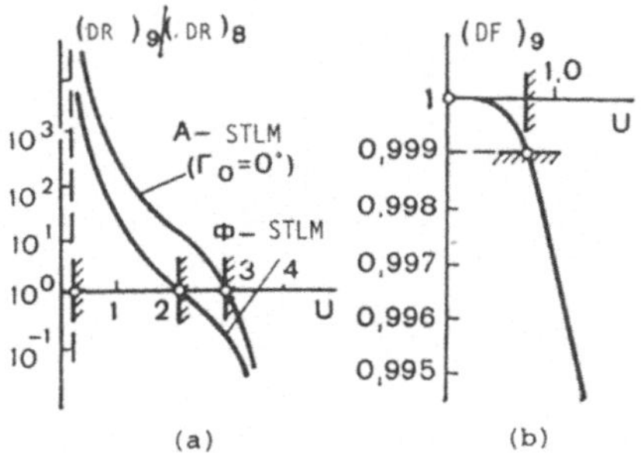

Figure 7.12: The fall of the (a) dynamic range; and (b) defocusing due to STLM non-linearities.

Taking into account the smallness of the higher orders of the Bessel function $J_n(U)$ with $U \approx 1$, from (7.48) and (7.49) we get the following evaluation (Fig. 7.12a)

$$DR_9 = \begin{cases} DR_8 \frac{2}{3}[J_1(U)/J_3(U)]^2 & \text{amplitude STLM} \\ DR_8 \frac{3}{8}[J_1(U)/J_2(U)]^2 & \text{phase STLM} \end{cases} \tag{7.50}$$

where $U = \frac{\pi}{2}\mathcal{E}/\mathcal{E}_{\lambda/2}$ = the slow-changing normalized amplitude of the space–time signal which controls the space–time light modulator. If $J_3(U)/J_1(U) \leq \sqrt{(2/3)}$ and $J_2(U)/J_1(U) \leq \sqrt{(3/8)}$ (and the coefficients of non-linear distortions of the amplitude and phase space–time light modulator, respectively), then $DR_9 \geq DR_8$. Therefore,

$$\mathcal{E}_{max} \leq \begin{cases} 1.85\mathcal{E}_{\lambda/2} & \text{amplitude STLM} \\ 1.34\mathcal{E}_{\lambda/2} & \text{phase STLM} \end{cases} \tag{7.51}$$

As can be seen, with identical $\mathcal{E}_{\lambda/2}$ amplitudes, amplitude space–time light modulators (with balanced modulation) are preferable to the phase space–time light modulators (see Fig. 7.11).

Requirement (7.51) could turn out to be insufficient for assuring $DR_9 >$ 70 dB the in case of a planar electrooptical array without taking special measures to suppress the adjacent images. In this case $DR_8 \approx e_{AA}$ and, for example, with magnitudes of the latter around 20 ... 30 dB to accomplish the given task, it is necessary that $\mathcal{E}_{max} \leq 0.1\mathcal{E}_{\lambda/2}$ for the amplitude as well as the phase space–time light modulator (see also Sections 7.3.4 and 7.3.5).

Aside from the noise, space–time light modulator nonlinearities also produce relevant order of diffraction defocusing $DF_9 < 1$ with several emis-

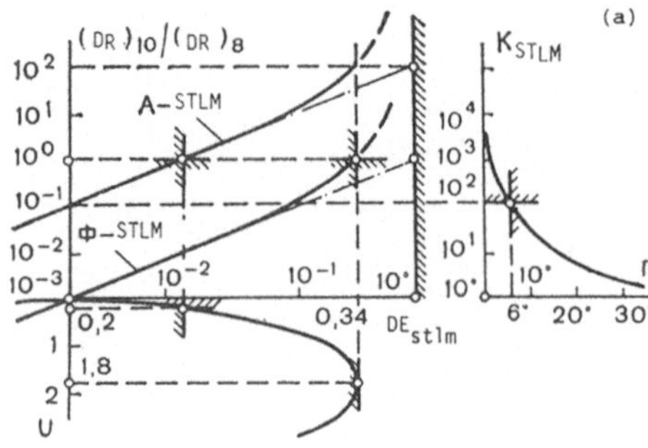

Figure 7.13: The effect of the zero order of diffraction on the CO processor dynamic range.

sion sources present in the image zone. Indeed, in this case the normalized amplitude $U(\mathbf{R},t)$ of the space–time signal received by an array antenna element at point $\mathbf{R}$, could vary within limits from zero to some value $U_{max} = \frac{1}{2}\pi\mathcal{E}/\mathcal{E}_{\lambda/2}$. In such a case, the numerator $2J_1(U)/U$ in (7.45) and (7.46) should be regarded as an random function of a spatial variable whose correlation radius is on the order of a minimum wavelength in the received signal spectrum. This circumstance allows us to use an evaluation of type (7.18), where

$$\alpha = \overline{\left[\frac{2J_1(U)}{U}\right]^2} - \overline{\frac{2J_1(U)^2}{U}} \le \overline{\left[1-\frac{U^2}{8}\right]^2}$$

$$- \overline{1-\frac{U^2}{8}}^2 = \frac{\overline{U^4}-\overline{U^2}^2}{64} \le \left(\frac{U_{max}}{4}\right)^2 \qquad (7.52)$$

is the deviation of the random wavelength $2J_1(U)/U$. If onto the array antenna falls a dominating planar wave, then the normalized space–time signal amplitudes fluctuate "in phase" $[U(\mathbf{R},t) \approx U(t)]$ and deviation (7.52) is minimum (approximately equal to zero) [180]. If a large number of comparable waves is incident on the array, the space–time spatial correlation weakens and the deviation is close to its maximum value (approximately equal to $U^4/4^4$). In this way for the final result we get (see Fig. 7.12,b)

$$DF_9 \ge 1 - (U_{max}/4)^4 \qquad (7.53)$$

where U_{max} is the maximum allowed normalized voltage at the receiver output. Assuming that $DR_9 \geq 0.999$, we obtain a requirement more rigid than (7.51):

$$\mathcal{E}_{max} \leq 0.45\mathcal{E}_{\lambda/2} \tag{7.54}$$

which should be carried out when choosing the receiver amplification or when setting its automatic gain control operating mode.

7.3.3 Zero order diffraction

This factor is described by addend $(J_\circ)$ in expansions (7.43) and (7.44). It should be regarded as parasitic noise $[DR_{10} < \infty]$, which, unlike the ones discussed previously (multiplicative factors), has an additive nature and is present at the processor output even in the absence of a relevant signal. Moreover, in the latter case it is at its maximum $[(J_\circ) = 1]$. This circumstance worsens not only the processor dynamic range, but also the electrooptical array threshold sensitivity (see Section 8.4). Analogous to (7.48) and (7.49) we have

$$DR_{10} = DR_8 \text{ctg}^2(\Gamma_\circ/2)[J_1(U)/J_\circ(U)]^2 \tag{7.55}$$

for an amplitude space–time light modulator; and

$$DR_{10} = DR_8[J_1(U)/J_\circ(U)]^2 \tag{7.56}$$

for a phase space–time light modulator.

With $\Gamma_\circ = 2\pi m$ $(m = 0, 1, 2, ...)$, that is with a balanced light modulation, $\text{ctg}^2(\Gamma_\circ/2) \longrightarrow \infty$. However, actual space–time light modulators do not provide full suppression of the light carrier $[\text{ctg}^2(\Gamma_\circ/2) < \infty]$ due to various sources of distortion (light modulation inhomogeneities along space–time light modulator channel pupils, shifts in operation point $\Gamma_\circ$ and so on). It is convenient to characterize the remaining light flux by the concept of contrast K_{STLM} [16, 106, 107] (Fig. 7.13a)

$$K_{STLM} \approx \text{ctg}^2(\Gamma_\circ/2) < 10^2...10^3 \tag{7.57}$$

When using input devices with electronic beam or optical addressing which perform the spatial sweep of the signals received by the array antenna elements, the contrast (7.57) concept must be replaced by a more exhaustive concept such as the one of frequency contrast characteristic (FCC) of the spatial time signal sweep photosensitive carrier as the contrast function depending on the resolving power [16, 107]. Sections 3.2.2 and 7.3.5. present the discussion of this circumstance.

If the concept of space–time light modulator diffractional efficiency, which is defined as the ratio of the power in the $\pm 1^{st}$ relevant order of diffraction to the power in the read beam at the space–time light modulator input, is to be used (not taking into account the coherent optical processor diffraction efficiency, see Chapter 5 and Fig. 7.13b)

$$DE_{STLM} = J_1^2(U)\left[J_0^2(U) + 2\sum_{n=1}^{\infty} J_n^2(U)\right] = J_1^2(U) \qquad (7.58)$$

then, keeping in mind that with a small $U < 1$; $J_0^2(U) \approx 1 - 2J_1^2(U) = 1 - 2DE_{STLM}$, (7.55) and (7.56) can be represented as (Fig. 7.13c)

$$DR_{10} = \begin{cases} DR_8 K_{STLM} \frac{DE_{STLM}}{1-2DE_{STLM}} & \text{amplitude STLM} \\ DR_8 DE_{STLM}/[1 - 2DE_{STLM}] & \text{phase STLM} \end{cases} \qquad (7.59)$$

Thus we may neglect the zero order, that is, $DR_{10} > DR_8 > 10^7$, if

$$DE_{STLM} \geq \begin{cases} K_{STLM}^{-1} \approx & 10^{-2} \\ 1/3 \approx & 0.338 \end{cases} \qquad (7.60)$$

for amplitude and phase space–time light modulators, respectively.

Requirement (7.60) for phase space–time light modulators is only executable within limits with $DE_{STLM} \approx 34\%$. This corresponds to the limiting diffractional efficiency using electro and acoustooptic effects. This limit is achieved when $U \approx 1.8$, which does not satisfy condition (7.54). Therefore phase space–time light modulators are less preferable than the amplitude ones if special measures to eliminate the zero order of diffraction are not taken.

7.3.4 The presence of distortion in coherent optical processors with AOM-based input devices

In the this review the emphasis is placed on the pattern-forming electrooptical array channel, whereby not reflecting upon the specificity of the coherent optical processors on space–time light modulators with spatial sweep of the time signal (see Chapter 3). When using a multichannel AOM, it is necessary to analyze Raman-Nath and Bragg diffraction processes separately [185,

186]. In the first case, with high levels of the space–time signal (3.1) at the acoustooptic light modulator piezoelectric transducer nonlinear distortions of type (7.44) emerge. Here, however, the first term is different from (3.2):

$$\begin{aligned}\tilde{E}^{(1)}(x,y,t) &= E_{\circ}J(m_x x)J_t(y)\\ &\times J_1\{U[m_x x; t+(y-y_{\circ})/v]\}\exp(-i2\pi\nu t)\end{aligned} \tag{7.61}$$

In the second case, the analogous term is written as

$$\begin{aligned}\tilde{E}^{(1)}(x,y,t) &= E_{\circ}J(m_x x)J_t(y)\\ &\times \sin\{\frac{1}{2}U[m_x x; t+(y-y_{\circ})/U]\}\exp(-i2\pi\nu t)\end{aligned} \tag{7.62}$$

Thus, when using an AOM, aside from the additive noise, which is defined by expansion terms (7.44) on the even frequencies, $n\Omega_{\circ}$, we observe amplitude distortions which, as evident from comparing (7.61) and (7.62), are expressed better in the Raman-Nath diffraction process. These distortions, apparently lead to the distortion of the diffraction pattern (3.4) at the processor output. Figure 7.14 shows the normalized responses of a spectro-analyzer with uniform amplitude-phase distribution J_t of a monochromatic signal with various modulation indices U [186]. Distortion of signals (7.61) and (7.62) can be defined using respective coefficients of non-linear distortions of the AOM amplitude characteristic

$$\kappa_{RN} = 2[0.5U - J_1(U)]/U \tag{7.63}$$

$$\kappa_{BR} = w[0.5U - \sin(0.5U)/U \tag{7.64}$$

The plots of the above are shown in Fig. 7.15.

As with electrooptical space–time light modulators, nonlinearity of the AOM input device causes not only amplitude signal distortions, which cause defocusing of the output picture, but also parasitic noise, which lowers the coherent optical processor radiation pattern and causes false signals.

The noise is described by frequencies with indices $n \geq 2$ in expansion (7.44), which bring about higher order of diffractions in the output picture

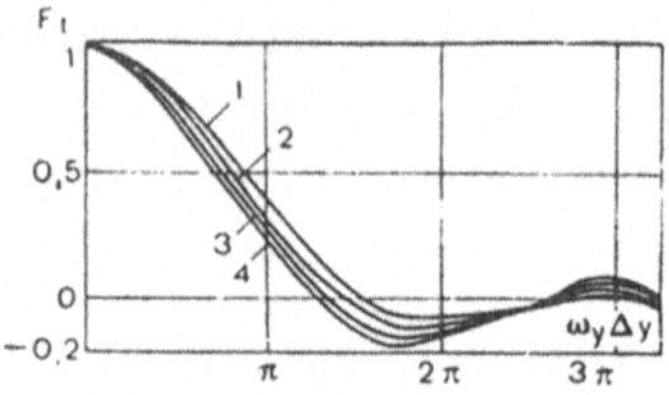

Figure 7.14: For optical spectro-analyzer output diffraction pattern distortion: (1) $U << 1$; (2) $U = 2$ (Raman-Nath Process); (3) $U = 2$ (Bragg Process); (4) $U = 4$ (Bragg Process).

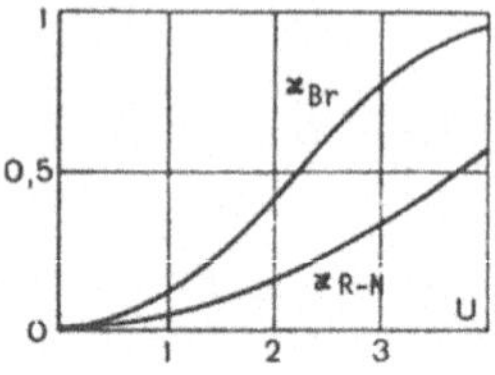

Figure 7.15: Non-linear distortion coefficient.

(3.4). With two or more independent signals, come about so called intermodulational distortions, which lead to practically total processor output noise. To evaluate the light intensity increase at the maximum of the relevant order of diffraction relative to the average intensity of the intermodulatory background we may use the following evaluations of partial dynamic range [185, 186]:

$$DR_{RN} = 20 \log[J_o(U)/J_2(U)] \tag{7.65}$$

$$DR_{BR} = 20 \log[J_1(U)/J_3(U)] \tag{7.66}$$

for the Raman-Nath and Bragg diffraction processes, respectively (Fig. 7.16).

An analysis of relations (7.63)–(7.66) allows conclusions to be made concerning the advantages of the Bragg diffraction process as compared to the Raman-Nath diffraction process. Particularly, in that, to obtain a signal-to-noise ratio equal to the one at the analyzer output with a acoustooptical light modulator using the Raman-Nath process, a modulation index $U \simeq 1$ radian is required. With a permissible non-linearity of amplitude characteristic at

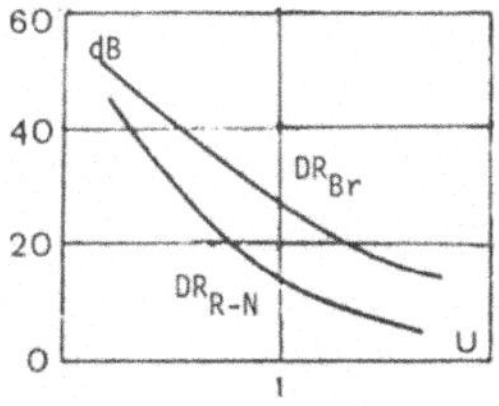

Figure 7.16: Spectro-analyzer dynamic range.

5%, the processor radiation pattern will be 36 dB. The same permissable nonlinearity in the Bragg mode allows us to obtain a radiation pattern of 40 dB.

7.3.5 Non-linear distortions in coherent optical processors with input devices based on space–time light modulators with electronic and optical addressing [107].

Linear operating mode of such processors is reviewed in Section 3.2. It is in many aspects similar to linear operation of processors on AOMs. It differs, however, by blurring the signal which is being recorded (3.20) in virtue of the electronic or optical recording beam end size as well as the continuity of the transmitting characteristic (frequency contrast characteristic—FCC[6]) of the electrical target. Both of these factors can be reflected in (3.20) by a convolution of the signal which is being recorded $U = \frac{\pi}{2}\mathcal{E}/\mathcal{E}_{\lambda/2}$ with a resulting space–time light modulator response function O_{STLM}. The response is to a recording end input signal with size $\delta_t = \delta_y/v_p$ [see (3.20)]:

$$\tilde{U}(m_x x, t) = U(m_x x, t) \otimes O_{STLM}(t) \tag{7.67}$$

If signal (7.67) is sufficiently large ($U \approx 1$), then during its reading from the electrooptical space–time light modulator we will observe nonlinear distortions of type (7.43) or (7.44). The first term of these is similar to (7.67):

$$\tilde{E}(x, y) = E_{read} J(m_x x)\{J(y) J_1 \left\{\tilde{U}\left[m_x; t_n + \frac{y - y_o}{v_p}\right]\right\} \tag{7.68}$$

[6]In the linear operating mode, FCC coincides with the dependence on spatial coefficient frequency of the modulator transmitting characteristic. This is a spatial space–time light modulator response spectrum to a spatial impulse [107: 198].

where all symbols are defined in the comments for (3.19).

In this way, if using a space–time light modulator with electronic or optical addressing two types of effects emerge: "blurring" (7.67) of the recording signal, and amplitude distortions during its reading, provided that the signal is sufficiently large. Each of these effects was actually already researched separately. The first leads to emergence of a corresponding envelope ($\hat{\mathcal{F}}_t\{O_{STLM}\}$) in the output picture; the second, to output diffraction pattern form distortion (defocusing) (see Fig. 7.14). The resulting influence of these factors has a complex character which could be studied (and evaluated quantitatively) using only concrete signal examples. Let us note that these factors often compensate for each other. Indeed, the blurring of signal type (7.67) causes a decline in its amplitude which leads to lesser amplitude distortions (7.68). Therefore the requirements formulated previously for the space–time light modulator and the signal, which limit the influence of the mentioned factors within given limits, retain their power. To these requirements we should simply add a limitation for nonuniformity of FCC to ω_y. It is convenient to tie this in with the corresponding decline of the space–time light modulator diffractional efficiency DE [107: 198]:

$$DE(\omega_y) = \frac{(FCC)^2(\omega_y/2\pi)T^2_{(read)}}{16} \tag{7.69}$$

where T_{FC} is the space–time light modulator transmission coefficient with respect to the reading light intensity. Assuming a 50% decline in the diffractional efficiency within limits of the space–time light modulator resolving power (3.23), we obtain

$$FCC(R_y)/FCC(0) \geq \frac{1}{\sqrt{2}} \tag{7.70}$$

Requirement (7.70) allows us to choose a space–time light modulator by the FCC having a recording signal maximum operating frequency $F_{max} = R_y v_p$.

7.3.6 Comments

Table 7.1 contains systemized formulated requirements for the processor and the space–time light modulator parameters. These guarantee undistorted reproduction of the electrooptical array radiation pattern (its main maximum) with accuracy not worse than the potential (about 1%); also, the output image parasitic noise level, caused by non-idealities in the processor and the space–time light modulator, of not higher than 60 dB (of the limiting dynamic range).

Table 7.1: STLM Parameters.

	Distortion factor	Evaluation of Contribution		Notes
		Defocusing	Dynamic Range (dB)	
1	STLM pupil effect	1	∞	When CO compensation is available
	STLM interaction effect	—	$DR_{COP} \geq 0.5\Sigma[(DR)_n]^{-1}$	$SU \geq 0.5$
2	Aberration of light beams	0.99*	∞ STBW $\leq 10^3$	Astronomical lenses, $f \approx 1$ m $\Delta x/f \leq 10^{-3}$,
3	Longitudinal misalignment of COP	0.999	∞	Element accuracy $\lvert\delta z\rvert \leq 100\mu$m
	Transverse misalignment of COP	0.999	∞	$\lvert\delta y\rvert \leq 1\mu$m for COP with masks
4	Spatial incoherence of light	0.999	∞	$\lvert\gamma_{12}\rvert \geq 0.9995$, $2r_k \leq 4.2\ \mu$m $\sigma_k \geq 10^3\sigma_{STLM}$
5	Time incoherence of light: chromatic aberration	0.999	∞	$l_k \geq 1$ mm, $V_k \geq$ cm^3
	Spatial frequency dispersion	1	70	$l_k \geq 0.03$ mm, $V_k \geq$ cm^3
6	Second order reflection	1	60**	Reflection coefficient $r \leq 10^{-2}$
7	Dispersion of light in discontinuities	1		Correlation radius, $r_\alpha \leq 10\ \mu$m dispersion, $\alpha \leq 4$, class of process $\bar{r}_\alpha \leq 0.2\lambda$
8	2-input STLM AM or PM STLM	1	70	Equivalent lens with $f_e \leq 10cm$ at array
9	STLM nonlinearity:	0.999	—	$\mathcal{E}_{max} \geq 0.45\mathcal{E}_{\lambda/2}$ ***
	AM–STLM	—	70	$\mathcal{E}_{max} \geq 1.85\mathcal{E}_{\lambda/2}$
	PM—STLM	0.999	—	$\mathcal{E}_{max} \geq 0.45\mathcal{E}_{\lambda/2}$ ****
		—	70	$\mathcal{E}_{max} \geq 1.34\mathcal{E}_{\lambda/2}$
10	Zero order Diffraction AM-STLM	1	70 216	$(DE)_{STLM} \leq 10^{-2}$ (1%) in a balanced modulated light system
	PM-STLM	1	=70	$(DE)_{STLM} = 0.34$ (34%) for reduced COP
	Overall	DF_{COP} $\approx$)0.99	$DR_{COP} \approx 60$dB	If all requirements restrictions, tolerances met

* Potential accuracy of CO processor operation.

** Maximum attainable dynamic range for CO processor.

*** $\mathcal{E}_{max} \geq 0.45\mathcal{E}_{\lambda/2}$ with balanced light modulation.

**** $\mathcal{E}_{max} \geq 0.45\mathcal{E}_{\lambda/2}$ not considering zero order diffraction.

Chapter 8

Electrooptical Array Sensitivity

Along with parasitic noise (additive spatial noise) and defocusing (multiplicative spatial noise) of the radio-range image as a result of processor imperfections at the actual electrooptical array output, noise (time) is unavoidable. This noise, in this case, is composed of thermal and quantum noise [187, 188]. The latter condition is due to the fact that electrooptical array antennas belong to both the radio (electro) and the optical ranges where thermal and quantum noise is prevalent, respectively [187, 188]. Within electrooptical array linearity limits time noise is often additive and does not distort the output space–time signal, and it does not degrade the gain or the actual electrooptical array antenna gain (limiting dynamic range). However, the additive power of time noise, converted to the electrooptical array input, determines its sensitivity, i.e. the minimum power or EMF of the signal received by the array at which the power or the voltage of the signal at the output of the output device reaches the set value or the required SNR [187]. This relation is, in turn, defined by the accuracy of signal reproduction or by qualitative indicators of its detection or by the probabilities of correct detection or false alarm [56, 57]. This chapter evaluates the influence of thermal and quantum noise on the electrooptical array sensitivity in whole during incoherent and coherent output of information at the output of the coherent optical processor. The chapter also compares the threshold sensitivities (with SNR at the output equal to one) of an electrooptical array, a phased array, and an active phased array antenna (APAA). Complex requirements are also formulated for electrooptical array parameters which provide for sensitivity not worse than that of an active phased array of comparable size.

8.1 Thermal Noise in Electrooptical Arrays

Thermal noise of a radio-receiving device is a combination of internal (in) and external (ex) noise [187].

8.1.1 External thermal noise

This noise is caused by thermal radiation in the radio band. The radiation is characterized by the sky temperature $T(\mathbf{K})_{ex}$ °K, which in general depends on direction $\mathbf{K}$. Except for directions toward point (quasi-point with respect to array resolution) thermal radiation sources (stars, sun, planets) sky temperature $T(\mathbf{K})_{ex}$ is distributed uniformly and in the centimeter range comprises [189]

$$T(\mathbf{K})_{ex} \sim \begin{cases} 30^\circ & \text{in the sky} \\ T_\circ = 290^\circ & \text{on the ground} \end{cases}$$

The antenna thermal noise component, caused by the external noise reception is defined by expression (8.1)

$$T(\mathbf{K}') = \frac{D\eta}{4\pi} \iint_{4\pi} T(\mathbf{K})_{ex} |F_\circ(\mathbf{K},\mathbf{K}')|^2 \mathrm{d}^2\mathbf{K} = \eta T(\mathbf{K}')_{ex} \tag{8.1}$$

where D and ν are the gain and the efficiency respectively; $|F_\circ(\mathbf{K},\mathbf{K}')|^2$ is the normalized power radiation pattern; it was also taken into account that $D = 4\pi / \iint_{4\pi} |F_\circ(\mathbf{K},\mathbf{K}')|^2 \mathrm{d}^2\mathbf{K}$, and that the sky temperature was distributed uniformly. Let the efficiency of this antenna coincide with the efficiency of the antenna feed of the electrooptical array. The latter efficiency, unlike the equivalent e_{EOAA} (7.2), is defined by ohmic losses. Then, (8.1) may be viewed as equivalent noise temperature of an ideal electrooptical array without corresponding internal noise (thermal and quantum). Therefore, a light beam of intensity (8.2)[1] coincides with the noise temperature (8.1) at the output of this electrooptical array.

$$I_{ex}(\mathbf{K}') = j_\circ k\eta T(\mathbf{K}')_{ex} \Delta F \tag{8.2}$$

[1] Let us assume that the blinding done to the electrooptical array by point sources is eliminated by methods of Chapter 6.

where $j_\circ$ is the proportionality coefficient defined below; $k = 1.38 \cdot 10^{-23} W/Hz \cdot K$ = Boltzmann's constant, ΔF is the electrooptical array transmission band, in Hertz.

To define coefficient $j_\circ$, let us assume that sky temperature $T(\mathbf{K})_{ex} = T_{ex}$ = const for the entire solid angle 4π. Then all noise intensities induced at the outputs of all antenna array elements will be uncorrelated [189]. Therefore, the light beam intensity at the electrooptical array output, defined by the processing of these potentials by algorithm (5.2) is:

$$I_{ex}(\mathbf{K}') = e_\circ^2 \langle \eta_n k T_{ex} \Delta F] [|\dot{J}(\mathbf{R}_n, \mathbf{K}')|^2 \rangle = j_\circ k \eta_\circ T_{ex} \Delta F \tag{8.3}$$

where $e_\circ$ is a normalizing coefficient similar to the coefficient in (2.7) which describes the coherent optical processor; η_n is the efficiency of the nth element of the electrooptical array, which, in case of identical elements is equal to $\eta_\circ$. From (8.3) we can see that

$$j_\circ = e_\circ^2 \sum_{n=1}^{N} |\dot{J}(\mathbf{R}_n, \mathbf{K}')|^2 \approx e_\circ^2 M_\circ \tag{8.4}$$

where $M_\circ$ is the average number of elements on the exposed side of the non-planar array (in antennas with uniform amplitude phase distribution, $M_\circ$ is the total number of elements)

8.1.2 Internal thermal noise in electrooptical arrays

A significant contribution to the electrooptical array noise temperature is made by internal noise. These, in general, are caused by temperature of the array antenna receiving elements $T_n = T_\circ = 290K$, by the noise factor $N_n = N$ of the each element's receiver-amplifier module, and by the multichannel space–time light modulator temperature, $T_{STLM} \approx T_\circ$. Other sources of noise such as the coherent optical processor and the readout device (photoreceiver) are less important since the first operates in the optical range (at temperature $T_\circ$ the brightness of the "heated" processor elements does not exceed 10^{-23} W/m^2 or about 3 quanta in 24 hours), and the second is characterized by more powerful "shot" (quantum noise, see below). Therefore, analogously to (8.3) we obtain [187].

$$I_{in}(\mathbf{K}') = e_\circ^2 \langle k T_{in_n} \Delta F] [|\dot{J}(\mathbf{R}_n, \mathbf{K}')|^2 \rangle = j_\circ k T_{in_n} \Delta F \tag{8.5}$$

Here

$$T_{in_n} = T_n[1-\eta_n] + T_\circ(N_n - 1) + T_{STLM}/K_{pr_n} = T_{in} \approx T_\circ[N_\circ - \eta_\circ] \quad (8.6)$$

where $K_{pr_n} = K_{pr_o} \gg 1$ is the gain for each electrooptical array receiver-amplifier module; $\eta_n = \eta_\circ$ is the efficiency of the nth element and its high frequency channel.

The signal-to-noise ratio (S/N) at the electrooptical array output taking into consideration (8.3), (8.5), and (5.2) is

$$(c/n_T)_{EOAA} = I_q(\mathbf{K}_q)/[I_{ex}(\mathbf{K}_q) + I_{in}\mathbf{K}_q)]$$

$$= e_\circ^2| < \dot{\mathcal{E}}_\Omega(\mathbf{R}_n)_q][\dot{J}(\mathbf{R}_n, \mathbf{K}'))|^2 \Delta F / j_\circ k\{\eta_\circ T_{ex} + T_\circ[N_\circ - \eta_\circ]\}\Delta F$$

$$\leq e_\circ^2 \eta_\circ A_\circ |\dot{S}_q(\Omega)|^2 \left| \sum_{n=1}^{N} |\dot{J}(\mathbf{R}_n, \mathbf{K}')|^2 \right|^2 / j_\circ k T_\circ[\eta_\circ t_{ex} + N_\circ - \eta_\circ]$$

$$= \eta_\circ A_\circ M_\circ |\dot{S}_q(\Omega)|^2 / kT_\circ[N_\circ + \eta_\circ(t_{ex} - 1)] \quad (8.7)$$

where $I_q(\mathbf{K}_q)$ is the response of electrooptical array with processing algorithm (5.2) to a plane wave incident from $\mathbf{K}_q$; $t_{ex} = T_{ex}/T_\circ$ is the relative noise temperature of the radio range; $A_n = A_\circ$ is the effective area of the element. With no internal thermal noise, i.e., with an ideal electrooptical array with $\eta_\circ = 1$ and $N_\circ = 1$, expression (8.7) reduces to

$$(c/n_T)_{ideal} = A_\circ M_\circ |\dot{S}_q(\Omega)|^2 / kT_\circ t_{ex} \geq (c/n_T)_{EOAA} \quad (8.8)$$

Let us introduce the electrooptical array thermal noise coefficient. We shall define it as the relative decrease in the signal-to-noise ratio at the output of an actual electrooptical array with respect to the ideal case [179].

$$N_T = \frac{(c/n_T)_{ideal}}{(c/n_T)_{EOAA}} = \frac{N_\circ + \eta_\circ(t_{ex} - 1)}{\eta_\circ t_{ex}} = 1 + \frac{1}{t_{ex}}(N_{T_\circ} - 1) \quad (8.9)$$

where $N_{T_\circ} = N_\circ/\eta_\circ$ is the noise factor at room temperature (when $t_{ex} = 1$).

The coefficients, N_T and $N_{T_\circ}$, are useful for comparing the electrooptical array noise characteristics of both active (APAA) and passive (PPAA) phased array antennas. Since modern phased array designs have efficiency

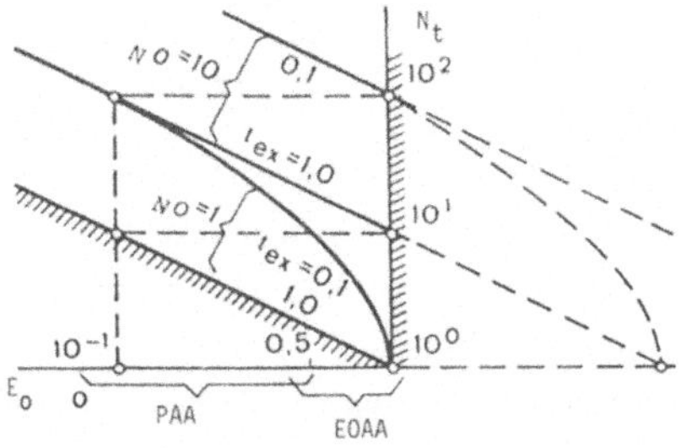

Figure 8.1: Thermal noise comparisons

values $e_{PAA} \approx 0.1...0.5$, and $e_{APAA} = \eta_\circ \approx 0.8...0.95$ [51-53], the electrooptical array noise characteristics caused by external and internal thermal noise are at least as good as that of a phased array, and are comparable to an active phased array. The latter is obvious since the electrooptical array high frequency receiving section is, in essence, an active phased array. Let us evaluate the maximum power transfer, G_m:

$$G_m = N_T^{PAA}/N_T^{EOAA,APAA} = \frac{1+(N_\circ\eta_\circ - 1)t_{ex}}{1+(N_\circ - 1)/t_{ex}}$$

$$= 1 + \frac{1-e_{PAA}}{e_{PAA}} \cdot \frac{T_\circ + T_{pr}}{T_{ex} + T_{pr}} \qquad (8.10)$$

where $T_{pr} = T_\circ(N_\circ - 1)$ and it is kept in mind that $e_{PAA} \approx 1$. Magnitude (8.10) is independent of the number $N_\circ$ of antenna array elements and agrees with an analogous result for a single antenna with a receiver [179].

The advantages of an electrooptical array antenna and an active phased array antenna compared to a phased array antenna are graphically represented in Fig. 8.1, where $N_\circ = 1$; 10 and $t_{ex} = 0.1$; 1 are used as parameters.

8.2 Quantum Noise in Electrooptical Array with Incoherent Photodetection

Information from the processor output is read by a multi-element matrix or by scanning photorecorders employing external or internal photo effects (see Section 1.3.2). The output information is presented as the following light intensity distribution

$$I_{out}(\mathbf{K}') = \sum_q I_q(\mathbf{K}') + \underbrace{I_{ex}(\mathbf{K}') + I_{in}(\mathbf{K}')}_{I_T(K')} + I^{(0)}(\mathbf{K}') \tag{8.11}$$

where $I^{(0)}(\mathbf{K}') = I_{10}$ is the zero order of diffraction (see Section 7.3.3). Expression (8.11) does not reflect the influence of other factors reviewed in Chapter 7, due to the assumption that the requirements for the coherent optical processor are met (see Table 7.1).

Let one of the photoregistration device elements be located at point $\mathbf{K}' = \mathbf{K}_q$ in the processor output plane. In this case, the light flux (8.11) generates the following photo-electron flux there [188].

$$i_{out} = i_q + i_T + i^{(0)} + i_d \tag{8.12}$$

Where

$$\begin{aligned} i_q &= \frac{e\varepsilon}{h\nu}\int_\sigma\int_\phi I_q(\mathbf{K}')\mathrm{d}^2\mathbf{K}' \\ &= \frac{e\varepsilon}{h\nu}\eta_{COP}P_L DE_{STLM} DE_{COP}\frac{D_{AA}}{4\pi}\int_{\sigma_\phi}\int F_\circ^2(\mathbf{K}',\mathbf{K})\mathrm{d}^2\mathbf{K}' \\ &= \frac{e\varepsilon}{h\nu}\eta_{COP}P_L\frac{U_q^2}{4}K_{su}^\delta\lambda_\circ = \frac{e_\circ P_\circ}{h\nu}K_{pr_\circ}q_\circ P_q \end{aligned} \tag{8.13}$$

where $e = 1.60 \cdot 10^{-19}$ C is the electron charge; $\varepsilon \leq 1$ is the photoregister quantum efficiency; $h = 6.62 \cdot 10^{-34}$ J/Hz is Planck's constant; σ_Φ is the photoregistering device photoemissive surface; ν_{COP} is the processor efficiency (generally keeps track of the collimator and optical system losses); P_L is the laser power; $DE_{STLM} = J_1^2(U_q) \approx U_q^4/4$ is the space–time light modulator diffraction efficiency [see (7.58)]; $DE_{COP} \approx K_{su}^\delta$ [see (5.32)]; $\lambda_\circ = D_{AA}/4\pi \int_\sigma \int_\phi |F_\circ(\mathbf{K}',\mathbf{K}_q)|^2\mathrm{d}^2\mathbf{K}'$ is the fraction of energy spot I_q within the limits σ_Φ which is approximately equal to $\lambda_\circ$ in (6.44) due to the proximity of the normalized radiation pattern $F_\circ$ to its own function $\Psi_\circ$ (see Section 6.3); $P_\circ = \varepsilon\lambda_\circ K^\delta \eta_{COP}$; $P_l/4 \approx (10^{-2}...10^{-3})P_l$; $q_\circ = U_q^2/P_q$ is the space–time light modulator quality [89]; $P_q = A_\circ/2\pi\int_{-\infty}^{\infty}|\dot{S}_q(\Omega)|\mathrm{d}\Omega$ is the power of the signal received by an array antenna element; and $\mathbf{K}_{pr_\circ}$ is the gain of the receiver amplification modules

$$i_T = i_q/(c/n_T)_{EOAA} \leq i_\circ\frac{kT_\circ}{h\nu}N_{T_\circ}\frac{K_{pr_\circ}}{M_\circ}q_\circ P_\circ \tag{8.14}$$

where (8.7) is considered and $i_\circ = e\Delta F \approx 10^{-11}A$ is the photon limit [188]

$$i^{(0)} \approx i_q/(DR)_{10} \leq \frac{e_\circ P_\circ}{h\nu}/(DR)_8 \approx 10^{-10}A \tag{8.15}$$

where $(DR)_8$ and $(DR)_{10}$ are partial dynamic ranges reviewed in Sections 7.3.1 and 7.3.3; $i_d \geq 10^{-9}$ A is the dark current. DC current components can be filtered out at the photoreceiver output. However, even in this case they make their contribution into the common, shot, (quantum) photoreceiver noise, which is caused by the irregular manner of photon arrivals and photoelectron emissions

$$i_\nu^2 = 2N_\phi i_\circ i_{out} \tag{8.16}$$

where N_ϕ is the excess noise factor of the photoreceiver ($N_\phi \leq 2$ for photoelectronic multipliers and is equal to 1 for solid state detectors). Adding to (8.16) thermal noise at the equivalent output resistance

$$i_{R_\phi}^2 = kT_\phi \Delta F / R_\phi \tag{8.17}$$

which is increased to the equivalent noise temperature T_ϕ, and taking into consideration the possible photomultiplying in the photoreceiver by M(Ω) times (e.g., if using avalanche photodiodes), we derive the following expression for the signal-to-noise ratio at the electrooptical array output in presence of thermal and quantum noise [188].

$$\begin{aligned}(c/n_{T,\nu})_{EOAA} &= \frac{M^2(\Omega)i_q^2}{M^2(\Omega)[(i_q+i_t)^2 - i_q^2] + M^2(\Omega)i_\nu^2 + i_{R_\phi}^2} = \\ &= \frac{(i_q/i_T)^2}{2(i_q/i_T) + 1 + (i_\nu/i_T)^2 + [i_{R_\phi}/M(\Omega)]^2/i_T^2} = \\ &= \frac{(c/n_T)^2_{EOAA}}{2(c/n_T)_{EOAA} + 1 + [i_\nu^2 + i_{R_\phi}^2/M^2(\Omega)]\ i_T^2}\end{aligned} \tag{8.18}$$

Let us introduce the overall electrooptical array antenna noise factor. It is defined, analogously to (8.9), as the ratio of signal to noise ratios at the ideal electrooptical array input without the internal thermal and quantum noise of a real electrooptical array (8.18):

$$N_{T,\nu} = (c/n_T)_{ideal}/(c/n_{T,\nu})_{EOAA} = N_T \cdot N_\nu > N_T \tag{8.19}$$

where

$$N_\nu = N_{sq.det.} + \frac{i_\nu^2 + i_{R_\phi}^2/M^2(\Omega)}{i_T^2} \Big/ \frac{i_t}{i_q} \tag{8.20}$$

where

$$N_{sq.det.} = 2 + 1/(c/n_T)_{EOAA} \tag{8.21}$$

is the noise factor of an ideal quadratic detector [188].

In this way, the overall noise factor of an electrooptical array antenna with incoherent (direct) photodetection is at least $N_sq.det$ times worse than that of the active phased array. If the ratio (8.7) is large and the second term in (8.20) is small, we can live with this circumstance. For this, two conditions must be observed. First, the thermal noise at the equivalent output resistance (8.17) must not be greater than the overall quantum noise (8.16) $i_{R_\Phi}^2 < i_\nu^2$. This, according to (8.16)–(8.18) is fulfilled if

$$M^2(\Omega)R_\phi > kT_\phi/2N_\phi e i_{out} \approx 10^9 \quad \text{ohms} \tag{8.22}$$

This limitation can be fulfilled only if photoelectronic multipliers are used. Second, the numerator of the second term in (8.20) must not be greater than its denominator; i.e., according to (8.12), (8.16), and (8.18) the following must be true:

$$\begin{aligned} (i_\nu^2/i_T^2)/(i_q/i_T) &= 2N_\phi i_\circ(i_q + i_T + i^{(0)} + i_d)/(i_T i_q) \\ &< 2N_\phi i_\circ\left[2/i_T + \frac{(i^{(0)} + i_d)}{i_T^2}\right] \\ &\approx 2N_\phi i_\circ \frac{i^{(0)} + i_d}{i_T^2} < 1 \end{aligned} \tag{8.23}$$

From here, keeping in mind (8.14), we obtain the second requirement

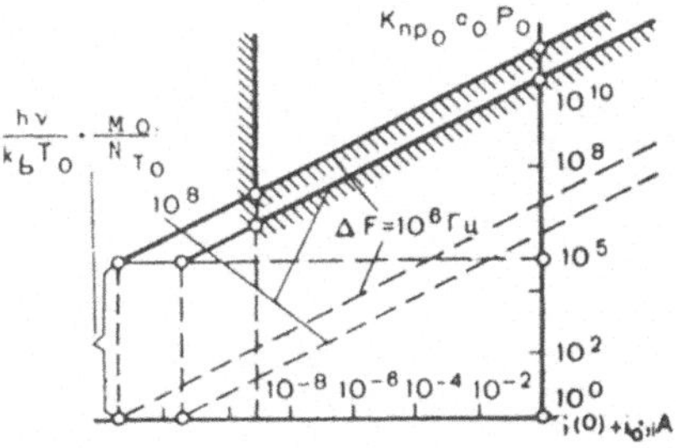

Figure 8.2: Requirements for electrooptical arrays with incoherent photodetection.

$$K_{pr_\circ} q_\circ P_\circ > \left[\frac{2N_\phi(i^{(0)} + i_d)}{i_\circ} \right]^{1/2} \frac{h\nu}{kT_\circ} \frac{M_\circ}{N_{T_\circ}} \tag{8.24}$$

Figure 8.2 shows limitation (8.24) ($M_\circ = 10^4$).

Thus, an electrooptical array with incoherent photodetection has the following shortcomings: at the photoreceiver output, information about the received signal frequency and phase is lost; the resulting noise coefficient is worse than the one of the equivalent active phased array antenna [see (8.9) and Fig. 8.1]. Whereupon, to attain a better case (that of an ideal quadratic detector), it is necessary to observe strict requirements (8.24) for the photoregister "quality" and the array antenna receiving module amplification as well as for the space-time light modulator "quality" and the laser power.

8.3 Quantum Noise in Electrooptical Array Antennas with Coherent Photodetection

Photodetectors devices transform optical signals into electrical current with an ideal square law characteristic and as such can serve as a mixer for a heterodyne photodetector.

Suppose that an information-bearing light flux (8.11) arrives at the output of a coherent optical processor simultaneously with a collimated reference beam from an optical oscillator $\dot{E}_h \exp\{i[2\pi\nu_h t + \pi_h(t)]\}$ (Fig. 8.3). Then, the interference will result in a light flux whose intensity is:

$$I_{out}(\mathbf{K}', t) = |\dot{E}_h \exp i[2\pi\nu_h t + \phi_h(t)] + \dot{E}_{out}(\mathbf{K}', t) \exp i[2\pi\nu t + \phi_h(t)]|^2$$

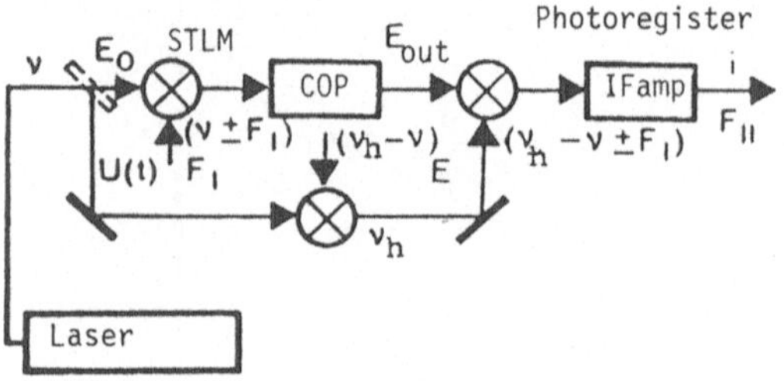

Figure 8.3: Structural diagram of the coherent optical processor for electrooptical arrays with coherent photodetection.

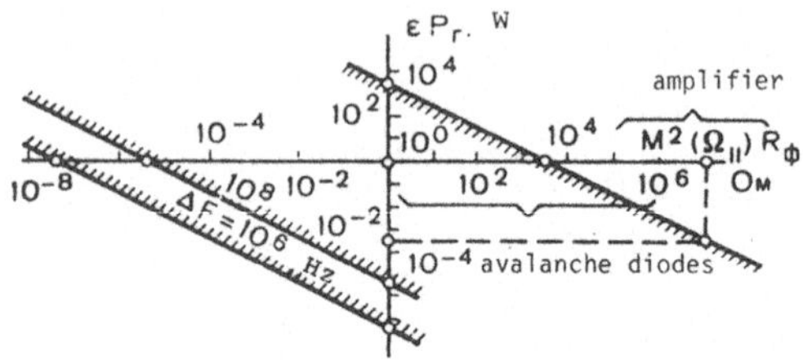

Figure 8.4: Optical oscillator power consumption.

$$= I_h + I_{out} + 2\Re e\{\dot{E}_h \dot{E}^*_{out}(\mathbf{K}', t) \exp i[2\pi(\nu_h - \nu)t + \phi_h(t) - \phi(t)]\} \quad (8.25)$$

Here,

$$\overset{*}{E}_{out}(\mathbf{K}', t) = \left[\sum_q \dot{E}_q(\mathbf{K}', t) + \dot{E}_h(\mathbf{K}', t)\right] \exp(i\Omega_I\, t) + \dot{E}^{(0)}(\mathbf{K}') \quad (8.26)$$

where Ω_I is the intermediate frequency of the signal at the receiver-output, and

$$\left.\begin{array}{rcl} |\dot{E}_{out}(\mathbf{K}', t)|^2 &=& I_{out}(\mathbf{K}') \\ |\dot{E}_q(\mathbf{K}', t)|^2 &=& I_q(\mathbf{K}') \\ |\dot{E}_h(\mathbf{K}', t)|^2 &=& I_h(\mathbf{K}') \\ |\dot{E}^{(0)}(\mathbf{K}', t)|^2 &=& I^{(0)}(\mathbf{K}') \end{array}\right\} \quad (8.27)$$

As can be seen, if $\phi_h(t) = \phi(t)$ (this is achieved through synchronous detection using a common laser, see Fig. 8.3), then (8.25) contains beat frequencies $\Omega_{II} = |2\pi(\nu_h - \nu) \pm \Omega_I|$, where $(\nu_h - \nu)$ is the laser frequency displacement, done by a corresponding single band light modulator. Without the single band light modulator: $\nu_h = \nu$ and $\Omega_{II} = \Omega_I$ a homodyne mode is realized [188].

As before (see Section 8.2), let one of the photoregistration device elements be located at point $\mathbf{K}' = \mathbf{K}_q$ in the processor output plane. Then, the light flux (8.25) generates in it a photocurrent

$$\imath_{out}(t) = \imath_h + \imath_{out} + \imath_q(t) + \imath^{(0)}(t) \tag{8.28}$$

where

$$\begin{aligned} \imath_h &= \frac{e\varepsilon}{h\nu}\int_{\sigma_\phi}\!\!\int I_h \mathrm{d}^2\mathbf{K}' \\ &= \frac{e\varepsilon}{h\nu}\lambda_\circ\sigma_\phi P_h \Big/ \iint_{4\pi} \mathrm{d}^2\mathbf{K}' \leq \frac{e\varepsilon}{h\nu}\lambda_\circ P_h / D_{AA} \end{aligned} \tag{8.29}$$

where $\lambda_\circ$ is the same as in the comments for (8.13) and is the essential of the energy in the light spot within limits of area σ_ϕ in the photodetector; $P_h = \iint_\infty I_h \mathrm{d}^2\mathbf{K}'$ is the optical oscillator power. Here, the equality in (8.29) takes place when the cross-sectional area of the oscillator light beam is coordinated with radioband at the processor output, and inequality, when when it does not exceed the latter; $\imath_{out}$ is defined by expression (8.12);

$$\begin{aligned} \imath_q(t) &= \frac{e\varepsilon}{h\nu}\int_{\sigma_\phi}\!\!\int_\phi 2\sqrt{I_h I_q(\mathbf{K}')}\mathrm{d}^2\mathbf{K}' \cdot \Re e\{\exp[i(\Omega_{II}t + \phi_q)]\} \\ &= 2\sqrt{\imath_h \imath_q}\cos(\Omega_{II}t + \phi_q) \end{aligned} \tag{8.30}$$

where $\phi_q = \arg\{|\dot{E}_q|/|\dot{E}_h|\}$ and expressions (8.25)–(8.27) are used:

$$\imath_T(t) \approx 2\sqrt{\imath_h \imath_T}\cos(\Omega_{II}t + \phi_T) \tag{8.31}$$

where $\phi_T = arg|\dot{E}_T|/|\dot{E}_h|$, and $\imath_T$ is defined in (8.14);

$$\imath^{(0)}(t) = 2\sqrt{\imath_h \imath^{(0)}}\cos[2\pi(\nu_h - \nu)t] \tag{8.32}$$

where $\imath^{(0)}$ is defined in (8.15).

The transmission band $\Delta\Omega$ of the intermediate frequency Ω_{II} (see Fig. 8.3) should be chosen so as to filter out all the currents in (8.28) except (8.30). However, current (8.31) and shot noise (8.16) penetrate into the indicated band. The shot noise is generated by summed current of (8.28). The shot noise, along with the thermal noise (8.13) lowers the signal-to-noise ratio at the electrooptical array output [compare to (8.18)]:

$$
\begin{aligned}
(c/n_{T,\nu})_{EOAA} &= \frac{M^2(\Omega_{II})\overline{i_q^2(t)}}{M^2(\Omega_{II})\overline{i_T^2(t)} + M^2(\Omega_{II})\overline{i^2\nu(t)} + i_{R_\phi}^2} \\
&= \frac{\overline{i_q^2(t) i_T^2(t)}}{1 + [\overline{i_\nu^2(t)} + i_{R_\phi}^2/M^2(\Omega_{II})]/\overline{i_T^2(t)}} \\
&= (c/n_T)_{EOAA} \cdot \left(1 + \frac{\overline{i^2\nu(t)} + i_{R_\phi}^2/M^2(\Omega_{II})}{\overline{i_T^2(t)}}\right)
\end{aligned}
\tag{8.33}
$$

From here, analogous to (8.20)

$$
N_\nu = 1 + \frac{\overline{i_\nu^2(t)} + \overline{i_{R_\phi}^2/M^2(\Omega_{II})}}{\overline{i_T^2(t)}} \tag{8.34}
$$

The quantum noise can be disregarded if ($N_\nu \approx 1$), if $\overline{i_\nu^2(t)} \geq \overline{i_{R_\Phi}^2}/M^2(\Omega_{II})$, which, in keeping with (8.33), (8.16), (8.17), and (8.19) is equivalent to requiring

$$
\varepsilon P_h > \frac{kT_\phi h\nu D_{AA}}{e^2 2N_\phi M^2(\Omega_{II}) R_\phi} \tag{8.35}
$$

where it was assumed that $M^2(\Omega_{II})R_\phi \approx 10^6$ [113,121]; and $D_{AA} \approx 10^5$. Unlike requirement (8.22), requirement (8.35) is not as strict on photoregister parameters (see Fig. 8.4). Second, for the numerator of the second term in (8.34) not to be greater than the denominator, it is necessary that

$$
\frac{\overline{i_\nu^2(t)}}{\overline{i_T^2(t)}} = \frac{2N_\phi i_\circ \sqrt{\overline{i_{out}^2(t)}}}{4 i_h i_T \cos^2(\Omega_{II} + \phi_h)} \approx \frac{N_\phi i_\circ i_h}{i_h i_T} > 1 \tag{8.36}
$$

From here, using (8.14) and (8.24), we derive

$$
K_{pr_\circ} q_\circ P_\circ > N_\phi \frac{h\nu}{kT_\circ} \cdot \frac{M_\circ}{N_{T_\circ}} \tag{8.37}
$$

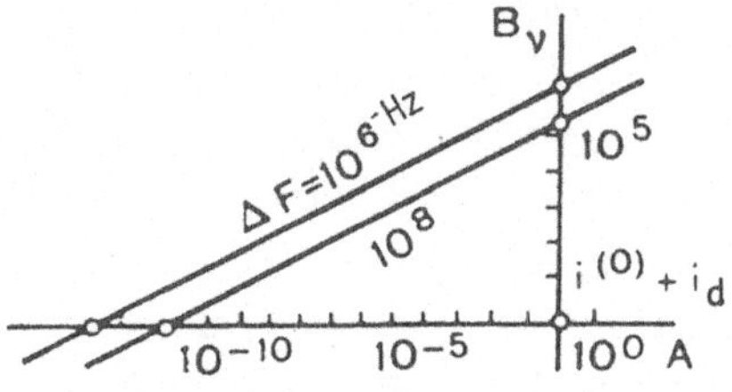

Figure 8.5: For comparison of the electrooptical array and coherent and noncoherent photodetection.

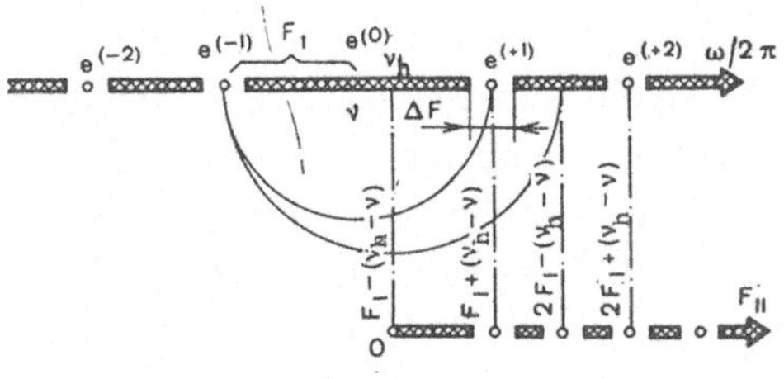

Figure 8.6: For choosing the filter pass-band for the intermediate-frequency amplifier of the photoregister in the electrooptical array with coherent photodetection.

As can be seen, limitation (8.37) is

$$B_\nu = \sqrt{\frac{2(i^{(0)} + i_d)}{N_\phi e \Delta F}} \tag{8.38}$$

times weaker than limitation (8.24) (see Fig. 8.5, where $M_\circ \approx 10^4$).

And so, electrooptical arrays with coherent heterodyne photodetection have the following advantages over processors which output by direct (incoherent) photodetection (see section 8.2): information about the radio signal frequency and phase is saved at the corresponding photoreceiver in intermediate frequency current form (8.30); the resulting noise factor (8.33) is close to the thermal noise factor of an equivalent active phased array when feasible requirements for optical oscillator power (8.33), the information reading (output) devices' "quality," which allows the use of photoregistration devices not only on photocell amplifiers, but also on the multielement solid state photoreceivers (photomatrices) are met; requirement (8.37) for the array receiving module amplification, space–time light modulator "quality" and laser power are weakened by an order of magnitude; a possibility of using an intermediate frequency $F_{II} = \Omega_{II}/2\pi$ filter or an intermediate frequency amplifier having a pass-band $F_I - (\nu_h - \nu) - 0.5\Delta F \leq F_{II} \leq F_I - (\nu_h - \nu) + 0.5\Delta F$ (see Fig. 8.6) to suppress all parasitic diffraction orders $e^{(m)}$ ($m = 0, -1, \pm 2, \pm 3,$ ...) exists, if $F_I > 2\Delta F$ and $0.5(F_I + \Delta F) < (\nu_h - \nu) < 0.5(2F_I - \Delta F)$. The latter allows us to eliminate uncertainty in determining the target coordinates in case a dual-sideband input device is used (see Section 5.2.4). It also makes it easier to reach the maximum coherent optical processor dynamic range.

The disadvantages of electrooptical arrays with heterodyne photodetection of information output from the coherent optical processor are complications in the coherent optical processor optical system (see Fig.9.1), the critical nature of the signal and oscillator light beam alignment [188], and additional noise caused by the optical oscillator.

Let us evaluate the contribution made by the latter. Let the laser oscillator relevant mode to noise ratio be finite: $I_n/I_h > 0$ (for example, if a He-Ne laser is used $I_n/I_h \approx 10^5$ [182]). Then a noise current is added into (8.28). It is caused by the oscillator noise background I_n

$$i_n(t) = \sqrt{i_h i_n} \cos[\Omega_{II} + \varphi_n(t)] \tag{8.39}$$

where $i_n = i_h(I_n/I_h)$ and (8.30) is used. Current (8.29) can be ignored if $i_h < i_T$, or if according to (8.36) and (8.29) the following requirement is fulfilled:

$$\varepsilon P_h \leq \frac{I_h}{I_n} N_\phi h\nu \Delta F D_{AA} \approx 0.1\,\mathrm{W} \tag{8.40}$$

This does not contradict requirement (8.35). Limitation (8.40) can be significantly weakened through using a balanced photodetector [188]

$$\varepsilon P_h \leq \left(\frac{2}{1-c}\right)^2 \frac{I_h}{I_n} N_\phi h\nu \Delta F D_{AA} \tag{8.41}$$

where c is the Δ and Σ photodetector shoulder coordination ($0 < c \leq 1$).

8.4 Electrooptical Array Sensitivity Evaluation

In accordance with the definition (see [187]), and also the introduction of Chapter 8, and the derived value of the overall electrooptical array noise factor $N_{T,\nu}$ (8.19), the sensitivity, or the minimum power of a detectable signal at the electrooptical array input, is defined by the expression

$$\begin{aligned} P_{min} &= kT_{ex}N_{T,\nu}\Delta F \approx kT_{ex}\left[1+\frac{1}{t_{ex}}(N_{T_\circ}-1)\right]\Delta F \\ &= kT_\circ(N_{T_\circ}+t_{ex}-1)\Delta F \leq kT_\circ N_{T_\circ}\Delta F \approx 10^{-12}\,\mathrm{W} \end{aligned} \tag{8.42}$$

where the detectability coefficient is assumed to equal one and it is taken into account that if requirements (8.35), (8.37), (8.40), or (8.41) for the electrooptical array with coherent detecting parameters are fulfilled, then its overall noise factor $N_{T,\nu}$ practically coincides with (8.9); $N_{T_\circ} \approx 10$; $\Delta F \approx 20$ MHz; $t_{ex} = 1$.

The minimum power (8.42) is consistent with the emission power flux falling on the array:

$$\Pi_{min} = P_{min}/A_e \leq \frac{kT_\circ N_{T_\circ}\Delta F}{M_\circ A_\circ} \tag{8.43}$$

where $A_e = M_\circ A_\circ$ is the effective aperture.

The electrooptical array with incoherent detecting sensitivity is at least two times worse than (8.42) and (8.43). To register a result of such power, the overall amplification at the radar station should be $10^9...10^{10}$ [56]. A fraction of this through amplification is done by the electrooptical array:

$$\begin{aligned} K_{EOAA} &= \frac{M^2(\Omega_{II})R_\phi\overline{i_q^2(t)}}{P_q} \\ &= M^2(\Omega_{II})R_\phi\frac{\lambda_\circ\varepsilon eP_h}{h\nu D_{AA}}\cdot\frac{eP_\circ}{h\nu}K_{pr_\circ}q_\circ \\ &\geq M^2(\Omega_{II})R_\phi\frac{\lambda_\circ e^2\varepsilon R_h}{(h\nu)^2M_\circ\pi}N_\phi\frac{h\nu M_\circ}{kT_\circ N_{T_\circ}} \\ &\approx M^2(\Omega_{II})R_\phi\lambda_\circ\varepsilon P_\circ \end{aligned} \tag{8.44}$$

where requirement (8.37) is kept in mind as well as the circumstance that amplification (8.44) is maximal when $P_h = 0.5P_o$ when the processor is constructed using the optic diagram given in Fig. 8.3 (synchronized detection diagram).

Chapter 9

Coherent Optical Processors for Electrooptical Array Antennas. Examples. Components. Research.

In this chapter we shall discuss some coherent optical processors for electrooptical array antennas which have different practical capabilities:

1. A coherent optical processor which forms the radiation pattern of a planar array antenna and which stores time data at the processor output for subsequent processing by traditional (including coherent-optical) methods (see Section 9.1 [93, 94, 124]);

2. A coherent optical processor which carries out the space–time processing of the linear array antenna signals, so that at the processor output a one-dimensional angular (only one angular coordinate) and frequency (magnitude) spectra (with the potential for correlation processing; (see 9.2.1 [128, 130]) are simultaneously reproduced;

3. A coherent optical processor for a planar array antenna with space–time signal processing, so that data on two angular coordinates and the frequency spectrum of the signal are present at the output (see 9.2.2 [13]);

4. A coherent optical processor for an interference suppression linear electrooptical array antenna which forms a radiation pattern with controlled nulls directed at the interfering signals [166, 167, 38]; and

5. Coherent optical processors which carry out the space–time processing of signals from circular array antennas so that at the processor output a

frequency-angular spectrum (one coordinate) and a frequency spectra from all targets in the azimuth sector 360° are simultaneously formed [80, 82].

The array antenna signals are input into the coherent optical processor by using a multichannel space–time light modulator employing the electroopti-cal effect, a combined space–time light modulator with multichannel optical addressing, and static signal rasters.

Coherent optical processors using a multichannel AOM and space–time light modulators with electronic signal addressing are treated in [63, 65, 71, 72, 75–77, 22, 84, 85, 88, 103]. The output devices may be photoelectric multiplier or a linear multielement photodiode matrix. The signals from the array antenna receiving elements, in most of the research, were modeled on an array antenna simulator which consisted of generators and a 16-channel amplifier with independent signal amplitude (from 0 to 50 V) and phase (from 0 to 360° with 3° samples) control. It allows modeling of any combination of signals from the 16-element array antenna on the intermediate frequencies F_1 and F_2.

The research was conducted on the Soviet optical benches "OSK-1" and "OSK-2" production measuring instruments and devices. The technique of photographic information recording was based on [190]. The holographic part of the research was carried out on the Soviet holographic stand "UIG-2M" [191].

9.1 Coherent Optical Processor for Forming the Radiation Pattern of Planar Array Antennas

9.1.1 The coherent optical processor and its components

A diagram of a linear array antenna processor that allows not only the formation of the continuous beam array of a radiation pattern but also the storage of the time information about the input signal at the processor output for subsequent processing is presented in Fig. 9.1. The coherent optical processor includes a He-Ne laser 1; a collimator which consists of a micro lens 2; a diaphragm 3; and a lens 4; a cylindrical lens 5 ($f = 0.5\,\mathrm{m}$) which compresses the light flux for the corresponding measurement; a 16-channel simulator of array 7; a multi-channel space–time light modulator which consists of electrooptical element 8; a polarizer 6; and an analyzer 9; a lens 10

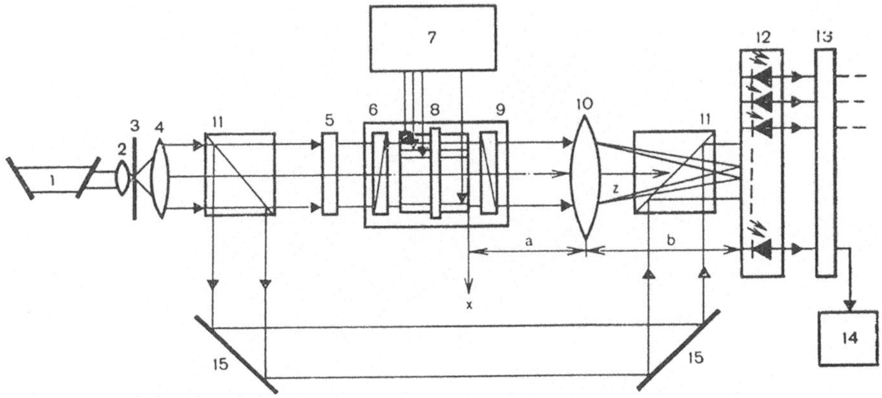

Figure 9.1: Schematic of the array antenna coherent optical processor: (1) He-Ne laser; (2,4) collimator lenses; (3) diffraction pattern (5) cylindrical lens; (6) polarizer, (7) array antenna simulator; (8) electrooptical element; (9) analyzer; (10) spherical lens; (11) light diffractor; (12) multielement photoreceiver; (13) multichannel amplifier; (14) registration device; (15) mirror.

(f = 0.6 m); light-separators 11; a multielement linear photodetector space–time light modulator 12; a multichannel amplifier 13; a recording device 14; and a mirror 15. The key elements of the processor are the multichannel electrooptical space–time light modulator which includes positions 6, 8, and 9, and the multielement photodetector 12.

The 32-channel space–time light modulator used the electrooptical Pockels effect in a 90° section of a lithium tantalate crystal. The electrooptical element consists of two identical $LiTaO_3$ plates with dimensions 16×8 ×0.5 mm. For thermal compensation of natural anisotropy, a half-wave plate was placed between crystals where the optical axes were opposed. The entire assembly was mounted on a copper base, which equalizes the crystal temperature and dampens the parasitic piezoelectric effects. The edges of the crystals were polished. The wide edges of the plates had 32 electrodes of chrome and gold, 250 μm wide and spaced by 400 μm, applied by the method of "vacuum plating"; the opposite edges of the crystal were completely covered with a thin layer of gold which served as the common electrode. In order to increase the contrast of the optical picture, at the modulator output near the edges of the crystals (near the input and the output of the modulator), masks were installed to cover the spaces between the electrodes. The electrooptical element was placed between the polarizer and analyzer. As the result, during the corresponding orientation of the crystal and polarization

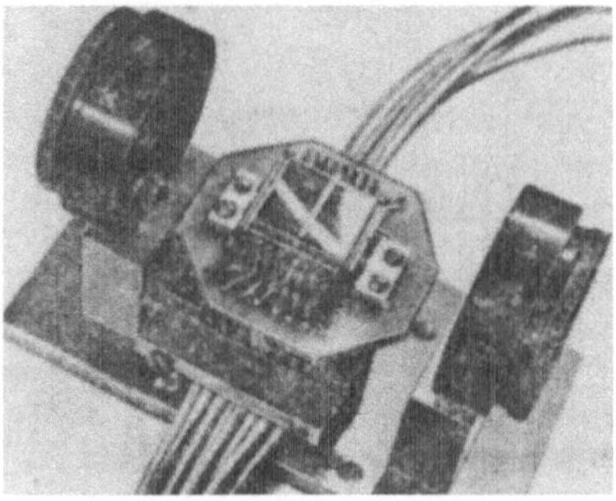

Figure 9.2: 32-channel dihedral electrooptical space–time light modulator on lithium tantalate monocrystal plates.

Figure 9.3: 59-element linear photodetector.

of light the transmission coefficient $\dot{T}_a$ is described by expression (1.6a) [89, 93, 95]. The half wave potential of the space–time light modulator $= 80 \pm 5$ V for all 32 channels ($\lambda = 0.63\mu$m). The modulator channel capacitance does not exceed 4.5 pF. The specific power of the modulator channel [89] is 13.4 mW/MHz. The interference between channels with frequencies of 10 to 70 MHz did not exceed 5% [coefficient $C_{01} = 0.05$ in (5.42)], contrast $K_{STLM} = 100$. The outside appearance of the 32-channel space–time light modulator is shown in Fig. 9.2. A similar space and time light modulator, which operates in the blue region of the visible spectrum, with a system of array antenna signal distribution designed on the basis of a polyamide film (the dependability of the instrument is increased and an assumption for a dual-channel unitizing is created) is used in the combined space–time light modulator with multichannel optical addressing (see Section 9.2).

The basis for the multielement linear photodetector was a common silicon p-n photodiode (with the help of boron diffusion in silicon cores, p-type regions were created). A row of 59 photodiodes spaced by 175 μm was arranged on a 5.6 x 10 mm crystal. The size of the photosensitive area of one element $\sigma_\phi = 70 \times 5000\,\mu\text{m}^2$. The spectral sensitivity of the multielement linear photodetector is equal to 0.18 Å/W on wavelength, $\lambda = 0.63\,\mu$m, where the spread in the elements' spectral sensitivity does not exceed $\pm 1\%$. Obscure currents of the photosensitive elements lie within $i_d = (5...8)10^{-9}$Å. The maximum threshold sensitivity is $0.3 \cdot 10^{-12}$ V Hz$^{-1/2}$. The high-speed response of the multielement linear photodetector is evaluated by the impulse front increase time and is 10 ns. with loading $R_\phi = 75\Omega$. The outside appearance of the multielement linear photodetector is shown in Fig. 9.3.

9.1.2 Forming the angular spectrum

In accordance with the results of Section 2.1, the output diffraction pattern (Fig. 9.4 and 9.5) in the multielement linear photodetector processor plane in Fig. 9.1 (see also Fig. 1.6) appears as a superposition of the zero order diffraction [parasitic illumination caused by the ending contrast of the space–time light modulator, see (7.57)] and $\pm 1^{\text{st}}$ order diffraction, complexly joined and centrally symmetrical with respect to the processor's optical axis. The intensity of the latter, in accordance with (2.12), is proportional to the two-dimensional convolution of the received radio phase spectrum whose power is $|\dot{s}\mathbf{K}_\perp, \Omega)|^2$ to the array antenna radiation pattern.

Whenever point-source emissions are received from far regions of the array antenna, the latter are represented at the processor output as diffraction spots whose intensity distribution coincides powerwise with the radiation pattern (2.19). The location of the spot maxima defines, in accordance with (2.15), the coordinates of the radio sources.

Figures 9.4 a–d show the output diffraction patterns (above) and the results of its photometry (below) with various positions of one (a–c) and two (d) radio sources, where elements 11 and 15 (see Fig. 9.1) were not used. As you can see, the output diffraction pattern is periodic, caused by the space–time light modulator channels being periodic and having step size $d_x \gg \lambda$ (see Fig. 5.3) and also due to the diffraction pattern being weighted by an envelope which appears as a radiation pattern of the space–time light modulator channels' pupil [see Fig. 9.4,a (dotted line)]. The horizontal axis is marked with a constant scale of $\cos\theta$, where angle θ is measured from the axis of the linear array antenna. The region where $|\cos\theta| \leq 1$ defines the area of "radio visibility" (observation area). The shape of the light spots coincides with the radiation pattern of the array as $\sin^2(N\Psi)/\sin^2\Psi$, where $\Psi = (Kd_x\cos\theta)/2$ ($N = 16$ is the number of array antenna elements, d_x

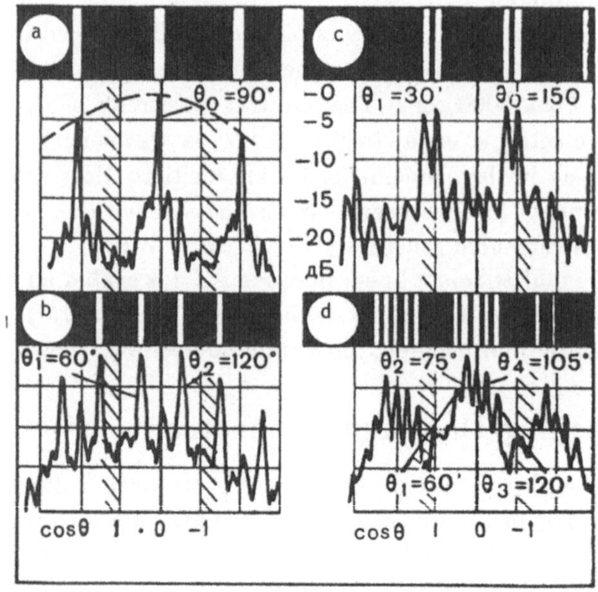

Figure 9.4: Examples of point sources at the coherent optical processor output shown in Fig. 9.1: (a) One source ($\theta_0 = 90°$); (b) one source ($\theta_1 = 60°$) and the conjugated image ($\theta_2 = 120°$); (c) one source ($\theta_1 = 30°$) and the conjugated image ($\theta_2 = 150°$); (d) two sources ($\theta_1 = 60°, \theta_2 = 75°$) and their conjugated images ($\theta_3 = 120°, \theta_4 = 105°$).

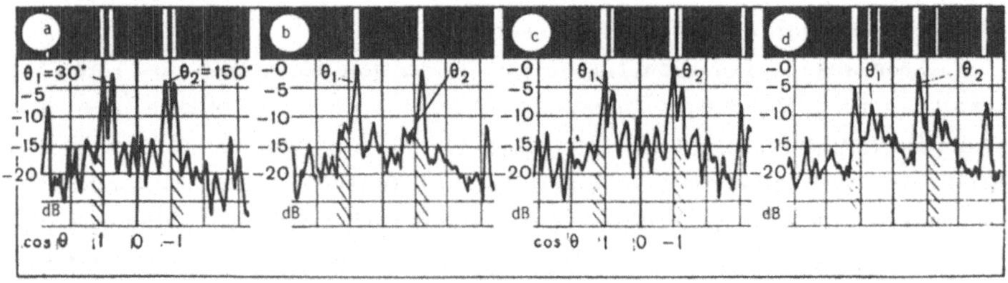

Figure 9.5: The equivalent lens method: (a) Radio object ($\theta_1 = 30°$) and the conjugated image ($\theta_2 = 150°$); (b) adjacent image suppression ($\theta_2 = 150°$) by a "convex equivalent" lens with $f_e = 12$ m; (c) image ($\theta_1 = 30°$) by a "concave equivalent" lens with $f_e = 24$ m; (d) image ($\theta_1 = 30°$) by a "concave equivalent" lens with $f_e = 6$ m.

$=$ array antenna step size, $K = 2\pi/\lambda$, $d_x/\Lambda = 0.50$). Since we used an amplitude (dual band) space–time light modulator, in Fig. 9.4 we also see the adjacent image. The zero order diffraction is described by the same expression as the radiation pattern and is significantly dampened due to the equilibrium operation of light modulation (all space–time light modulator channels are darkened when there is no signal).

9.1.3 Defocusing (suppressing) the adjacent image using an equivalent lens

The presence of adjacent images leads to ambiguity in determining the coordinates of the radio sources (θ or $180 - \theta$). It may be removed in one of the ways noted in Section 5.2.4. The most effective method appears to be the method of equivalent lens (see Fig. 5.2.1), constructed on the array antenna level from pieces of the feed, phase rotators, and so on. According to (5.2) the equivalent lens can be made to focus the ± 1 and to diffuse the ± 1 diffraction order. The smaller the focal length f_e, the greater the diffusion of the adjacent image [see (5.29) and Fig. 5.2]. When the space–time light modulator size is $\Delta x = 16$ mm, $\lambda = 0.63\,\mu$m, and $f_e = 1$ m it is suppressed by 30 dB. However, according to Section 5.2.2 the effect of the space–time light modulator pupil of a lensless processor leads to the necessity of fulfilling requirement (5.35), which guarantees that the distance diagrams of the channel pupils intersect on the level which assures distortion-free reproduction of the radio phase spectrum in the entire image area (see Fig. 5.5). In this case $\delta x = 100\mu$m (defined by the mask window size),$\Delta x = 8$ mm, $\lambda = 0.63\,\mu$m; therefore it follows from requirements (5.5) that $f_e \geq 12$ m.

Fig. 9.5a depicts the main and adjacent images at the output of unreduced Fourier processor. When a convex equivalent lens with $f_e = 12$ m, the first order, according to (5.35) and Fig. 5.6 is suppressed by 12 dB (see Fig. 9.5b). In order to shorten the length of the group (12 m) an auxiliary length was used ($f = 0.6$ m). Lengths f_e, f, a, and b (see Fig. 9.1) were chosen according to $1/(f_e - a) + 1/f = 1/b$. Figs. 9.5 c and d depict the diffraction patterns, obtained using concave equivalent note that when $f_e = 6$ m and the above conditions are fulfilled the adjacent lens with $f_e = 24$ mm (suppression of the image by -9dB) and $f_e = 6$ m respectively. Note that when $f_e = 6$ m and the above requirements are met, image suppression would reach -15 dB, however due to the effect of the pupil [requirement (5.35) is contradicted] we cannot expect to attain this level. If a greater adjacent image suppression is needed it is necessary to decrease f_e, which in turn decreases δx sizes.

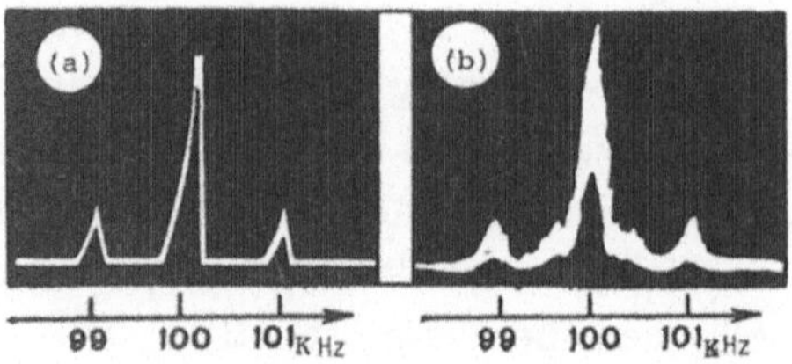

Figure 9.6: Signal spectrum oscillograms.

9.1.4 Coherent (heterodyne) photodetection

During coherent optical processing of complex radio (acoustic) signals by the processor in Fig. 9.1 storage of all the information about its amplitude, frequency, and phase requires the coherent removal of of the optical signal from the processor output (see 8.3). The simplest operation of coherent photodetection is carried out in the coherent optical processor (see Fig. 9.1) with the help of additional elements 11 and 15. Fig. 9.6 a and b shows light beam recordings of signal spectrum at the space–time light modulator output (Fig. 9.6a) and at the output of the multielement linear photodetector which is performing the heterodyning (Fig. 9.6b). During noncoherent photodetection according to (8.13), the output signal does not contain the information about spectral composition.

In order to illustrate the possibility of recording spatial information using the multielement linear photodetector in Fig. 9.7, we show the result of taking a light reading of the output diffraction pattern from a single source (see Fig. 9.4a) that was made by a recorder with aid of "FAY28" and a sample of this diffraction pattern obtained from the output of the multielement linear photodetector elements.

9.2 Coherent Optical Processors for Space–Time Processing of Linear and Planar Array Antenna Signals

A coherent optical processor for linear or planar array antennas, discussed in 9.1, is, in general, an optical pattern-forming circuit, which forms a continuous one- or two-dimensional radiation pattern that permits panoramic spatial scanning. When a linear arrays are used with a coherent optical processor to perform spacial scanning of the time signal, it is possible to do some additional processing (spectrum and correlation analysis). By complicating

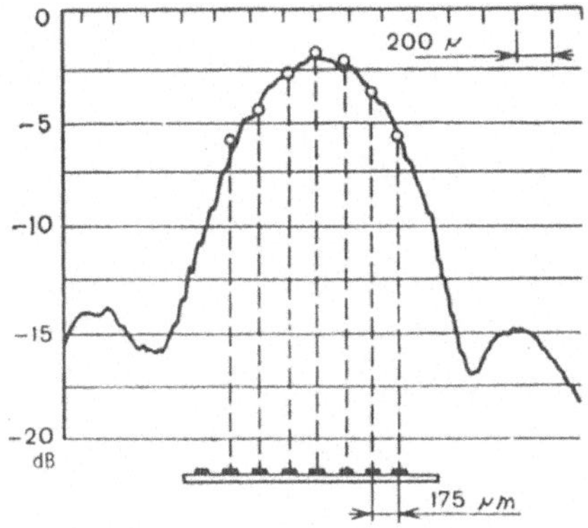

Figure 9.7: Reading optical signals from the processor output with a multi-element photodetector.

the format of the recorded space–time signal we can also derive information about two angular coordinates and the spectrum of the signals received from the object at the coherent optical processor's output.

9.2.1 Coherent optical processor for linear array antennas which forms frequency-phase signals with the space–time light modulator with multichannel optical addressing

The given processor, which actualizes the schematic in Fig. 1.7 and discussed in Section 1.2.2, is shown in Fig. 9.8. The key element of the processor is the combined space–time light modulator with a multichannel optical address system which consists of the multichannel light modulator 15, an optically controlled mask (OCM) 18, and a sweeping device 4. The coherent optical processor uses a 16-channel space–time light modulator exhibiting the transverse electrooptical effect in a lithium tantalate crystal with a 90° cut which differs from the construction described in Section 9.1.1 in that it operates in the blue region of visible spectrum ($\lambda = 0.44\,\mu$m), and the separation system is based on polyamide film. The half-wave voltage in the modulator channels is 60 ± 5 V.

The OCM 18 used was a PROM (Pockels Readout Optical Modulator) [105–108] which is a multilayer structure of metal, dielectric, electrooptical, photoconducting, crystal, dielectric, and metal (Fig. 9.9). The device used a plate of $Bi_{12}SiO_2$ crystal section (001) 300 μm thick with an aperture $20 \times 20\,\mu$m, the surface of which is covered by a film of Boron-Silicate glass 500

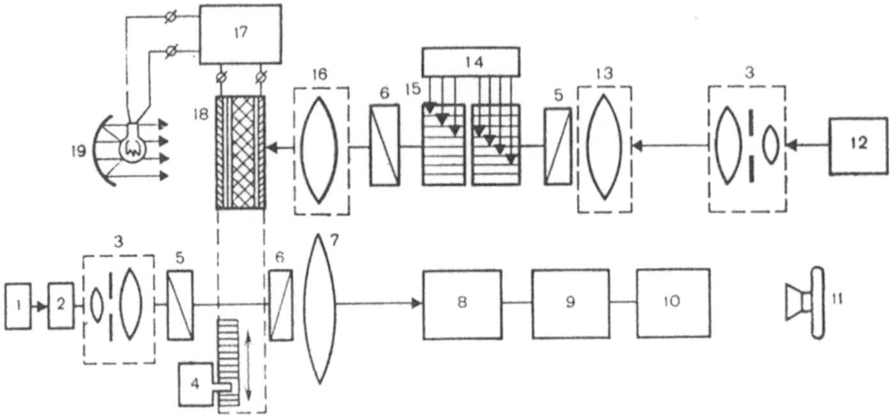

Figure 9.8: A linear electrooptical array antenna coherent optical processor with multichannel optical addressing: (1) He-Ne laser; (2) mechanical modulator; (3) collimator; (4) scanning device; (5, 6) polarizer, analyzer; (7) spherical lens; (8) photoelectric amplifier; (9) amplifier; (10) printer; (11) camera; (12) He-Cd laser; (13, 16) cylindrical lenses; (14) array antenna simulator; (15) multichannel space-time light modulator ($LiTaO_3$); (17) power supply; (18) optically controlled mask (PROM); (19) flash bulb.

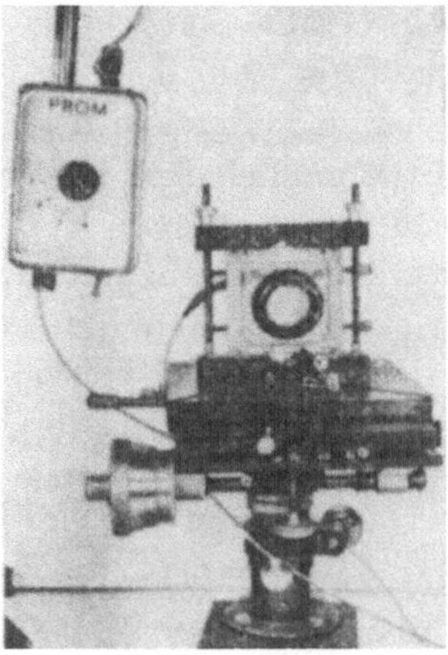

Figure 9.9: Optically controllable mask-PROM device.

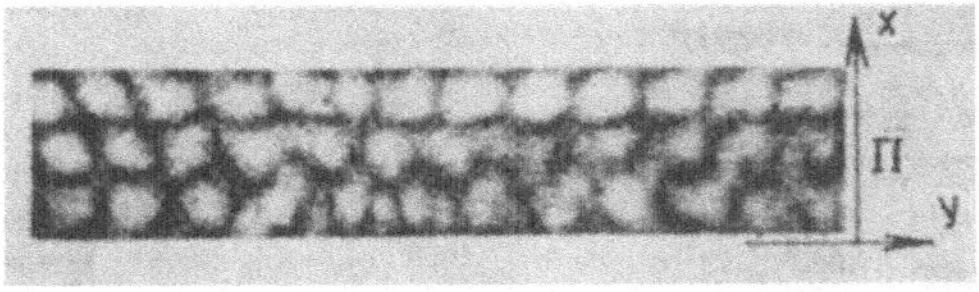

Figure 9.10: Array input signal format (on an optically controllable mask).

Å thick. The plate is glued between two glass bases, onto which transparent electrodes of In_2O_3 were dusted beforehand, and then a film of borosilicate glass 3 μm thick. Before recording the projected image onto the PROM with lens 16 of modulator channel 15, a voltage of 1.5 kV was applied to the electrodes, and then the device was uniformly illuminated by a short intensive light flash. Subsequently, the voltage was inverted and a recording was made. This recording was made by the light from a He-Cd laser 12 ($\lambda = 0.44\,\mu$m), reading by the beam of He-Ne laser 1 ($\lambda = .63\mu$m) crossed polarizers 5, 6. The instrument sensitivity when the contrast is 2:1 or 50:1, clearance on the order of 5 mm makes 10^{-5} $(1.5^{-4})\mathrm{J/cm}^2$; the number of clearance elements in row Δy, $R_y = 100$ with 100 rows in orthogonal measuring; this allows one to form signals from a 100-element linear array antenna. An optically controlled mask was placed on a mechanically movable alignment table 4, whose speed of movement v_p is governed by conditions (3.72, 3.23, 3.32).

Fig. 9.10 shows a fragment of the array antenna input signal with frequency Ω_o. The signal came from a point source located at an angle $\theta_o = 40°$. The fragment is recorded by PROM 18 by depictions of channels ($\delta y = 80\,\mu$m) of multichannel light modulator 15.

Fig. 9.11a shows a diffraction pattern in the output plane of the processor and the results of its photometry which allow an analysis of light distribution in diffracted image of the target ($e^{(+1)}$) and to determine the angular coordinate $\Omega_Z = \Omega_o \cos\Omega_o/c$ as well as frequency Ω_o (frequency spectrum).

The result of zero order diffraction $e^{(0)}$ photometry allows us to judge the optical homogeneity of the OCM within the aperture fragment.

Fig. 9.11f shows a diffraction pattern, at the processor output, from two sources of the same frequency Ω_o with different angular coordinates, $\theta_1 = 40°$ and $\theta_2 = 60°$.

Let us note that during recording through each optical light modulator channel 15 (see Fig. 9.8) of light flux of $1.2^{-3}\,\mathrm{V/cm}^2$, $\lambda = 0.44\,\mu$m was passed. Moreover, no phenomena which could be linked with degradation of the $LiTaO_3$ crystal were observed.

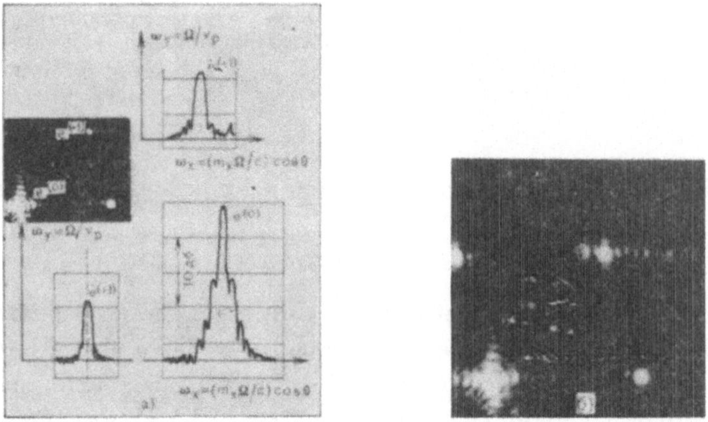

Figure 9.11: The processor output plane for a monochromatic point source: (a) One object ($\theta_\circ = 40°$); (b) two objects ($\theta_1 = 40°$, $\theta_2 = 60°$).

9.2.2 Coherent optical processor for planar array antennas with complex signal recording

In Section 3.3.1 we discussed a coherent optical processor which performed the space–time processing of signals from planar array antennas and which also allow the extraction of information at the output on about the two angular coordinates and the frequency of the received signals. This can be shown with the coherent optical processor shown in Fig. 9.8. Here a two-dimensional array antenna consisted of $M \times N = 4 \times 4$ receiving elements. To relate them and the multichannel light modulator channel 15 (see Fig. 9.8) an addressing law such as (3.32)(see Fig. 3.3) was chosen. This law allowed one to condense the two-dimensional information on linear format.

Figure 9.12a shows the diffraction pattern in the output plane of the processor as well as the result of its photometry (Fig. 9.12b). These permit us, using the methods in Section 3.3.1 (see Fig. 3.5), to determine the directivity cosines $\cos\alpha_q$ (on the "rough") and $\cos\beta_q$ (on the "fine") diffraction-pattern structures, and the frequency (spectrum) of the signal. The values of $\cos\alpha_q$ and α_q are determined by angles Φ_q and ψ_q on (3.41) and (3.43), respectively. The diffraction pattern in Fig. 9.12c and the result of its photometry (Fig. 9.12d) have an auxiliary character for visual description of the "rough" light field structure. They were formed during operation of four element array antenna bar which was modeled by four space–time light modulator channels whose size was $\Delta\tilde{x}$ (see Fig. 3.3). Fig. 9.12 describes the

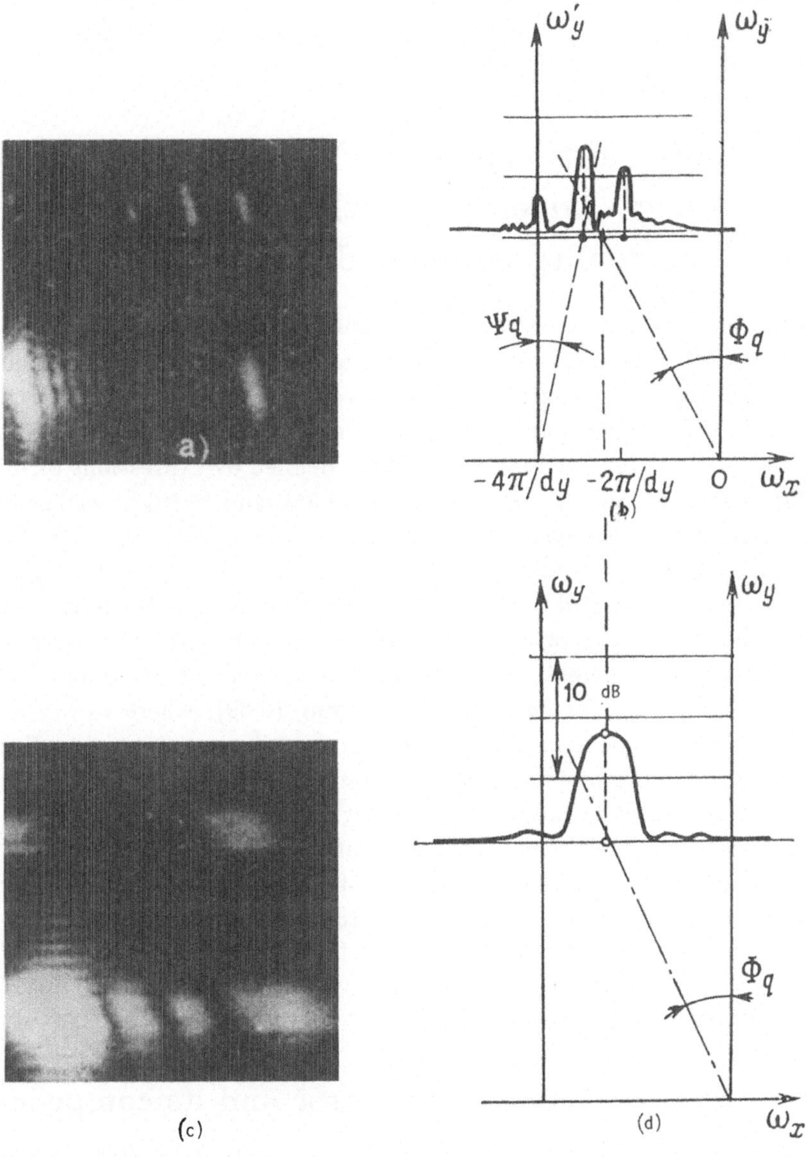

Figure 9.12: Coherent optical processor output plane of a planar array (4×4) with complex signal recording format.

light field modulation at the processor output which corresponds to a point object with frequency Ω_o and coordinates $\theta_q = 60°$, $\phi_q = 60°$, ($\cos\alpha_q = 0.75$, $\cos\alpha_q = 0.43$).

9.3 Coherent Optical Processor for Planar Array Antennas with Rejection of Interfering Spatial Signals

In Chapter 6 we discussed the method and synthesized the structure of the coherent optical processor of electrooptical arrays which reject the interfering spatial signals by controlled nulls directed at the interfering signals [38, 164, 166, 167]. When planar electrooptical array antenna's with rectangular apertures are involved the algorithm is defined by expressions (6.78) and (6.89) [taking into account (6.80), (6.83), (6.85), and (6.89) as well as results of Appendix C]. The schematic of this processor appears in Fig. 6.6 and in a fuller perspective in Fig. 9.13. The transmission function τ_Ψ (6.69), reproduced with the aid of OCM (Fig. 9.13), is chosen in the following way: within the entire observation area $\tau_\Psi(\mathbf{K}_\perp^S)| = 1$, and in the characteristic region $\sigma_x \times \sigma_y$ in the neighborhood of each q interference $\arg\tau_\Psi$ differs by 180°. There the size σ_x, y is determined according (6.89), where $c_{X,Y} = 0.5402$ when a linear array is used and 0.847 for two-dimensional array. Such an interference suppression processor is characterized by the formation of a continuous array antenna radiation pattern with ideal gaps directed $\mathbf{K}_q$ toward the interfering signals; the localization of gaps near directions $\mathbf{K}_q$ at the interfering signals (minimal suppression of the true signals); accuracy of realizable radiation patterns in the sense of root-mean-square criterion for ideal interference suppression radiation pattern (6.1); by the possibility of realizing it through Fourier optics; by the simplicity of its design; and by maximum OCM diffraction efficiency.

9.3.1 Coherent optical processor and its components

The coherent optical processor for interference suppression linear electrooptical array antenna is shown in Fig. 9.14. The signal $\mathcal{E}(X)$, which is the superposition of the array's far field of the useful radio signal C_o and interfering signal Π_o is modeled by an array simulator 7 and is controlled by the transmission function of the 16-channel space–time light modulator 8 on lithium tantalate (see 9.1.1). This creates an optical model of the received space wave field at the space–time light modulator, which is weighted by function $J_\Psi(x)$ with the help of the apodization mask 4 (the function $J_\Psi(x)$

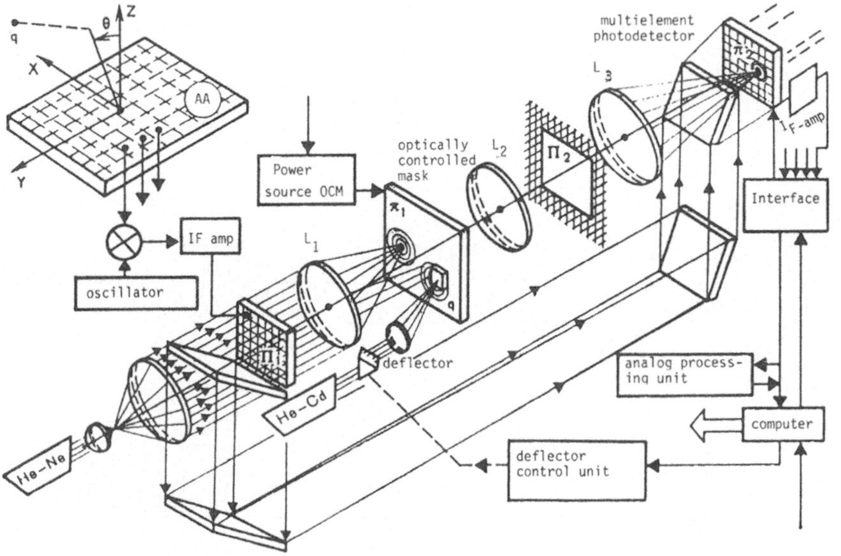

Figure 9.13: Interference protected coherent optical processor for a planar electrooptical array antenna.

is a spheroidal zero order function with parameter c_X. The function was realized by the silhoutte function method [158]). Next, the optical signal is Fourier-transformed by lens 10 and multiplied by the transmission function with the help of OCM 13. The most ready for practical use are the structures on liquid crystals, and also instruments such as "Phototitus" and PROM [106–108, 176].

The processor uses an OCM with a liquid crystal light valve in the read light reflecting mode. Such an OCM is characterized by high sensitivity to controlling light (10^{-5}....10^{-7} J/cm^2 with 100% modulation of light beam), good spacing possibilities (up to 200/mm on "half-decline" of specific-contrast characteristic), low controlling voltage (below 10 V) and small phase distortion. The on-time, τ_{on}, of the electrooptical response during normal supply operation is on the order of tens of milliseconds, and the off-time τ_{off} runs from tens of milliseconds to seconds. Better processing speeds ($\tau_{on} = 1\,\mu$s and $\tau_{off} = 20$ to $30\,\mu$s) and a higher sensitivity (10^{-8}...10^{-9} J/cm^2) is achieved in OCM of metal-dielectric-liquid crystal light valve-type. Further improvements in time response of the liquid crystal OCM are achieved by using a dual-power-supply method which permit us to carry out the switching cycle in tens to hundreds of microseconds and to

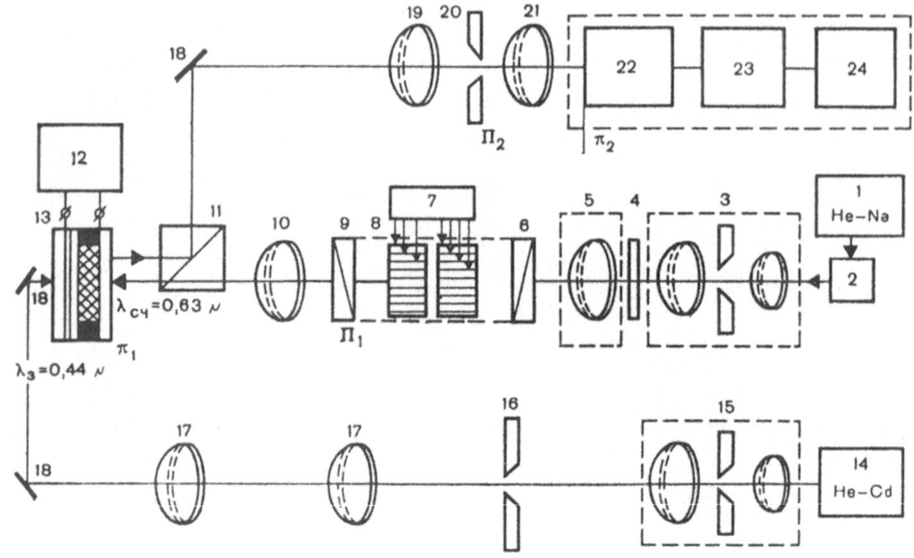

Figure 9.14: The coherent optical processor for an interference protected electrooptical array antenna: (1) He-Ne Laser; (2) mechanical modulator; (3, 15) collimator; (4) apodising mask; (5) cyllindrical lens; (6, 9) polarizer and analyzer; (7) array antenna simulator; (8) multichannel modulator on $LiTaO_3$; (10) spherical lens (f = 1 m); (11) diffractor; (12) optically controlled mask power supply; (13) optically controlled mask; (14) He-Cd laser; (16) filter mask; (17) display system; (18) mirrors; (19, 21) spherical lenses (f = 0.6 m); (20) limiting diaphragm; (22) photoelectric amplifier; (23) amplifier; (24) printer.

Figure 9.15: Optically controlled mask (OCM) with liguid crystal light valve.

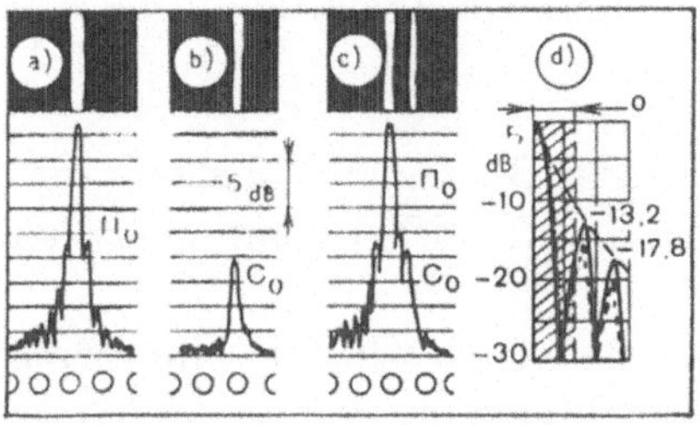

Figure 9.16: Initial distribution of emission sources.

have a relaxation controlled memory up to 2 s [176].

Structurally the OCM, the appearance of which can be seen in Fig. 9.15, consists of two glass bases with two transparent metal electrodes plated on them on In_2O_3 film, between which the liquid crystal light valve is located. Between PS and LC (larger thickness = $5\,\mu$m aperture 2.0×2.0 cm) is a dielectric mirror.

A control voltage (of less than 10 V) is supplied to the metal electrodes, and the recording of the filter onto the OCM is performed by the He-Cd laser (λ = 441 nanometers) 14, imaging system 17, and filter mask 16. The reflected optical signal from OCM 13 undergoes additional processing by a diffraction limiting system 19, 20, and 21.

9.3.2 Experimental results and discussion

The initial phase distribution of radiation sources (the diffraction patterns and photometric results at processor output area π_2 in Fig. 9.13) is presented in Figs. 9.16 a, b, and d where the direction from which the relevant signal C_o arrived coincides with the maximum of the second side lobe of the interfering signal Π_o, and the level of the relevant signal is equal to the magnitude of

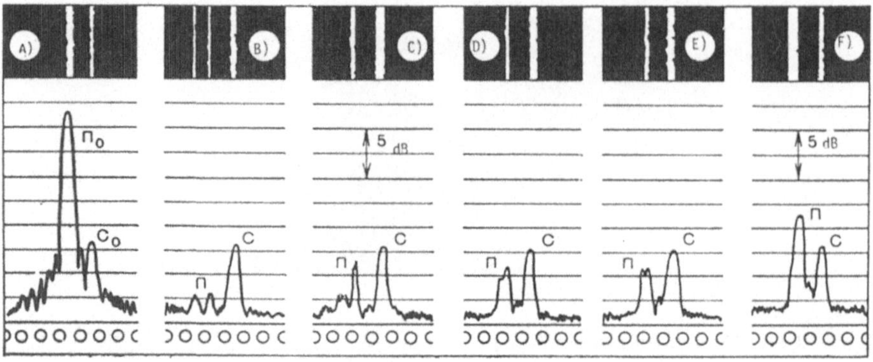

Figure 9.17: Interference rejection using a phase filter.

the first side lobe $\Pi_\circ$ (−14 dB). Fig. 9.16d shows the theoretical radiation pattern type $F = F_\circ = \mathrm{sinc}(\Omega_X/\delta\Omega_X)$ (continuous) and $F = \Pi_\circ$ (spherical function – dotted line).

Fig. 9.17f shows the way in which the interference is dampened during optimal damping conditions ($c_X = c_{opt} = 0.54$; $F = \Pi_\circ$). Damping of about 20 dB is achieved (theoretically complete damping of the interference should be observed see Figs. 6.11 and 6.32). In Fig. 9.17c optimal damping conditions were interrupted ($F = \mathrm{sinc}(...)$, the apodization mask 4), and the degree of damping was 14.8 dB (theoretical evaluation [see Fig. 6.8] yield the value of 12 dB). Figs. 9.16 d and e allow one to evaluate how critical was the degree of interference damping to the filter width σ_x (6.89), which is determined by the parameter c_x (compared to Fig. 9.17b).

Looking from the point of view of coming up with a self-adjusting reverse link loop which control the position of the filter with transition coefficient, τ_Ψ (see Fig. 9.13) in the OCM plane, one of the chief characteristics is the dependence of the interference damping degree upon the precision of the Δ filter placement in the direction of the interference. This is partially illustrated in Fig. 9.17f (the degree of damping is worsened by 10 dB with Δ around 0.1 of the radiation pattern width compared to the "ideal" case where $\Delta = 0$ in Fig. 9.17b).

Fig. 9.18 shows results of interference signal rejection with aid of nonideal amplitude filters $\tau_\Psi(\Omega_X) = \tau(\Omega_X)$, where $\tau(\Omega_X) = 0$ with $\Omega_X \in \sigma_x$ and $\tau(\Omega_X) = 1$ with $\Omega_X \ni \sigma_x$, depending on the width of the latter, given by parameter c_X. As can be seen with small c_X, a small damping is achieved, so with $c_X = 1$ ($\sigma_x = 100\,\mu$m a given coherent optical processor) it corresponds exactly to the half-width of the radiation pattern at the processor output, the degree of damping (Fig. 9.18c) is 10 dB, then the approximate evaluation

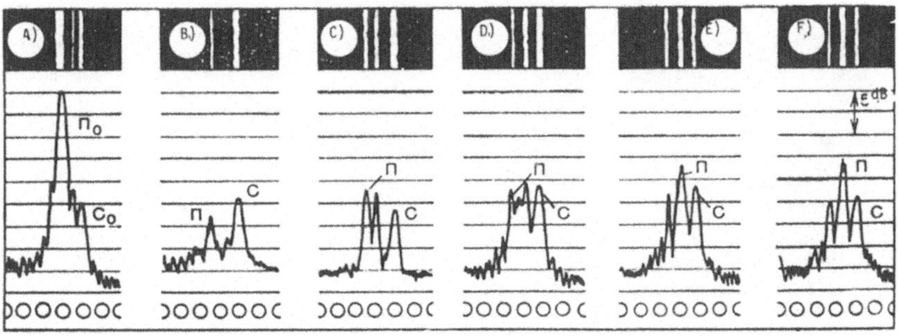

Figure 9.18: Interference rejection using an amplitude filter: F=sinc(...).

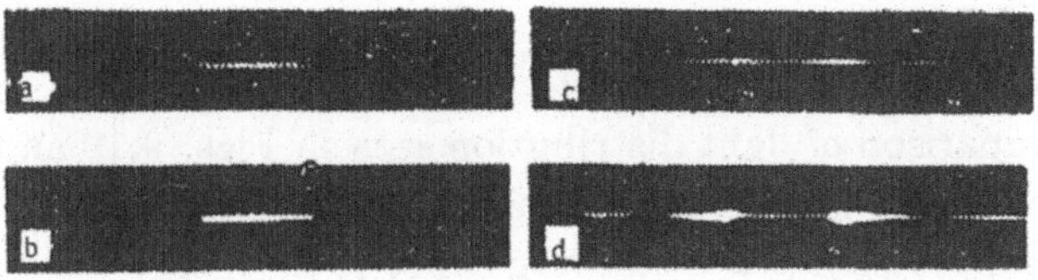

Figure 9.19: Light intensity distribution in various coherent optical processor planes.

yields 8 dB (see Fig. 6.8). An increase in c_X inevitably leads to shadowing the relevant signals. A merit of interference rejection via the use of an amplitude filter is that it is less critical to the mounting in the direction of the interference precision (Figs. 9.18 e and f).

The light intensity distribution at the linear 16-channel space–time light modulator output 8 (see Fig. 9.14; or Π_1 plane in Fig. 9.13) with an apodization mask 4 during formation of the interference $Pi_\circ$ at the coherent optical processor input is shown in Fig. 9.19a as a diffraction pattern and in Fig. 9.20 together with photometric results. In the case where there is no suppression, the diffraction pattern in the diaphragm plane 20 (see Fig. 9.14 or in plane Π_2 in Fig. 9.13) is shown in Figs. 9.19b and 9.20b. Here the light distribution, as opposed to Figs. 9.19a and 9.20a, is continuous due to the fact that in the OCM 13 plane (see Fig. 9.14; plane π_1 in Fig. 9.13) there is a diaphragm which separates out only one period of the radiation pattern which is periodic due to the discreteness of the array antenna and the space–time light modulator. In forming the filter with transmission function τ_Ψ the light distribution in the diaphragm 20 plane (see Fig. 9.14; π_2 plane in Fig. 9.13) is transformed into the distribution shown in Figs. 9.19d and 9.20d (for optimal phase filter in Fig. 9.17b) and in Figs. 9.19c and 9.20c

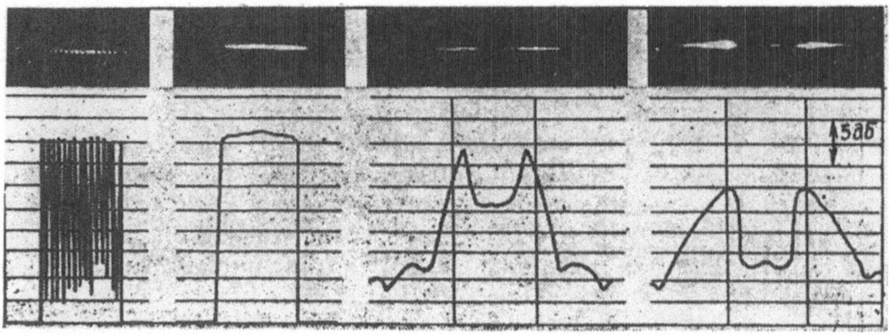

Figure 9.20: Evaluating the mechanism of interference suppression.

(for amplitude filtering, Fig. 9.18d).

A comparison of light distribution seen in Figs. 9.19 and 9.20 graphically illustrates the interference suppression "mechanism". It is apparent, that in the suppression process, a large part of its energy is distributed outside the limiting diaphragm 20 (see Fig. 9.14) as result of the disturbance caused by the filter; diaphragms in the π_2 plane are in Fig. 9.13) and are blocked by it. Therefore, in the case described in Figs. 9.19d and 9.20d, the measured ratio of energy lilluminating the diaphragm 20 to the full interference energy is 20 dB, which corresponds to the degree of interference suppression for the case presented in Fig. 9.17b. These results demonstrate the effectiveness of this method of spatial interference rejection.

The achieved suppression of 20 dB is not the limiting one, and is explained first of all by the OCM phase inhomogeneity, the imprecise filter positioning in the direction of the interference, the filter edge distortion due to finite resolving power of the OCM, and the distortions in amplitude phase distribution at the level of the array antenna's receiver-amplifier modules and space–time light modulator (the array antenna terminator in the experiment). Evaluations, given in Section 6.6 show that when the coherent optical processor dynamic range is 40 dB, using a more perfected element base we can attain a more than 30 dB suppression.

Let us note that the given coherent optical processor with this modification (see Section 6.5.2) could be used for linear electrooptical array antennas with a space–time light modulator to perform spatial scanning for time signals (a multichannel AOM, for example). Here, although in the case of broad-band interference in the sense of criterion (2.23) it is, in principle, impossible to achieve a theoretically full suppression in terms of the approach discussed in Chapter. 6; however, evaluation of (6.101) (see Fig. 6.19) is sufficiently optimistic.

Recall that Sections 6.5.1 and 6.5.3 discuss the development of this approach to interference suppression by coherent optical processors for electrooptical arrays with respect to sector interference, and also for cylindrical (circular) array antenna.

9.4 The Circular Array Antenna Coherent Optical Processor

The circular array antenna coherent optical processor allows panoramic scanning of the azimuth and frequency simultaneously when determining the elevation angle.

9.4.1 The coherent optical processor and mask recording

A schematic of this processor is given in Fig. 9.21 and the details are given in Fig. 9.22. We can separate out three functional parts: the signal recording channel and the input into the coherent optical processor (elements 18–23); the coherent optical processing channel (elements 1–4 and 6–17) with angular-frequency spectrum formation at the output plane ($\Omega' - \phi'$); the reference channel (elements 1–8), which are used only during analog holographic recording of the spatial pattern-forming filter $\dot{T}(\Omega_\phi, \Omega_{T_\theta})$.

The methods for recording mask $\dot{T}$ is based on the classical schematic of Vander Lugt and is described in general in Section 4.3.2. An auxiliary signal beam is formed as a Fourier image of a reference signal $\mathcal{E}_e(t, \phi) = \Re e\{J(t,\phi)\exp[i(\Omega_o t - \Omega_o R_o \sin\theta_o \cos\phi/c)]\} + E_o$ (E_o is the base level necessary for recording negative signals). The reference signal is input into the coherent optical processor in the corresponding recording scale (4.39), so that

$$\mathcal{E}_e(t,\phi) \longrightarrow \mathcal{E}_e(x,y)$$

$$= \Re e\{J_t(x)J_\phi(y) \times \exp\{i[\Omega_o t + \Omega_o x/v_p - \Omega_o R_o \sin\theta_o \cos(m_y y)/c]\}\} + E_o$$

The input is carried out either through standard signal format 9 (of a standard raster), synthesized by digital holography (Fig. 9.24), or a space–time light modulator (see Fig. 9.22. elements 20–23).

During the recording of mask $\dot{T}$ for broad-band signals $\Delta\Omega = \Omega_{max} - \Omega_{min}$, it is necessary to reconstruct in succession the signal of frequency band $\Delta\Omega$ with sample determined by the resolution of the spectrum analyzer

($\delta F = v_p/\Delta x$), while isolating the corresponding region on the frequency axis with the help of a slotted diaphragm (see Fig. 4.10). In the lens 10 focal plane a photo plate (with a resolution of $10^3\,\text{mm}^{-1}$) on which the hologram was recorded as the result of interference of the reference and supporting beams $-R = R_\circ \exp\,(-if\omega_y \sin\alpha_\circ)$. (The angle $\alpha_\circ$ is determined by the condition of three-dimensional separation of output signals corresponding to various numbers of the transmission coefficient T_A—see Fig. 9.23, and the relationship between the supporting and the signal beam levels was selected using neutral light filters 7). According to (4.41) the transmission coefficient for linearly recorded hologram-mask (9.1) is

$$T_A \sim [\overset{+}{F}_t\, \dot{T} + \bar{F}_t\, \overset{*}{T} + F_\circ + \dot{R}][\overset{+}{F}_t + \bar{F}_t\, \overset{*}{T} + F_\circ + \dot{R}]^* \tag{9.1}$$

where

$$\begin{aligned} \overset{+-}{F}_t &= \overset{+-}{F}_t\,(\omega_x \mp \Omega_\circ/v_p) = \mathcal{F}_x\{J_t(x)\exp(\pm i\Omega_\circ x/v_p)\} \\ F_\circ &= \mathcal{F}\{E_\circ\} \\ \dot{T}, \overset{*}{T} &= \frac{1}{2}\mathcal{F}_y\{J_\varphi(y)\exp[\mp\Omega_\circ R_\circ \sin\theta_\circ \cos(m_y y)/c]\} \end{aligned}$$

$*$ denotes complex conjugation.

Multiplier $\overset{+-}{F}_t$ "cuts out" $\dot{T}(\overset{*}{T})$ in the corresponding place on the frequency axis of mask 11 (plane π in Fig. 9.22). This place is limited by the slotted diaphragm within limits of which, in order to avoid the target image diffusion $\overset{+-}{F}_t \sim$ const (for calculations for the optical spectrum analyzer resolution element [see Appendix A.5]). Mask (9.1) performs a key role in forming the radiation pattern in the electrooptical array coherent optical processor.

Let us further assume that a signal from the far field is incident on the array, then according to Section 4.2.3 a signal of the form $E(x,y) = \Re e\{E(x,y)\}$ is recorded at the processor object plane π, and the mask 11 receives the light flux

$$\begin{aligned} \dot{e}_\pi &\sim \mathcal{F}\{E(x,y)\} = \frac{1}{2}\mathcal{F}_y\{\bar{F}_t \dot{\mathcal{E}}_{\Omega_\circ}(m_y y) + \overset{+}{F}_t \overset{*}{\mathcal{E}}_{\Omega_\circ}(m_y y)\} + F_\circ \\ &\sim \overset{+}{F}_t \overset{*}{T} + \bar{F}_t \dot{T} + F_\circ \end{aligned} \tag{9.2}$$

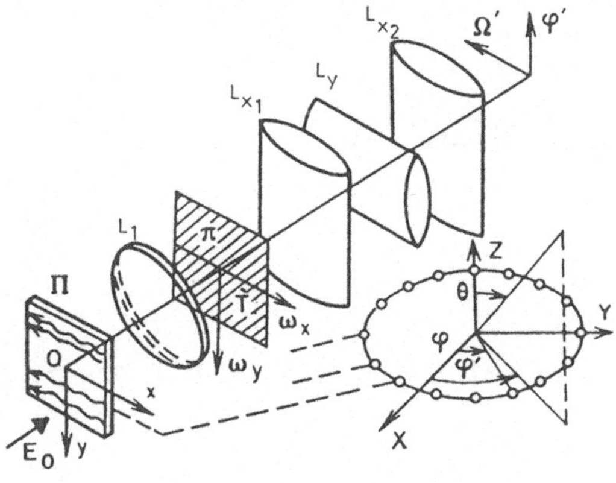

Figure 9.21: Schematic of the circular array antenna coherent optical processor.

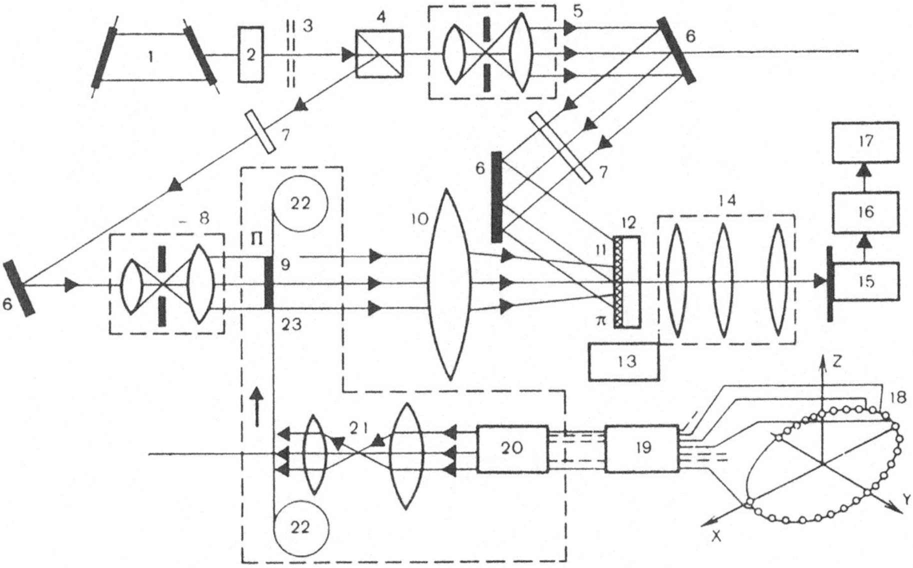

Figure 9.22: Circular array antenna coherent optical processor: (1) He-Ne laser; (2) shutter; (3) mechanical modulator; (4) light diffractors; (5) collimator; (6) mirrors; (7) neutral light filters; (8) collimator; (9) input array antenna signal format channel (standard signal scanning pattern); (10) Fourier lens; (11) mask; (12) adjusting device; (13) "spare" mask cassette; (14) astigmatic lens system; (16) amplifier, (17) printer; (18) circular array antenna (array antenna simulator); (19) amplifiers; (20) linear photodiode matrix; (21) imaging system; (22) tape mask mechanism; (23) photocarrier.

The light distribution in the output plane of the processor generally consists of 48 components:

$$\dot{e}_{out} \sim \dot{\mathcal{F}}^{-1}\{T_A \dot{e}_\pi\} = \dot{e}_{out}^{(+1)} + \dot{e}_{out}^{(0)} + \dot{e}_{out}^{(-1)} + \dot{e}_{out}^{(\pm1)} \tag{9.3}$$

where

$$\begin{aligned}
\dot{e}_{out}^{(+1)} \sim \hat{\mathcal{F}}^{(-1)} \Big\{ & \overset{[1]}{\overset{+}{F}{}_t^2 \overset{*}{\dot{R}} \overset{*}{\dot{T}} \dot{T}} + \overset{[2]}{\overset{+}{F}_t F_\circ \overset{*}{\dot{R}} \overset{*}{\dot{T}}} + \overset{[3]}{\overset{+}{F}_t F_\circ \overset{*}{\dot{T}}} + \overset{[4]}{\overset{+}{F}{}_t^2 F_\circ \overset{*}{\dot{T}} \dot{T}} \\
& + \overset{[5]}{\overset{+}{F}{}_t^2 F_\circ \overset{*}{\dot{T}} \dot{T}} + \overset{[6]}{\overset{+}{F}{}_t^3 |\dot{T}|^2 \overset{*}{\dot{T}}} + \overset{[7]}{\overset{+}{F}_t |\dot{R}|^2 \overset{*}{\dot{T}}} + \overset{[8]}{\overset{+}{F}_t F_\circ \dot{R} \overset{*}{\dot{T}}} + \overset{[9]}{\overset{+}{F}_t F_\circ \dot{R} \overset{*}{T}{}^2} \\
& + \overset{[10]}{\overset{+}{F}_t F_\circ \overset{*}{\dot{R}} \dot{T}} + \overset{[11]}{\overset{+}{F}_t F_\circ \dot{R} \overset{*}{\dot{T}}} + \overset{[12]}{\overset{+}{F}_t F_\circ^2 \overset{*}{\dot{T}}} + \overset{[13]}{\overset{+}{F}_t F_\circ^2 \dot{T}} + \overset{[14]}{\overset{+}{F}{}_t^2 F_\circ \overset{*}{T}{}^2} \Big\}
\end{aligned}$$

$$\dot{e}_{out}^{(0)} \sim \hat{\mathcal{F}}_y^{-1} \Big\{ \overset{[15]}{F_\circ^2 \overset{*}{\dot{R}}} + \overset{[16]}{F_\circ |\dot{R}|^2} + \overset{[17]}{F_\circ^2 \dot{R}} + \overset{[18]}{F_\circ^3} \Big\}$$

$$\begin{aligned}
\dot{e}_{out}^{(\pm)} \sim \hat{\mathcal{F}}^{(-1)} \Big\{ & \overset{[19]}{\overset{+}{F}_t \overset{-}{F}_t \overset{+}{F}_t \overset{*}{\dot{T}}{}^3} + \overset{[20]}{\overset{+}{F}_t |\dot{T}|^2 \dot{T}} + \overset{[21]}{\overset{-}{F}_t |\dot{T}|^2 \overset{*}{\dot{T}}} + \overset{[22]}{F_\circ \overset{*}{\dot{T}} \dot{T}} \\
& + \overset{[23]}{\overset{-}{F}_t \dot{T}^3} + \overset{[24]}{F_\circ \overset{*}{T}{}^2} + \overset{[25]}{F_\circ \dot{T}^2} + \overset{[26]}{F_\circ \overset{*}{T}{}^2} + \overset{[27]}{F_\circ \dot{T}^2} + \overset{[28]}{\overset{+}{F}_t |\dot{T}|^2 \dot{T}} + \overset{[29]}{\overset{-}{F}_t |\dot{T}|^2 \overset{*}{\dot{T}}} \\
& + \overset{[30]}{F_\circ \overset{*}{\dot{T}} \dot{T}} + \overset{[31]}{\dot{R} \overset{*}{\dot{T}} \dot{T}} + \overset{[32]}{\dot{R} \overset{*}{\dot{T}} \dot{T}} + \overset{[33]}{\overset{*}{\dot{R}} \overset{*}{T}{}^2} + \overset{[34]}{\overset{*}{\dot{R}} \dot{T}^2} \Big\};
\end{aligned}$$

$$\dot{e}_{out}^{(-1)}(\Omega) = \overset{*}{e}{}_{out}^{(+1)}(-\Omega)$$

is the complex-conjugate component of the output image, formed in the

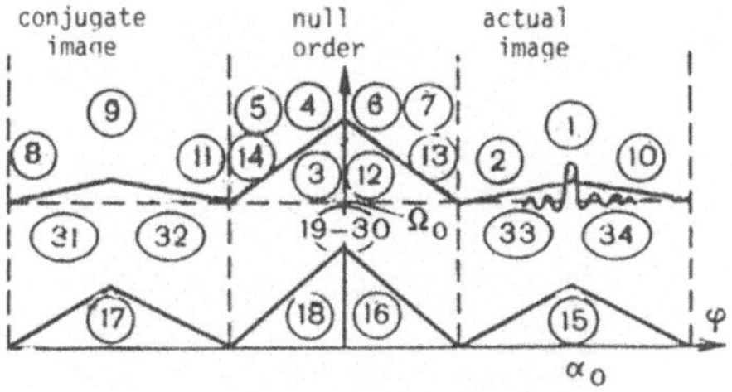

Figure 9.23: Restoring the frequency-angular spectrum for a circular array antenna with an analog holographic mask (the circles denote the number of the components in Section 9.2).

lower half-plane ($\Omega < 0$ see Figs. 9.23 and 9.25)

The distribution of composing parts $\dot{e}_{out}^{(+1)}$, $\bar{e}_{out}^{(0)}$, $\dot{e}_{out}^{(\pm 1)}$ in the processor output half plane ($\Omega > 0$) is presented schematically in Fig. 9.23 where the component numbers are marked in the circles. The first term $\hat{\mathcal{F}}_y^{(-1)}\{\overset{+}{F}_t^2 \overset{*}{R} \overset{*}{T}\, \dot{T}\}$ is informational and is relevant response of the system with circular array antenna signal space–time processing algorithm (4.41). Unless special measures are taken, this signal is distorted by the defocused 2nd, 10th, and 15ths terms, and also significantly weakened through multiplier $\overset{+}{F}_t \overset{-}{F}_t$ in terms 33 and 34.

The output image components to the left of the zero order diffraction region are determined by the specific holographic mask recording T_A. Particularly, term 9 is a convolution of the relevant signal with defocused circular array antenna radiation pattern with respect to ϕ noted in Section 4.3.3. The rest of the terms that are localized in the central regions of the output plane, determine the angle $\alpha_\circ$ of the support wave. Note that generally the terms containing $F_\circ$ and caused by the presence of the reference energy level $E_\circ$ and also by the dual-band input of the space–time light modulator are eliminated at the recording stage, as well as at the reproduction stage using the appropriate filtration schemes [21, 23]. Terms $\dot{e}_{out}^{(-1)}$ and $\dot{e}_{out}^{(+1)}$ (not shown on Fig. 9.23) symmetrical to the ϕ axis are due to the double-band signal input into the coherent optical processor with the help from the space–time light modulator (see Fig. 9.26). The circular array antenna phasing, to different angles $\theta_\circ$, is done with help of junction 13 (see Fig. 9.22). Thus, the junction changes masks recorded for different angles $\theta_\circ$.

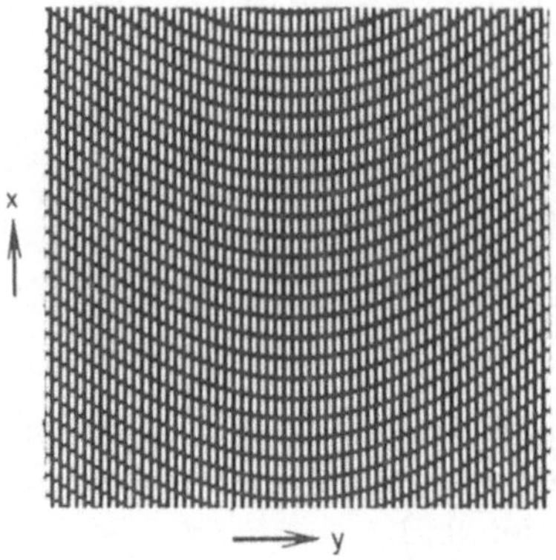

Figure 9.24: Circular array input signal frequency format Ω_o.

9.4.2 Experimental results and discussion

The experimental results, given in Figs. 9.26 and 9.25, show the following initial data: signal of harmonic frequency Ω_o; electrical radius of the circular array antenna $K_oR_o = 33.5$; number of receiving elements on the illuminated side of the array antenna ($|\Delta\varphi| \leq \pi/2$) $N = 61$; step size in the circular array antenna $d_\phi R_o/\lambda = 0.55$; $\theta_o = \pi/2$ the space–time light modulator sizes and the rest of the initial data are given in Section 9.4.3.

Fig. 9.24 shows the enlarged output signal format from the source with frequency Ω_o located in the far field of the circular array antenna with angle $\phi_o = 0$ (this specific format was obtained through calculations and subsequent reproduction with help of a precision synthesizer). The phase front delay of the wave equivalent to the space–time signal received by the circular array antenna elements was tracked in contour.

The initial format serves as the basis for recording masks by the analog method discussed above. This is presented in Fig. 9.25 (the critical and the horizontal axis are Ω and Ω_ϕ, respectively). The information (relevant) part of the recorded holographic mask T_A, that performs the main pattern-forming function during the coherent optical processor's operation in real time is the structure $\dot{T}_{inf} \approx \overset{*}{\dot{F}}_t \overset{*}{\dot{R}}\, \dot{T}$ in (9.1) (the horizontal format is the upper part of the figure) whose maximum intensity along the vertical axis determines the frequency of signal Ω_o. In the $\dot{T}_{inf}$ structure we can see interference bands formed by the signal and the support beams. When a mask is recorded from a corresponding standard signal whose frequency is different from Ω_o, the $\dot{T}_{inf}$ structure moves vertically according to the frequency change. The difficulty in creating a pattern-forming mask becomes apparent in the case

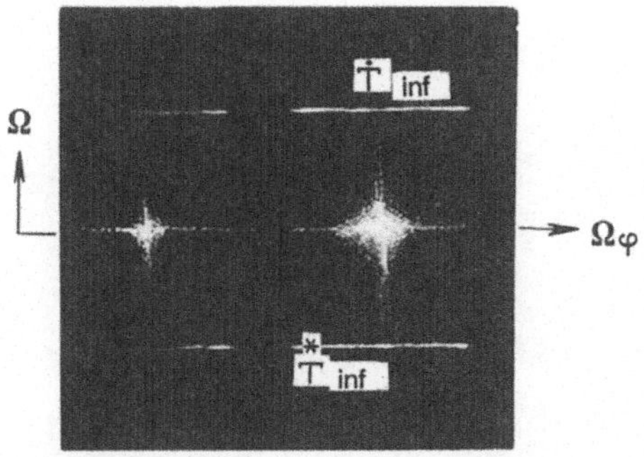

Figure 9.25: Holographic mask recorded by analog methods.

of a broad-band space–time signal. The $\overset{*}{T}_{inf}$ structure, which is symmetrical with respect to the Ω_ϕ axis is determined by the dual-band signal input into the coherent optical processor.

Fig. 9.26 shows the diffraction pattern at the output plane of the coherent optical processor $|\dot{e}_{out}|^2$ for a harmonic signal with frequency Ω_o. The signal incident from a far field onto the circular array antenna at an angle $\phi = 0°$ (the supporting channel in Fig. 9.22 is disconnected). The diffraction pattern coinciding with the location scheme of the diffraction terms in (9.3) appears in Fig. 9.23. The optical images in the upper right corner are a relevant operational order of diffraction whose vertical location, in accordance with (4.45), determines the frequency of the signal Ω_o and the horizontal azimuth of the target. The optical image in the upper left hand corner, in correspondence with (9.3) is a convolution along ϕ (along y) of the circular array antenna Π signal with a defocused radiation pattern, the convolution being displaced in the direction opposite to image 1. In the lower part of the diffraction pattern of Fig. 9.26 is a grouping of terms that are defined by the dual-band mode of the processor operation. These do not cause uncertainty in determining the azimuth coordinates of the objects (Section 3.1.1). As we mentioned earlier the noninformational images do not significantly affect the reading of the relevant information. However, they might limit the processor's dynamic range. It is therefore wise to filter them out (Fig. 9.27).

Figs. 9.28 a and b show a fragment of a diffraction pattern in the output plane of the processor along with the results of its photometry in the section of the relevant order for an instance of two objects having the same frequency, Ω_o, but located at different azimuth angles $\phi_1 = 0°$ and $\phi^2 = 30°$ $(\Omega_o\tau_\theta = 33.5)$. The optical target reconstruction at the coherent

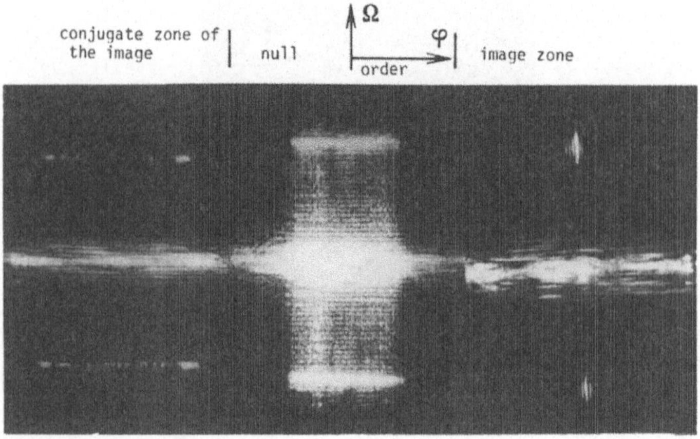

Figure 9.26: Diffraction pattern at the input plane of a circular array coherent optical processor.

optical processor output occurs in the parallel mode where the light intensity distribution on the ϕ-axis coincides with the circular array antenna radiation pattern as far as the power, and along Ω with the square of the optical spectrum analyzer kernel. The dotted line in Fig. 9.28a shows the theoretical circular array antenna radiation pattern for $\phi_1 = 0°$ with the assumption of uniform excitation of the array antenna receiving elements and their isotropy [133].

Figs. 9.29 a and b illustrate a fragment of a diffraction pattern along with the result of its photometry in sections of the relevant orders for cases with two objects with different frequencies, Ω_1 and Ω_2 and having different azimuth angles $\phi = -30°$ ($\Omega_1 \tau_\theta = 41.875$) and $\phi_2 = +30°$ ($\Omega_2 \tau_\theta = 33.5$). In this case the holographic mask T_A is recorded into the coherent optical processor sequentially with the help of two standard signal formats with frequencies Ω_2 and $\Omega_1 = 1.25\Omega_2$ shown in Fig. 9.24.

9.4.3 A Coherent optical processor for a circular array antenna employing a pattern-forming mask synthesized by digital holography

The experimental research was conducted on an experimental model of a coherent-optical processor shown in Fig. 9.22. Its support channel was disengaged. The pattern-forming mask 11 was synthesized according to the digital holography methods described in Appendix A. All necessary data pertaining

Figure 9.27: Diffraction pattern with the zero order filtered out in the scanning zone.

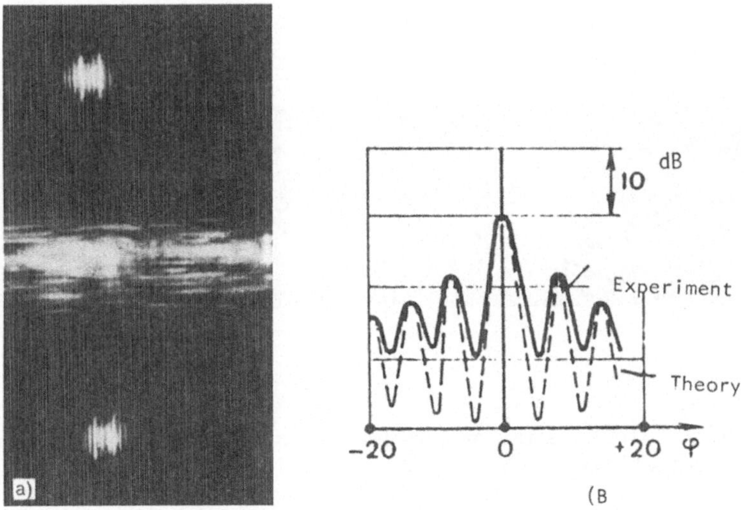

Figure 9.28: (a) Diffraction pattern; and (b) the photometric result of two emission sources $\Omega_2 = \Omega_1$, $\phi_2 \neq \phi_1$. (Same frequencies).

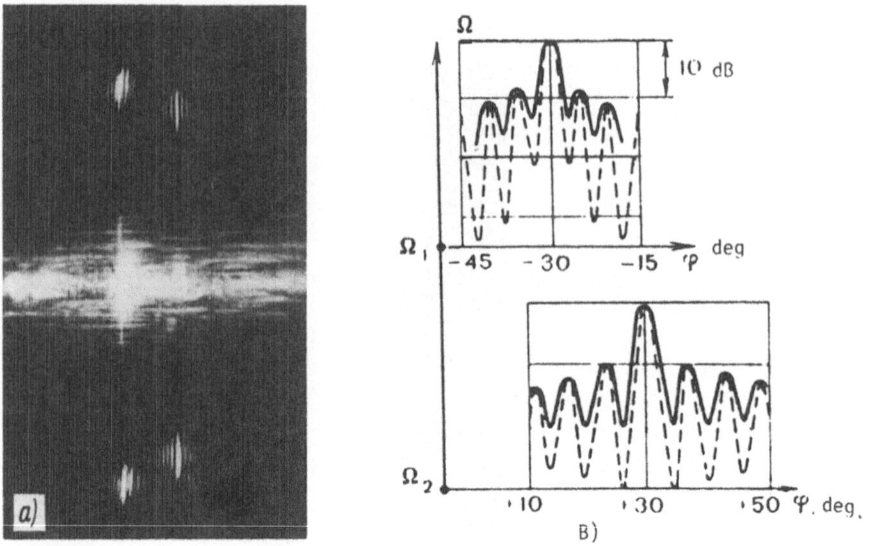

Figure 9.29: (a) Diffraction pattern; and (b) the photometric result of two emission sources $\Omega_2 = \Omega_1$, $\phi_2 \neq \phi_1$. (Different frequencies).

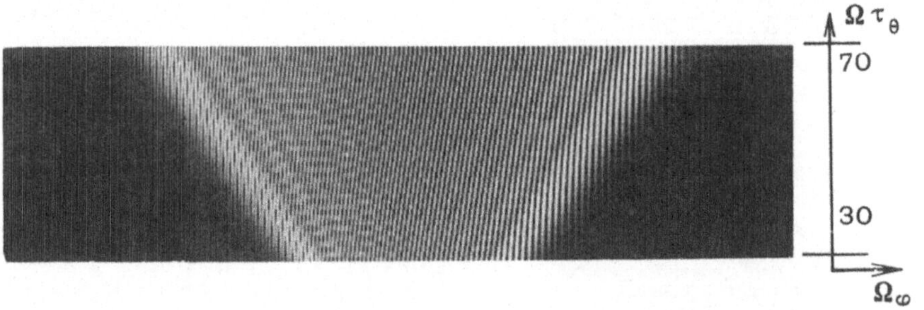

Figure 9.30: Circular array pattern-forming mask, recorded by digital holographic methods (see Section 9.4.3).

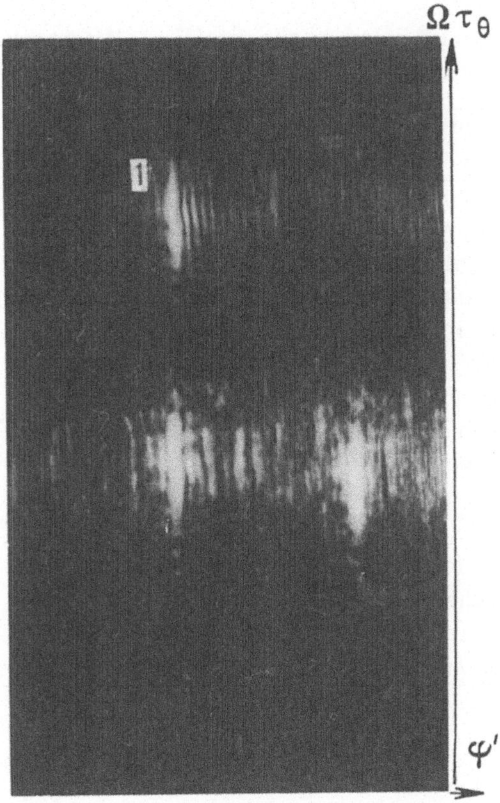

Figure 9.31: The diffraction pattern in the output plane of the circular array antenna coherent optical processor (useful diffraction orders) with three sources: 1- Ω_1 $\varphi_1 = -45°$; 2-$\Omega_2 = \Omega_1/1.75$ $(\Omega_2\tau_\theta = 33.5)$, $\varphi_1 = \varphi_2$; 3 -$\Omega_3 - \Omega_2$ $(\Omega_3\tau_\theta = 33.5)$, $\varphi_3 = +45°$.

to the circular array antenna, space–time light modulator, coherent-optical processor, and precision photoplotter used during the mask recording, are given in Appendix A.

The structure of the obtained mask is presented in Fig. 9.30. Figure 9.31 shows a fragment of a diffraction pattern of relevant diffraction orders in the output plane of the coherent-optical processor in a case where the circular antenna receives harmonic signals from the three sources located in the far field and differing in frequency, as well as in the azimuthal direction.

Table 9.1: The parameters of the precision photosynthesizer and mask.

The dimensions of the square recording aperture, $\delta_x \times \delta_y\, \mu m^2$	12.5×12.5
The recording step size, Δx by $\Delta y\, \mu m^2$	12.5×12.5
Reproducible body	$4.1 \cdot 10^4$
The number of quantizing levels $Q = 2^n$	256
The pattern-forming mask size $Dx \times Dy$, m^2	2.0x12.5
Mask recording time with magnetic tape reading, s	1.42
The step size error, μm, not over	1
The level of normalizing brace α_v	$0.05\mathcal{T}^{IV}(\Omega_\phi, \Omega_{\tau_\phi})\max$
Mask reading time:	
Simpson's quadrature formula,	
Asymptote, min	120
Using the fast Fourier transform	10
The linearity signal—degree of film blackening, %, not worse than	10
Mask diffraction efficiency, %	0.35

Table 9.2: The initial data on the array antenna, space–time light modulator, and coherent-optical processor necessary for pattern-forming mask realization using digital holography methods.

The array antenna electrical radius	
$\Omega_{r_0} = KR_\circ \sin\theta_\circ = \frac{2\pi F}{c}(R_\circ \sin\theta_\circ)$	30–70
The frequency band F_{max}/F_{min}	2.3
The angle coordinate $\theta_\circ$, with which the array antenna is phase synchronized, radians.....................	$\pi/2$
The array antenna excitation sector $\lvert\Delta_\phi\rvert$,radians........	$3\pi/4$
The number of the array antenna receiving elements $(d_\phi R_\circ = 0.55\,\lambda_{min})$	91
The signal scanning speed in the space–time light modulator v_p, m/s	5
The wave distribution speed in the medium c, m/s......	1,500
The space–time light modulator dimensions:	
Pupil, μm^2	25×100
Scale grating dy, μm..........................	200
apertures Δx by Δy, mm^2..........................	12.5×12.5
The light wavelength λ, μm	0.638
Reproducing coefficients:	
a, m^{-1}	$2.22 \cdot 10^4$
b, m^{-1}	$1.25 \cdot 10^4$
The Fourier lens focal length , f, m	0.6
The image forming system reproduction coefficient $M = f_2/f_1$	1 : 2
The optical spectrum analyzer resolution δF, Hz	4

Chapter 10

Conclusions. Trends in Electrooptical Array Antenna Theory and Development

An electrooptical array system is an extremely complex device which unites an active receiving device—more precisely, a coherent-optical processor—and electronic (and this includes computers) hardware (see Fig. 1.1). Serious structural-technological problems have arisen during the development of such devices. Here, there are a large number of problems, including processor element configuration, decreasing the effect of mechanical vibration and thermal influences, decreasing the weight and size and so on. Figures 10.1 and 10.2 give some definite ideas on possible configurations for array antenna coherent optical processors.

Figure 10.1 shows a coherent optical processor spectrum analyzer with controlling and operating devices [90]. The processor is mounted on a vertical plate (part of the elements are located on the back). The signal input to the space–time light modulator is accomplished using an electron-ray tube with a thermoplastic target (see Fig. 1.5).

Figure 10.2 shows a compact coherent image correlator with filtering in the frequency plane (cf. Figs. 1.8, 4.4, 4.9) and [192]. The definitive characteristic of the device is its portability, achieved by using laser diodes and lens 3 simultaneously as a collimator and a Fourier lens and also by bending the light beam trajectory inside the processor (Fig. 3.2a).

Instead of a helium-neon laser in the correlator, the device uses four laser diodes. The correlation analysis is therefore carried out in parallel for the four laser beams, thus increasing processor's information capacity. A lens which decreases the divergence of each laser beam from $20 \times 40°$ to $20 \times 6°$ is placed in front of each diode (body diameter $= 6$ mm, wavelength 830 nm).

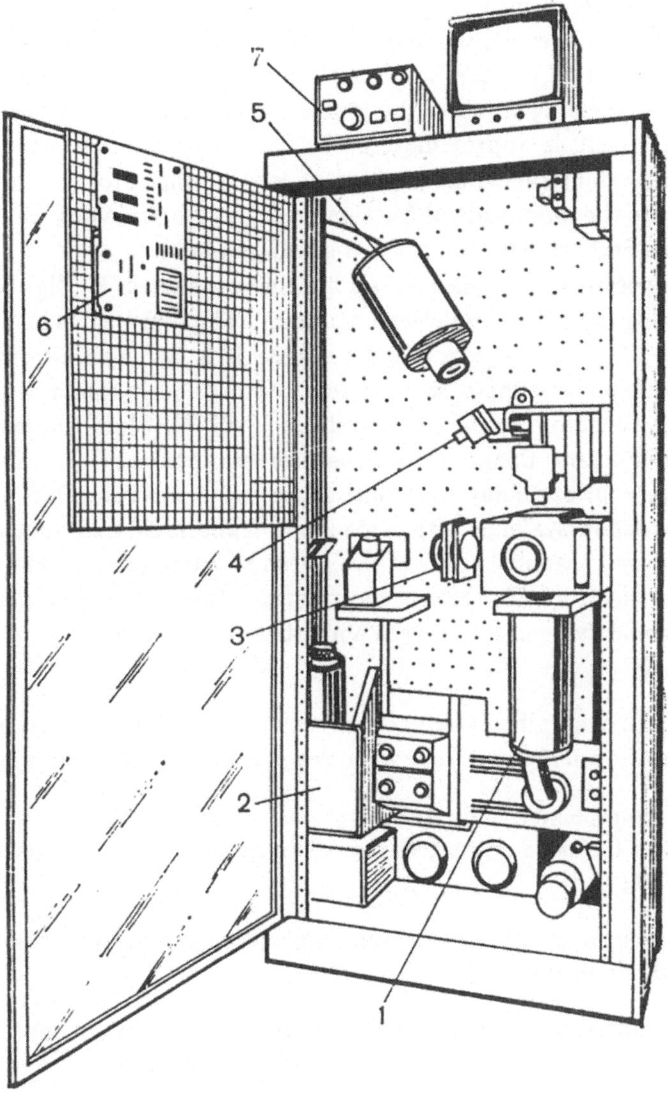

Figure 10.1: The cutaway view of a coherent optical spectrum-analyzer: (1) Electronic ray tube with thermoplastic target; (2) He-Ne laser; (3) collimator; (4) Fourier lens; (5) TV monitor; (6) microprocessor which controls the processor operation; (7) STLM operation controlling device.

A fiber optic cable is used to join each diode with a phototransistor which is connected to the laser diode current supply control. This control maintains the output power at a 5 mW level at 70 mW operation.

The processing speed of the image transformer 2 which uses a liquid-crystal light valve is approximately 100 ms [86, 106, 176]. Four matched filters 7 located in the Fourier plane of lens 3 whose focal length is 250 mm consitute the amplitude diffraction gratings, fabricated by an analogous method (see Section 9.4.1) on a 5×5 cm glass plate.

The correlator, in essence, is a hybrid optoelectronic (digital) processor, consisting of optical and electronic parts. The dimensions of the optical portion are $15 \times 23 \times 42$ cm^3, and its mass is 8 kg. The electronic unit, together with power supply, is $18 \times 28 \times 35$cm^3, and its mass is 8 kg. The total power consumption of the correlator is 55 W, while the laser diodes and an liguid crystal image transformer use less than 1 W.

Electrooptical array antennas represent a new class of wide band receiving antenna systems with wide-angle, panoramic scanning and some other additional practical applications. Despite their apparent and potential advantages, electrooptical arrays have not yet taken their place in modern radar and sonar systems. There are a number of reasons for this, particularly the following:

- The lack of a unified approach which would combine the fast processing times of optical processors with the processing system as a whole;
- Their complexity, poor technical characteristics and lack of development in the production techniques for coherent optical processor components; and
- The lack of experience in applying optoelectronic processors, which would widen their versatility and use.

The following trends in the development of the theory and practice of electrooptical array antennas are now being investigated:

1. Further development of the components, improvements in their technical characteristics, expansion of their practical applications in radar and sonar systems, development of hybrid optoelectronic processors.

   Hybrid optoelectronic processors rationally combine the advantages of coherent-optical devices with that of digital devices [37, 193]. In this case a multichannel acoustooptical light modulator, used for inputing radio signals into the coherent optical processor, and a charge-coupled device for information output are beyond competition. Multichannel input devices for sonars are being developed using the magnetooptical

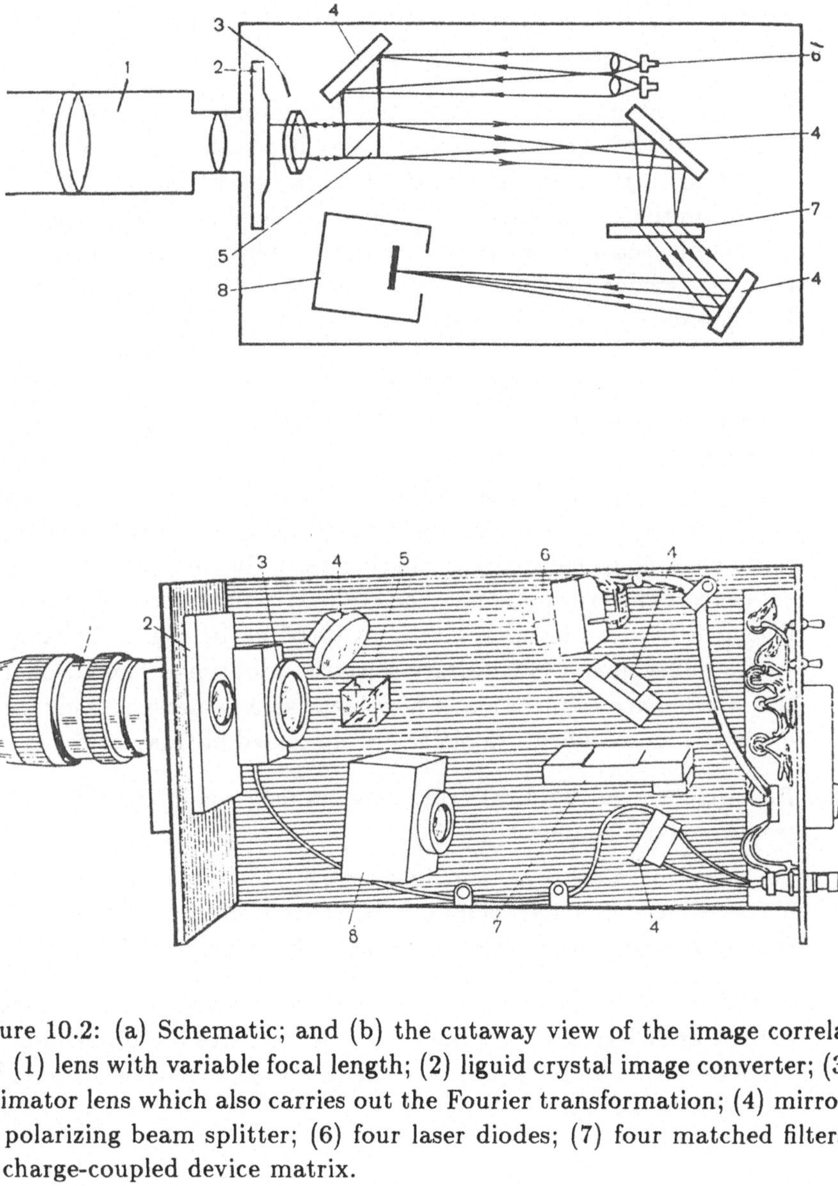

Figure 10.2: (a) Schematic; and (b) the cutaway view of the image correlator: (1) lens with variable focal length; (2) liguid crystal image converter; (3) collimator lens which also carries out the Fourier transformation; (4) mirror; (5) polarizing beam splitter; (6) four laser diodes; (7) four matched filters; (8) charge-coupled device matrix.

crystals and devices such as the liquid crystal charge-coupled device, which can be used also as a dynamic mask.

2. Development of analog optoelectronic processors for adaptive array antennas. Adaptive array antennas makes it possible to solve the problem of distinguishing the signals from the background noise and to maximize the signal-to-noise ratio under changing ambient conditions [24]. Optoelectronic processors, employing iterative vector-matrix procedures, remove the limitations on the volume of calculations and on the operating speed [193, 194].

3. Development of universal methods for coherent optical processing of the signals from nonplanar arrays and improvement in interference suppression.

   Solving these types of problems is necessary for the dynamic (adaptational) compensation of amplitude-phase distortions which appear as a result of varying deformations and changes in the three-dimensional orientation of an array antenna. The key role in this type of problem will belong to pattern-forming masks, synthesized by digital holography. At the same time it is necessary solve the problem of how to improve the interference suppression of nonplanar array antennas. This task is obviously more difficult than the one described in Chapter 6.

4. Research on electrooptical array antenna focuseds in the Fresnel zone (see Appendix D and [95]).

   Such electrooptical array antennas have broader practical applications in retrieving a maximum of information (about polar coordinates, range, spectrum, and object shape) from the received signal [26].

5. The development of a statistical method for processing space–time signals by optoelectronic processors in order to solve the problem of optimal detection and evaluation of the incoming signals on background interference [24, 26].

Appendix A

Circular Electrooptical Array Antenna Mask Synthesis by Digital Holography Methods

Operational filter-mask synthesis by digital methods allows significantly more coherent optical processor capabilities [12, 40]. Digital synthesis is free of some limitations which are present while transcribing masks by optical holography methods. The presence of these limitations is explained by partial light incoherence, imperfect optical components, vibration, and other disturbances. Computer aided mask synthesis can be accomplished in consecutive order and consists of a few steps:

- Computation of mask's complex transmission function;
- Code representation of real and imaginary parts as nonnegative functions;
- Possible transmission function modifications in order to improve the mask diffraction efficiency and to lower zero order diffraction;
- Space quantization with selection of grating, pitch, and size of aperture references; and
- Evaluation of quantization effects, nonlinearity of recording medium as well as the limited sampling quantity.

A.1 The Complex Mask Transmission Function

According to Section 4.2.3, it is possible to carry out space–time processing, permitting 360° azimuthal scanning, while simultaneously performing spectral analysis of the signals received by elements of the circular array antenna with the aid of coherent optical processors of the types shown in Figs. 1.4 and 4.7. The multichannel acoustooptical modulators, shown in these figures, can be used as input devices for such processors, but other space–time light modulators, described in Section 1.2.2, which perform three-dimensional space sweep of the command signals can be used as well. Coordinated masks, installed in spectral π-plane (see Fig. 4.7) are the main part of the processors mentioned above. According to (4.46), their complex transmission function is

$$\begin{aligned}\dot{T}\left(\Omega_{\varphi}, \frac{\Omega}{c}\cos\theta_{\circ}\right) &= \dot{T}\left[\Omega_{\varphi}, \sqrt{1-(\Omega\tau_{\phi}/R_{\circ})^2}\right] \\ = \dot{T}(\Omega_{\varphi}, \Omega\tau_{\theta}) &= \int_{-\infty}^{\infty} T_{\varphi}(\varphi)\exp[-i(\Omega\tau_{\theta}\cos\varphi + \Omega_{\varphi}\varphi)]\mathrm{d}\varphi \qquad \text{(A.1)}\end{aligned}$$

where $\Omega\tau_{\theta} = (\Omega/c)R_{\circ}\sin\theta_{\circ} = KR_{\circ}\sin\theta_{\circ}$ is a dimensionless quantity ($KR_{\circ}$ is the effective radius of the circular array antenna with physical radius $R_{\circ}$). Linear mask dimensions (A.1) in the π-plane, according to (4.43) and (1.2) are related to the dimensionless parameters Ω_{ϕ} and $\Omega\tau_{\theta}$ as follows:

$$\Omega\tau_{\phi} = v\omega_x\tau_{\theta} = v\frac{k}{f}x\frac{R_{\circ}}{c}\sin\theta_{\circ} = ax$$

$$\Omega_{\varphi} = \omega_y/m_y = \frac{k}{f}y/(\Delta\varphi/\Delta y) = by \qquad \text{(A.2)}$$

where $a = (v/c)(2\pi\sin\theta_{\circ})(R_{\circ}/\lambda f)$, $b = (2\pi/\Delta y)(\Delta\varphi/\lambda f)$; v is the three-dimensional sweep rate; f is the focal length of the lens L_1 (see Fig. 4.7); λ is the light wavelength; and $\Delta\phi$ is the sector where the array antenna receiving elements are housed ($\Delta\varphi = m_y\Delta y$, and Δy are the space–time light modulator dimensions).

Since $\dot{T} = |\dot{T}|\exp(i\arg\dot{T}) = \Re e\dot{T} + i\,\Im m\dot{T}$, the following ways of complex mask realization are possible. The first consists of the independent reproduction of two masks: one of the amplitude, $|\dot{T}|$, and one of the phase, exp(i arg$\dot{T}$), as, for example, a photomask and a cineform, respectively. This approach does not make the structure more complicated and does not worsen its pattern capabilities but is very difficult to implement. The second method, recording two uniform photomasks $\Re e\dot{T}$ and $\Im m\dot{T}$, is easier to produce, but complicates the processor optical design. Finally, there is a third way. This is to use only one mask, either the $\Re e\dot{T}$ or the $\Im m\dot{T}$ in a spectral plane of the processor, as shown in Fig. 4.7. This method is successful in spite of the fact that it causes a worsening of the processor's pattern and energy characteristics (see, for example, Fig. 4.12a), because it minimizes these disadvantages by modifying its transmission function structure (A.1).

A.2 The Mask Structure

Let us list some initial suppositions in order to be able to design the mask (A.1) structure. For convenience, let us assume, that the amplitude–phase distribution is uniform on the illuminated half of the circular array and is zero elsewhere $T_\varphi(\varphi) = \text{rect}(\varphi/\pi)$. Then, for (A.1) we obtain

$$\begin{aligned}\dot{T}(\Omega_\phi, \Omega\tau_\theta) &= \int_{-\pi/2}^{\pi/2} \exp[i(\Omega\tau_\theta \cos\varphi + \Omega_\varphi \varphi)]\mathrm{d}\varphi \\ &= \pi \exp\left(i\frac{\pi}{2}\Omega_\varphi\right)[A_{-\Omega_\varphi}(\Omega\tau_\varphi) + iV_{-\Omega_\varphi}(\Omega\tau_\theta)] \qquad \text{(A.3)}\end{aligned}$$

where A and V are the Anger and Weber functions. The term, $\exp\left(i\frac{\pi}{2}\Omega_\varphi\right)$, only causes the output image to shift by $\pi/2$ along the y-axis and therefore, is omitted everywhere below. Functions A and V with $|\Omega\tau_\varphi| \gg 1$ ($|\Omega\tau_\theta| \gg |\Omega_\varphi|$) are characterized by their asymptotic representation

$$\begin{aligned}A_{-\Omega_\varphi}(\Omega\tau_\theta) &\approx \sqrt{\frac{2}{\pi\Omega\tau_\theta}} \cos\left(\Omega\tau_\theta + \frac{\pi}{2}\Omega_\varphi - \frac{\pi}{4}\right) + O(\Omega\tau_\theta)^{-3/2} \\ V_{-\Omega_\varphi}(\Omega\tau_\theta) &\approx -\sqrt{\frac{2}{\pi\Omega\tau_\theta}} \sin\left(\Omega\tau_\theta + \frac{\pi}{2}\Omega_\varphi - \frac{\pi}{4}\right) + O(\Omega\tau_\theta)^{-3/2} \qquad \text{(A.4)}\end{aligned}$$

As can be seen, where the argument, $|\Omega\tau_\theta|$, values are large (where the array antenna effective radius is large) the mask is weighted by a multiplier $\sqrt{2/\pi\Omega\tau\theta}$. The latter is defined by the proportionality of the received signal power to the array antenna effective radius. It is obviously expedient, from the energy view point, to compensate for it. This will increase the specific weighting of the high frequency part of the signal spectrum with respect to the low frequency part.

The cosinusoidal array (A.4) has a transmission function of zero along parallel straight lines $\Omega_{T_\theta} + \frac{\pi}{2}\Omega_\varphi - \pi/4 = \pi/2 + n\pi$ $(n = 0, \pm 1, \ldots)$. The sinusoidal array has its maxima along these lines. According to (A.2) these lines are at angle α with respect to the x-axis

$$\alpha = \arctan(b/a) = \arctan\left[\frac{2\Delta\varphi}{\pi} + \frac{R_\circ v \sin\theta_\circ}{c\Delta y}\right] \tag{A.5}$$

The distance between the adjacent lines is

$$\begin{aligned} d &= \frac{\pi}{a}\cos\alpha = \frac{\pi}{a}\cos\left(\arctan\frac{b}{a}\right) \\ &\approx \frac{\pi/a}{\sqrt{1+(b/a)^2}} = \frac{\pi}{\sqrt{a^2+b^2}} \approx \pi/b = \frac{\lambda f}{\Delta y}\frac{\Delta\varphi}{2} \end{aligned} \tag{A.6}$$

since $a \ll b$ with $v \ll c$.

Since the distance (A.6) is not a function of the angle $\theta_\circ$, the readjustment of the coherent optical processor during an angle change may be accomplished by rotating the diffraction grating through the corresponding angle, α.

In a general case of wide-band signal reception, an asymptotic representation of type (A.4) is made in a limited frequency band. Let us describe the general approach to the synthesis of antenna pattern controlling masks which are free of these limitations with the use of digital holography.

A.3 The Mask Diffraction Efficiency

The conditions for producing masks with a transmission function of $\Re e\mathcal{T}$ on a precision photosynthesizer are the requirements of it being passive and nonnegative. Therefore, let us modify this function as follows:

$$0 \leq \mathcal{T}^I(\Omega_\varphi, \Omega_{T_\theta}) = \frac{\Re e\{\dot{\mathcal{T}}(\Omega_\varphi, \Omega_{T_\theta}) - \dot{\mathcal{T}}(\nu_{min}, z_{min})\}}{\Re e\{\dot{\mathcal{T}}(\nu_{max}, z_{max}) - \dot{\mathcal{T}}(\nu_{min}, z_{min})\}} \leq 1 \tag{A.7}$$

where ν_{min}, z_{min} (ν_{max}, z_{max}) are the values of Ω_φ and Ω_{T_θ}, respectively, where $\Re e\dot{\mathcal{T}}$ reaches an absolute minimum (maximum) on a given interval of

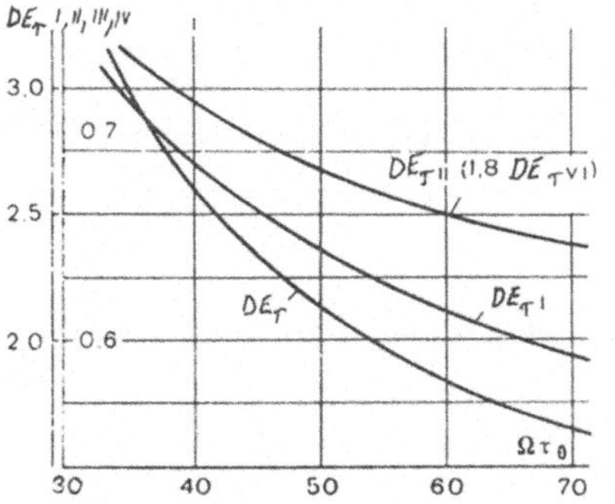

Figure A.1: Diffraction efficiency of a mask for a circular array.

Ω_φ and $\Omega_{\mathcal{T}\theta}$ values. Similar expressions ought to be written for the imaginary part of the mask (A.1).

The diffraction effectiveness, (DE), of the mask with a transmission function of (A.7) is

$$\mathrm{DE}_{\mathcal{T}I} = \frac{1}{2}\mathrm{DE}_{\dot{\mathcal{T}}}/\Re e\{\dot{\mathcal{T}}(\nu_{max}, z_{max}) - \dot{\mathcal{T}}(\nu_{min}, z_{min})\}]^2 \qquad \text{(A.8)}$$

where the multiplier 0.5 is caused by the existence of an adjacent image. $\mathrm{DE}_{\mathcal{T}I}$ is the diffraction efficiency of the complex mask (A 1.1). It is defined as

$$\begin{aligned}\mathrm{DE}_{\mathcal{T}I} &= \int_{-\infty}^{\infty}|\dot{E}_\pi\dot{\mathcal{T}}|^2 d\Omega_\varphi \Big/ \int_{-\infty}^{\infty}|\dot{E}_\pi|^2 d\Omega_\varphi \\ &= \int_{-\infty}^{\infty}|\dot{\mathcal{T}}|^4 d\Omega_\varphi \Big/ \int_{-\infty}^{\infty}|\dot{\mathcal{T}}|^2 d\Omega_\varphi \qquad \text{(A.9)}\end{aligned}$$

where $\dot{E}_\pi$ is the light distribution in a spectral plane at the input of the mask, $\mathcal{T}$. Also, it is assumed that $\dot{E}_\pi \sim \overset{*}{\mathcal{T}}$ (for maximum electrooptical array antenna power effectiveness). Let us note that the mask (A.3) which exists during an even amplitude distribution $\mathcal{T}_\varphi\varphi = \mathrm{rect}(\varphi/\pi)$, is characterized by the maximum value of DE (A.9) shown in Fig. A.1.

The value of DE (A.8) is quite low. Moreover, the mask has a constant part which causes parasitic exposure of the processor output. This illumination appears as the space–time light modulator image with dimensions of $\pi = m_y y$ and the image of the space–time signals' frequency spectrum with

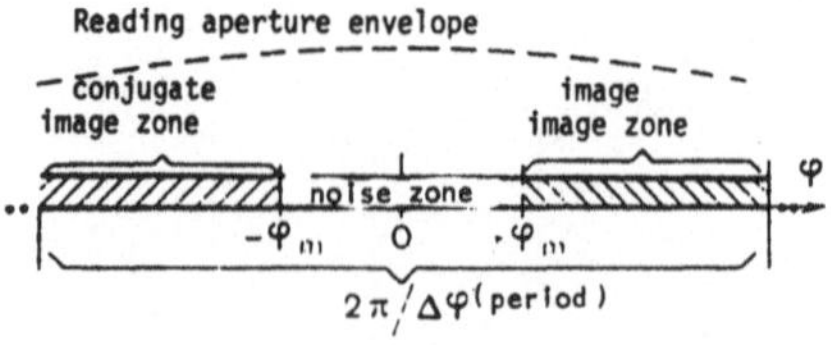

Figure A.2: Choosing the digitization step and the spatial subcarrier frequency for the mask.

dimensions $\Omega = v\omega_x$. Therefore, let us modify the transmission function (A.7). First, let us compensate for the decrease in the absolute value of the transmission function $\mathcal{T}^I$ during the argument $\Omega\tau$ increase by inverse of the local $\mathcal{T}^I$ maxima envelope. Such a multiplier, according to (A.4), in case of mask (A.3) and large $\Omega\tau_\theta$ values can be approximated by $\sqrt{\pi\Omega\tau_\theta/2}$. In general, due to the analyticity of function $\dot{\mathcal{T}}$, $\Re e\dot{\mathcal{T}}$ is the envelope of $\dot{\mathcal{T}}$. Secondly, let us substitute component $|\dot{\mathcal{T}}(\Omega_\varphi, \Omega\tau_\theta)|$ which is unevenly distributed along Ω_φ into (A.7) for the constant component $\Re e\dot{\mathcal{T}}(\nu_{min}, z_{min})$. As the result of this modification, we obtain the mask:

$$\mathcal{T}^{II} = \Re e\{\dot{\mathcal{T}}(\Omega_\varphi, \Omega\tau_\theta) + |\dot{\mathcal{T}}(\Omega_\varphi, \Omega\tau_\theta)|\}/2|\dot{\mathcal{T}}[\Omega_\varphi(\Omega\tau_\theta)\Omega\tau_\theta]| \qquad \text{(A.10)}$$

where $\Omega_\varphi = \Omega_\varphi(\Omega\tau_\theta)$ is the curve along which $|\dot{\mathcal{F}}|$ reaches the local maxima with a given $\Omega\tau_\theta$. The diffraction effectiveness, just as in (A.8), (see Fig. A.1) is

$$\mathrm{DE}_{\mathcal{T}^{II}} = \mathrm{DE}_{\dot{\mathcal{T}}}/8\dot{\mathcal{T}}[\Omega_\varphi(\Omega\tau_\theta), \Omega\tau_\theta]|^2 \qquad \text{(A.11)}$$

The modified mask $\mathcal{T}^{II}$ diffraction efficiency is somewhat higher than that of the mask $\mathcal{T}^I$. This provides for a lower exposure level, which in this case is approximately equal to $2\mathrm{DE}_{\mathcal{T}^{II}} \cdot 100\%$ (approximately 1.5%) of the light flux power at the mask inlet. In the case of $\mathcal{T}^I$, the exposure level reaches the value of $|\Re e\{\dot{\mathcal{T}}(\nu_{min}, z_{min})\}|^2/[\Re e\{\dot{\mathcal{T}}(\nu_{max}, z_{max}) - \dot{\mathcal{T}}(\nu_{min}, z_{min})\}]^2$ (approximately 25%).

At last, one more possibility can be used to eliminate the parasitic exposure effect.

A.4 Shifting to the Spatial Subcarrier Frequency

Since the undiffracted diffraction order occupies the azimuthal observation sector $|\varphi|^2 \leq \varphi_m$ ($\varphi_m = 3\pi/2$ with sesqui-entry addressing [137]), in order to eliminate the parasitic exposure of the $F^{(\pm 1)}$ diffraction order, it is necessary to shift the latter by at least $2\varphi_m$, by introducing a spatial subcarrier frequency into mask $\mathcal{T}^{II}$. Therefore, the expression

$$\dot{\mathcal{T}}^{III} \exp(i2\varphi_m \Omega_\varphi)\dot{\mathcal{T}}(\Omega_\varphi, \Omega\tau_\theta) \tag{A.12}$$

should be substituted into the transmission function (A.10) for (A.1). Since $|\dot{\mathcal{T}}^{III}| = |\dot{\mathcal{T}}|$, the mask structure does not change and is characterized by the same diffraction efficiency and minimum parasitic exposure.

A.5 Mask Recording by Digital Holography

The precision photosynthesizer creates a mask in discrete points $x = p\Delta_x$, $y = q\Delta_y$ ($p, q = 0; \pm 1; \pm 2; ...; \Delta_x, \Delta_y \gg 10^{-6}$m) in squares with dimensions $\delta x \delta y$. Therefore the mask

$$\begin{aligned} \mathcal{T}^{IV}(\Omega_\varphi, \Omega\tau_\theta) &= \sum_p \sum_q \mathcal{T}^{II}(p\Delta_\varphi, q\Delta_\Omega)\mathrm{rect}[(\Omega_\varphi - p\Delta_\varphi)/b\delta y] \\ &\times \mathrm{rect}(\Omega\tau_\theta - q\Delta_\Omega)/a\delta_x] \end{aligned} \tag{A.13}$$

where $\Delta_\varphi = b\Delta_y$, $\Delta_\Omega = a\Delta x$; a and b are defined by equation (A.2).

According to (4.46), mask (A.13) has the following amplitude phase distribution:

$$\begin{aligned} J_\varphi^{IV}(\varphi, \Omega\tau_\theta) &= \hat{\mathcal{F}}_\varphi^{-1}\{\mathcal{T}^{IV}(\Omega_\varphi, \Omega\tau_\theta)\} \\ &= J_\varphi^T(\varphi) \sum_{k=-\infty}^{\infty} J_\varphi^{II}(\varphi + k2\pi/\Delta_\varphi, \Omega\tau_\theta) \end{aligned} \tag{A.14}$$

where the Poisson's summation formula was used along with the equations:

$$J^{II}_{\varphi}(\varphi, \Omega\tau_{\theta}) = \hat{\mathcal{F}}^{-1}_{\varphi}\left\{\mathcal{T}^{II}(\Omega_{\varphi}, \Omega\tau_{\theta})\right\}$$

$$J^{T}_{\varphi}(\varphi) = \hat{\mathcal{F}}^{-1}_{\varphi}\left\{\sum_{q} \mathcal{T}^{II}(0, q\Delta_{\Omega})\mathrm{rect}(\Omega_{\varphi}/b\delta_{y})\mathrm{rect}[(\Omega\tau_{\theta} - q\Delta_{\Omega})/a\delta x]\right\}$$

The amplitude-phase distribution (1.14) appears to be periodic with respect to φ (the period is equal to $2\pi/\Delta_{\varphi}$). In virtue of this, the processor output image becomes periodic also. Therefore, in order to eliminate the periodic parasitic illumination effect, $2\pi/\Delta_{\varphi} \geq 2(2\varphi_m) + 2\varphi_m = 6\varphi_m$; thus, Fig. A.2.

$$\Delta_{\varphi} \leq \pi/3\varphi_m \qquad \text{(A.15)}$$

Condition (A.15) defines the step size in meters by calculating (A.2) with the requirement that:

$$\Delta_{y} \leq \pi/3\varphi_m b \qquad \text{(A.16)}$$

Let us explore the effect of the envelope J^{T}_{φ}, defined by the finite width of the precision photosynthesizer reading aperture:

$$\begin{aligned} J^{T}_{\varphi}(\varphi) &= \sum_{q} \mathcal{T}^{II}(0, q\Delta_{\Omega})\mathrm{rect}[(\Omega\tau_{\theta} - q\Delta_{\Omega})/a\delta x]\hat{\mathcal{F}}^{-1}_{\varphi}\{\mathrm{rect}[\Omega_{\varphi}/b\delta y]\} \\ &= \sum_{q} \mathcal{T}^{II}(0, q\Delta_{\Omega})b\delta y\,\mathrm{rect}[(x - q\Delta_{x})/\delta x]\mathrm{sinc}(\varphi b\delta y/2\pi) \qquad \text{(A.17)} \end{aligned}$$

The envelope, $\mathrm{sinc}(\varphi b\delta y/2\pi)$, weights the output image. Since the first period of the $\pm 1^{\mathrm{st}}$ diffraction orders $F^{(\pm 1)}$ (see Fig. A.2) falls within the sector of angles $\varphi_m \leq |\varphi| \leq 3\varphi_m$, the envelope J^{T}_{φ} nonuniformity can be neglected supposing that sinc(...) ≥ 0.9, i.e. $3\varphi_m b\delta y \leq \pi/2$ or

$$\delta y \leq \pi/6\varphi_m b \qquad \text{(A.18)}$$

The factor $\text{rect}[(x - q\Delta_x)/\delta x]$ with $\delta x < \Delta x$ causes the formation of dark strips in the mask, thus decreasing its diffraction efficiency. Therefore it is expedient to select a step size $\Delta_x = \delta x$, in which case $\Delta_\Omega = a\delta x$. On the other hand, according to the Kotel'nikov's theorem, undistorted mask reproduction is also possible with the step size $\Delta_\Omega = \lambda f a / 2\Delta x$, where Δx is the dimension of the space–time light modulator along the x-axis, and $2\pi/\Delta x$ is the diffraction spot size in the spectral plane π (see Fig. 4.7). The step size Δ_Ω selection can be tied with the radiation pattern defocusing (the decrease of the directivity diagram maximum with respect to power) at the coherent optical processor output, due to the discrete nature of the mask reproduction along the frequency axis. The step size is defined as $\Delta_\Omega = \tau_\theta \delta\Omega / M$, where $\delta\Omega = 2\pi\nu / \Delta x$ is the spectrum analyzer resolution, M is the number of intervals in the mask quantization per spectrum analyzer resolution element. Having been given the defocusing value of (DF) ≤ 0.9, from (4.49), let us derive the following for the worst-case situation:

$$
\begin{aligned}
M &\geq \text{Entier}\left[1.1 K R_\circ \sin\theta_\circ \frac{\delta\Omega}{\Omega_\circ}\right] \\
&\sim \text{Entier}[1.1 K R_\circ \sin\theta_\circ \Delta x R_x] \sim \text{Entier}[6.9\tau_\theta / \Delta\tau] \qquad \text{(A.19)}
\end{aligned}
$$

where $\Delta\tau = \Delta x / v$ is the signal recording (filling in) onto the space–time light modulator aperture time; $\tau_\theta = R_\circ / c \sin\theta_\circ$) is the array antenna aperture fill in time. As a rule, the step size Δ_Ω, obtained in accordance with Kotel'nikov's theorem or (A.13), is larger than $a\delta x$. In order to decrease the amount of mask (A.13) data in order to reduce computer time, it is advisable to use the following method. The mask (A.13) is computed with a step size of Δ_Ω, but during the mask reproduction the precision photosynthesizer keeps the value of $a\delta x$ along the entire range Δ_Ω (by maintaining a constant mask value or by introducing an appropriate extrapolation).

And so, points $p\Delta_\varphi$ and $q\Delta_\Omega$ where the function (A 1.10) must be computed are defined. The precision photosynthesizer quantizes the functions which are being recorded with respect to their number of graduations, $Q = (2)^n$. Therefore, the transmission function which is being realized is described by the following expression:

$$
\mathcal{T}^V = [\mathcal{T}^{IV}(\Omega_\varphi \Omega \tau_\theta) \cdot Q] / Q \qquad \text{(A.20)}
$$

where [...] is the integer part of the number.

The quantized mask (A.20) causes additional parasitic background at the processor output. This is similar to the background defined by the switching errors with (caused by) discrete phasing [54].

The mask $\mathcal{T}^V$ recording on a photocarrier which usually has a nonlinear amplitudinal transmission dependency on the exposure is the most distinguishing feature of the problem at hand. Therefore, the data base, which describes the transmission function $\mathcal{T}^{IV}$, should be the subjected to the nonlinear predistortion, which weakens the nonlinearity of the photo carrier [140], while the precision photosynthesizer has to reproduce the mask with the following transmission function:

$$\mathcal{T}^{VI} = [(2)^n - 1]\text{Entier}\{\log[\mathcal{T}^{IV}(\Omega_\varphi, \Omega\tau_\theta) + \alpha_\nu]/(\mathcal{T}^{IV}_{min} + \alpha_\nu)] : \ln[(1+\alpha_\nu)/(\mathcal{T}^{IV}_{min} + \alpha_\nu)]\} \quad (A.21)$$

where $\mathcal{T}^{IV}_{min}$ is the minimum value of the transmission function $\mathcal{T}^{IV}$ with respect to Ω_φ with fixed $\Omega\tau_\theta$ values; $0 < \alpha_\nu << 1$ is an additional substitute which eliminates the zeros of $\mathcal{T}^{IV}$ (see Section 9.4.3).The $DE_{\mathcal{T}^{VI}}$ value, defined similarly to $\text{DE}_{\dot{\mathcal{T}}}$ (A.9) with a substitution of $\dot{\mathcal{T}}$ for $\mathcal{T}^{IV}$ is lower than $\text{DE}_{\mathcal{T}^{II}}$ value which is defined by the periodicity of the mask $\mathcal{T}^{IV}$.

A.6 Numerical Evaluation of the Mask

In solving the practical problems, the range of $\Omega\tau_\theta$ values may be wide enough when real problems are being solved, for example, in sonar of $\Omega\tau_\theta = 30$ to 400. Therefore, taking into consideration the two-dimensional nature of the mask $\mathcal{T}(\Omega_\varphi, \Omega\tau_\theta)$ whose reproduction sample was evaluated above, and also the calculation of $\dot{\mathcal{T}}$ from the integral equation (A.l), some difficulties, namely large information volume and high computer time consumption arise. Therefore, considering the form of the mask transmission function (A.1) as a Fourier transform of the amplitude-phase distribution of $\mathcal{T}\exp(-i\Omega\tau_\theta\cos\varphi)$ type, we used the widely known method [18, 19, 50] of converging (A.1) with the help of quadrature formulas with a known error to a discrete Fourier transform and by calculating the latter using a fast Fourier transform [58].

Conclusions in Section 9.4.3 contain data which are used during circular electrooptical array antenna (A.3) mask synthesis with $F_{max}/F_{min} \approx 2.3$ ($KR_o = 30...70$) by methods of digital holography, conducted as described in this appendix. The calculations were performed on the computer type "M-4030" (operating speed 1.2×10^4 operations/s, memory size of 236 kilobytes) with subsequent data base loading (41 lines along $\Omega\tau_\theta$ with reading step size of $\Omega_\varphi = \pm 5 \times 10^2$ in each line) onto magnetic tape and reproduction by a precision photosynthesizer [140, 141]. The calculations were performed using a fast Fourier transform as follows: expressions (A.3)—the rectangular interpolation formula—fast Fourier transform (the average number of sampling points on the interval $[-\pi/2, \pi/2]$ is approximately 10^3), as well as, taking into account the special appearance of the mask (A.3), by a combined method. In the latter case, function decomposition into Anger and Weber series [144, 145] was used for the values $|\Omega\tau_\theta| \gg |\Omega_\varphi|(|\Omega\tau_\theta| \gg 1)$ and $|\Omega\tau_\theta| \ll |\Omega_\varphi|$. The region joining was done with the help of Simpson's quadrature formula [the number of points on the interval $[-\pi/2, \pi/2] \approx 3(\Omega\tau_\theta + \Omega_\varphi)$].

The accuracy of calculations by both methods was maintained at approximately 0.2%. Mask (A.21), recorded in accordance with the data in Section 9.4.3, is presented in Fig. 9.30.

Appendix B

Evaluation of the Diffraction Efficiency of Coherent Optical Processors

The coherent optical processor diffraction efficiency was previously defined as a ratio of the light power in the radio spectrum to the light power in the optical spectrum at the processor output. Assuming that the overall efficiency is equal to one (that is neglecting light losses in the coherent optical processor, lenses, masks, etc.) we have:

$$
\begin{aligned}
\mathrm{DE}_{COP} &= \iint_{4\pi} |\tilde{\hat{\mathcal{L}}}_{AA}\{\langle \dot{\mathcal{E}}_{\Omega}(\mathbf{R}_n)]\}|^2 \mathrm{d}^2\mathbf{K}' \\
&\quad \Big/ \iint_{\infty} |\tilde{\hat{\mathcal{L}}}_{AA}\{\langle \dot{\mathcal{E}}_{\Omega}(\mathbf{R}_n)]\}|^2 \mathrm{d}^2\mathbf{K}' \\
&\approx \frac{\iint_{4\pi} |\langle \dot{E}_{\Omega}(\mathbf{R}_n][\tilde{\mathcal{J}}(\mathbf{R}_n, \mathbf{K}')\rangle|^2 \mathrm{d}^2\mathbf{K}' / ||\hat{\mathcal{L}}_{AA}||}{\iint_{\Sigma} |\langle \dot{\mathcal{E}}_{\Omega}(\mathbf{R}_n)][\mathcal{J}_{\delta}(\mathbf{R} - \mathbf{R}_n)\rangle|^2 \mathrm{d}^2\mathbf{R}}
\end{aligned}
\tag{B.1}
$$

where expressions (4.5) and (5.8) were used along with the conservation of energy. According to this law, the light power flux at the processor output is the same at the processor input, provided that the efficiency is equal to 1; $||\hat{\mathcal{L}}_{AA}||$ is the normalized operator (5.2) (in this case $||\hat{\mathcal{L}}_{AA}|| \approx \Sigma/N$ is the average array antenna mesh area). According to (5.9)

$$
\tilde{\mathcal{J}}(\mathbf{R}_n, \mathbf{K}') \approx \dot{\mathcal{J}}(\mathbf{R}_n, \mathbf{K}') \iint_{\Sigma} \mathcal{J}_{\delta}(\mathbf{R} - \mathbf{R}_n) \mathrm{d}^2\mathbf{R}
$$

because the space–time light modulator channel pupil function is considerably more narrow than the amplitude phase distribution, (B.1) can be changed to:

$$\begin{aligned}
\mathrm{DE}_{COP} &= \frac{||\hat{L}_{AA}|| \int\int_{2\pi} \hat{L}_{AA}\{\langle \dot{\mathcal{E}}_{\Omega}(\mathbf{R}_n)] \int\int_{\Sigma} J_{\delta}(\mathbf{R}-\mathbf{R}_n)\mathrm{d}^2\mathbf{R}\}|^2 \mathrm{d}^2\mathbf{K}'}{||\hat{L}_{AA}||^2 \langle |\dot{\mathcal{E}}_{\Omega}(\mathbf{R}_n)|^2][\int\int_{\Sigma}[|J_{\delta}(\mathbf{R}-\mathbf{R}_n)|^2 \mathrm{d}^2\mathbf{R}\rangle} \\
&\approx \frac{\left|\int\int_{\Sigma} J_{\delta}(\mathbf{R})\mathrm{d}^2\mathbf{R}\right|^2 \int\int_{4\pi} |\hat{L}\{\dot{\mathcal{E}}_{\Omega}(\mathbf{R}\}|\mathrm{d}^2\mathbf{K}'}{(\Sigma/N)\int\int_{\Sigma}|J_{\delta}(\mathbf{R})|^2\mathrm{d}^2\mathbf{R}\int\int_{\Sigma}|\dot{\mathcal{E}}_{\Omega}(\mathbf{R})|^2\mathrm{d}^2\mathbf{R}} = K^{\delta}_{su}/\gamma_{AA} \quad \text{(B.2)}
\end{aligned}$$

where

$$K^{\delta}_{su} = \left|\int\int_{\Sigma} J_{\delta}(\mathbf{R})\mathrm{d}^2\mathbf{R}\right|^2 \Big/ (\Sigma/N)\int\int_{\Sigma}|J_{\delta}(\mathbf{R})|^2\mathrm{d}^2\mathbf{R} \quad \text{(B.3)}$$

is the coefficient of the space–time light modulator surface usage

$$\gamma_{AA} = \frac{\int\int_{\Sigma}|\dot{\mathcal{E}}_{\Omega}(\mathbf{R})|^2\mathrm{d}^2\mathbf{R}}{\int\int_{4\pi}|\hat{L}_{AA}\{\dot{\mathcal{E}}_{\Omega}(\mathbf{R})\}|^2\mathrm{d}\mathbf{K}'} \quad \text{(B.4)}$$

is the coefficient of the array antenna reactivity [52].

The expression for $(\mathrm{DE})_{COP}$ may be verified by the simplest example of a planar array antenna coherent optical processor. According to the definition and (5.30), the diffraction efficiency is the ratio of the light energy in the radio spectrum ($|\mathbf{K}_{\perp}| \leq K$) to the energy in the optical spectrum ($|\mathbf{k}_{\perp}| \leq |m\mathbf{K}_{\perp}| \leq k$), which in this case is:

$$\begin{aligned}
\mathrm{DE}_{COP} &= \frac{\int\int_{|\mathbf{K}_{\perp}|\leq K} F^2_{\delta}(\mathbf{K}'_{\perp})F^2_{AA}(\mathbf{K}_{\perp}-\mathbf{K}'_{\perp})\mathrm{d}^2\mathbf{K}'_{\perp}}{\int\int_{|\hat{m}\mathbf{K}_{\perp}|\leq k} F^2_{\delta}(K'_{\perp})F^2_{AA}(\mathbf{K}_{\perp}-\mathbf{K}'_{\perp})\mathrm{d}^2\mathbf{K}'_{\perp}} \\
&\geq F^2_{\delta min}\frac{\mathrm{circ}(\mathbf{K}_{\perp}/K)\otimes\otimes F^2_{AA}(\mathbf{K}_{\perp})}{F^2_{\delta}(\mathbf{K}_{\perp})\otimes\otimes F^2_{AA}(\mathbf{K}_{\perp})} \quad \text{(B.5)}
\end{aligned}$$

where $F_{\delta\mathrm{min}}$ is the minimum envelope of (5.31) in the radio band; $\mathrm{circ}(\mathbf{K}_{\perp}/K) = 1$ with $|\mathbf{K}_{\perp}| \leq K$ and 0 with $|\mathbf{K}_{\perp}| > K$.

It is convenient to evaluate (B.5) for an equidistant electrooptical array antenna (see Fig. 1.6) whose radiation pattern (5.1) may, according to (1.11) and (1.5), be written as

$$
\begin{aligned}
F_{AA}(\mathbf{K}_\perp) &= \sum_{n=1}^{N} J(\mathbf{R}_{\perp n}) \exp(-i\,\mathbf{K}_\perp \mathbf{R}_{\perp n}) \\
&= \hat{\mathcal{F}}\{J(\mathbf{R}_\perp)\mathrm{comb}(\mathbf{R}_\perp/\mathrm{d}_X\mathrm{d}_Y)\} \\
&= \frac{1}{(2\pi)^2}\hat{\mathcal{F}}\{J(\mathbf{R}_\perp)\} \otimes \otimes \hat{\mathcal{F}}\{\mathrm{comb}(\mathbf{R}_\perp/\mathrm{d}_X\mathrm{d}_Y)\} \\
&= \frac{\mathrm{d}_X\mathrm{d}_Y}{(2\pi)^2} F(\mathbf{K}_\perp) \otimes \otimes \mathrm{comb}(\mathbf{K}_\perp/\mathrm{d}_{\Omega X}\mathrm{d}_{\Omega Y})
\end{aligned}
\tag{B.6}
$$

where

$$
\begin{aligned}
\mathrm{comb}(\mathbf{R}_\perp \mathrm{d}_X\mathrm{d}_Y) &= \mathrm{comb}(X/d_X)\mathrm{comb}(Y/d_Y) \\
&= \frac{1}{\mathrm{d}_X\mathrm{d}_Y} \sum_{m=-\infty}^{\infty} \sum_{n=-\infty}^{\infty} \delta(X - md_X)\delta(Y - nd_Y)
\end{aligned}
$$

is the impulse train or Dirac "comb"; $d_{\Omega X} = 2\pi/d_X$, $d_{\Omega Y} = 2\pi/d_Y$ (see Fig. 5.3). Substituting (B.5) into (B.6) and considering that $[F(\mathbf{K}_\perp) \otimes \otimes(\mathbf{K}_\perp/d_{\Omega X})]^2 = F^2(\mathbf{K}_\perp) \otimes \otimes \mathrm{comb}(\mathbf{K}_\perp/d_{\Omega X}/d_{\Omega Y})$ we get

$$
\begin{aligned}
\mathrm{DE}_{COP} &\geq F^2_{\delta min} \frac{\mathrm{circ}(\mathbf{K}_\perp/K) \otimes \otimes F^2(\mathbf{K}_\perp) \otimes \otimes \mathrm{comb}(\mathbf{K}_\perp/d_{\Omega X}d_{\Omega_Y})}{F^2_\delta(\mathbf{K}_\perp) \otimes \otimes F^2(\mathbf{K}_\perp) \otimes \otimes \mathrm{comb}(\mathbf{K}_\perp/d_{\Omega X}d_{\Omega Y})} \\
&\geq \frac{F^2_{\delta min}}{\gamma_{AA}} \frac{\mathrm{circ}(\mathbf{K}_\perp/K) \otimes \otimes \mathrm{comb}(\mathbf{K}_\perp/d_{\Omega X}d_{\Omega_Y})}{F^2_\delta(\mathbf{K}_\perp) \otimes \otimes \mathrm{comb}(\mathbf{K}_\perp/d_{\Omega X}d_{\Omega Y})} \\
&= \frac{F^2_{\delta min}}{\gamma_{AA}} \frac{\sum_m \sum_n \mathrm{circ}\left[(\mathbf{K}_\perp - \mathbf{n}_X\, md\,\Omega_X - \mathbf{n}_Y - nd_{\Omega Y})/K\right]}{(2\pi)^2 d_X d_Y \hat{\mathcal{F}}\{\hat{\mathcal{F}}^{-1}\{F^2_\delta(\mathbf{K}_\perp)\}\mathrm{comb}(\mathbf{R}/d_X d_Y)\}}
\end{aligned}
\tag{B.7}
$$

where (B.4) and $d_X d_Y \mathrm{comb}(\mathbf{K}_\perp/d_{\Omega X \Omega Y}) = \hat{\mathcal{F}}\{\mathrm{comb}(\mathbf{R}_\perp/d_X d_Y)\}$ were used. Since $d_X = d_Y = \Lambda/2$ we have $d_{\Omega X} = d_{\Omega Y} = 2K$, the sum in the numerator of (B.7) is normalized in the radio band because the biased circ functions do not intersect each other. Therefore

$$
\mathrm{DE}_{COP} \geq \left|\frac{F_{\delta min}}{F_\delta(0)}\right|^2 \frac{F_\delta(0)}{\gamma_{AA} d_X d_Y \hat{\mathcal{F}}\{[J_\delta(\mathbf{R}_\perp) \otimes \otimes J_\delta(\mathbf{R})]\mathrm{comb}(\mathbf{R}_\perp/d_X d_Y)\}}
$$

$$= \left|\frac{F_{\delta min}}{F_\delta(0)}\right|^2 \frac{\left|\int\int_{-\infty}^{\infty} J_\delta(\mathbf{R}_\perp)\mathrm{d}^2\mathbf{R}_\perp\right|^2}{\gamma_{AA} d_X d_Y \int\int_{-\infty}^{\infty} |J(\mathbf{R}_\perp)|^2 \mathrm{d}^2\mathbf{R}_\perp}$$

$$= \left|\frac{F_{\delta min}}{F_\delta(0)}\right|^2 K_{su}^\delta/\gamma_{AA} \qquad \text{(B.8)}$$

where (5.30) and (B.3) were taken into consideration.

Appendix C

Notes on Calculating Extended Spheroidal Functions and Related Functionals

For obtaining numerical results, scientific papers [174, 175], show that the fundamental function $\Psi_\circ^X(\Psi_\circ^Y)$ and the corresponding eigenvalue $\lambda_X(\lambda_Y)$ are expressed by semi-integral index Bessel functions of the first kind.

Keeping in mind that in this case the parameter $c_{X,Y}$ is small ($c_{X,Y} < 1$), the second approximation for the extended spheroidal function [175] was used for calculations:

$$\Psi_\circ^X(\Omega_X, c_X) = \left\{A + \frac{B}{2}\left[1 - \left(\frac{\Omega_X \Delta X}{c_X}\right)^2\right]\right\}$$

$$\times \left[\mathrm{Si}\left(\frac{\pi}{2}c_X + \Omega_X\frac{\Delta X}{2}\right) + \mathrm{Si}\left(\frac{\pi}{2}c_X - \Omega_X\frac{\Delta X}{2}\right)\right]$$

$$-\frac{B}{2\left(\pi c_X/2\right)^2}\left\{\left[2\sin\left(\frac{\pi}{2}c_X\right) - \pi c_X\cos\left(\frac{\pi}{2}c_X\right)\right]\right.$$

$$\left.\times \cos(\Omega_X\Delta X/2) - \Omega_X\Delta X\sin\left(\frac{\pi}{2}c_X\right)\sin(\Omega_X\Delta X/2)\right\} \qquad \text{(C.1)}$$

where

$$A = a_\circ(1-\Gamma)\sqrt{2\pi}$$

$$B = 3a_\circ\Gamma\sqrt{2/\pi}$$

$$\Gamma = 5\mu_1 P_{0,2}/(1 - 5\mu_1 P_{2,2})$$

$$\mu_1 = \frac{\left(P_{2,2} + \frac{1}{5}P_{0,0}\right) - \sqrt{\left(P_{2,2} - \frac{1}{5}P_{0,0}\right)^2 + \frac{4}{5}P_{0,5}^2}}{2\left(P_{0,0}\, P_{2,2} - P_{0,2}^2\right)}$$

$$P_{0,0} = \frac{2}{\pi}\mathrm{Si}(\pi c_X) - J_{1/2}^2\left(\frac{\pi}{2}c_X\right)$$

$$P_{0,2} = \frac{\pi c_X}{12}[J_{3/2}(\pi c_X/2) J_{1/2}(\pi c_X/2)$$

$$- \quad J_{5/2}(\pi c_X/2) J_{-1/2}(\pi c_X/2)] - \frac{1}{3} J_{5/2}(\pi c_X/2) J_{1/2}(\pi c_X/2)$$

$$P_{2,2} = \frac{2}{5}\left[\frac{1}{\pi}\mathrm{Si}(\pi c_X) - J_{1/2}(\pi c_X/2) - J_{3/2}^2(\pi c_X/2) - \frac{1}{2} J_{5/2}^2(\pi c_X/2)^2\right]$$

$a_\circ$ is the normalizing constant. It depends on C_x and is chosen by the condition (6.64); Si is the sine integral. Similar relationships exist for $\Psi_\circ^Y$ as well. The eigenvalue which corresponds to the function $\Psi_\circ^{X,Y}(\Omega_{X,Y}, c_{X,Y})$ is defined as

$$\lambda_{X,Y}(c_{X,Y}) = 1/\mu_1(c_{X,Y}) \tag{C.2}$$

Let us note that the third approximation which is not shown here, differs from the second by less than 1% (with $c_{X,Y} = 1.5$, the difference reaches 4%).

Since in case of a rectangular (linear) electrooptical array antenna expression (6.79) holds true, the amplitude phase distribution is defined by expression (6.80). From this expression it can be seen that its form coincides with that of the radiation pattern (C.1) on the interval $|X| \leq \Delta(X/2)$, $|Y| \leq \Delta(Y/2)$. But on this interval considering that $c_{X,Y} < 1$, it is advisable to use a relationship, simpler than (C.1), namely:

$$J_\Psi(X) = \hat{\mathcal{F}}^{-1}\{\Psi_\circ^X(\Omega_X e_X)\} = \frac{a_\circ}{\sqrt{\pi X c_X/\Delta X}}$$

$$\times \quad J_{1/2}(\pi X c_X/\Delta X) + \Gamma J_{5/2}(\pi X c_X/\Delta X)]\mathrm{rect}(X/\Delta X) \tag{C.3}$$

To evaluate the degree of interference (6.66) suppression it is necessary to evaluate the $b_\circ$ function which is defined by the expression (6.56):

$$b_\circ(\Omega_X - \Omega_{X_q}) = \int_{-\infty}^{\infty} \Psi_\circ(\Omega_X - \Omega_{X_S})\Psi_\circ(\Omega_{X_S} - \Omega_{X_q})d\Omega_{X_S} \qquad \text{(C.4)}$$

Substituting (C.1) into expression (C.4) using the convolution theorem and the fact that $\Psi_\circ$ function is finite, we get

$$b_\circ(\delta\Omega_X \varepsilon_X) = b(\varepsilon_X)/b(0) \qquad \text{(C.5)}$$

where

$$\begin{aligned}
b(\varepsilon_X) &= [1 + (\pi c_X/2)^2 d_1 + (\pi c_X/2)^4 d_2] \sin(c_x \varepsilon_X)/c_X \varepsilon_X \\
&+ \left[2d_1 + 4(\pi c_X/2)^2 d_2\right] \cos(\pi c_X \varepsilon_X/2)/\varepsilon_X^2 \\
&- 24 d_2 \cos(\pi c_X \varepsilon_X/2)/\varepsilon_X^4 \\
&- \left[2d_1 + 12(\pi c_X/2)^2 d_2\right] \sin(\pi c_X \varepsilon_X/2)/\frac{\pi}{2} c_X \varepsilon_X^3 \\
&- 24 d_2 \sin\left(\frac{\pi}{2} c_X \varepsilon_X\right)/\frac{\pi}{2} c_X \varepsilon_X^5
\end{aligned}$$

where

$$\begin{aligned}
d_1 &= -\frac{1}{3} + \frac{2}{15}\Gamma \\
d_2 &= \frac{2}{45} + \frac{2}{65}\Gamma + \frac{1}{225}\Gamma^2 \\
\delta\Omega_X &= 2\pi/\Delta X \\
\varepsilon_X &= (\Omega_X - \Omega_{X_q})/\delta\Omega_X \\
b(0) &= 1 + d_1(\pi c_X/2)^2/3 + d_2(\pi c_X/4)^4
\end{aligned}$$

Appendix D

Electrooptical Arrays Focused in the Fresnel Region

If the objects are located in the near (Fresnel) field of the array antenna, an additional degree of information freedom, enclosed in the phase fronts of waves falling on the array antenna detector arises. In this case additional opportunity to extract information about the distance as well as to a certain extent about the shape of the radiation (scattering) objects [26, 195–198].

The problem of regenerating the information about the objects located in the array antenna's Fresnel region can be formulated in a way similar to that of the far field (see Chapters 2–4). It is required, using multielement array antennas with known parameters, to form a light distribution at the coherent optical processor output. This distribution's amplitude should be proportional to the emitting objects radiobrightness distribution. The distribution should also produce a geometric likeness of the emitting objects. The initial information for the above requirements is the results of the wave front reception. Let us not the peculiarities of the problem in question.

Let us recognize the range of "elementary-planar waves" [197]

$$\mathrm{d}^2/\Lambda < |\mathbf{R}| < D^2/\Lambda \tag{D.1}$$

as the Fresnel region, where d is the array antenna element "diameter"; λ is the wave length of the radiation being received; R is the vector radius of the radiation abject points in the array antenna coordinate system; D is the antenna array "diameter."

In general, from an electrodynamic point of view, it is expedient to characterize the emitting object by the electric $\jmath^e(\mathbf{R})$ and magnetic $\jmath^m(\mathbf{R})$ current distributions because, being enclosed in the object they unambiguously characterize its coordinates and, to a certain degree, its geometric shape. Furthermore, without significant loss of generality, let us limit the review of emitting objects to the electric currents $j_e(\mathbf{R})$ which flow in its ideally conducting surface.

According to [52], currents $j_e(\mathbf{R})$ induce voltages at the nth array antenna element output (Fig. 1.2) at point $\mathbf{R}_n$ with normalized vectoral three-dimensional element $\mathbf{n}(\text{grad}|\mathbf{R}-\mathbf{R}_n|)$ directivity characteristic, with gain, G_n, input impedance z_{nn}, and load impedance z_n. These voltages are

$$\begin{aligned} \dot{U}_n &= i\Lambda z_n(z_{nn}+z_n)^{-1}\sqrt{G_n \Re e z_{nn}/2\pi W}\iiint_{V_R} d\mathbf{E}(\mathbf{R},\mathbf{R}_n)\mathbf{f}_n(\text{grad}|\mathbf{R}-\mathbf{R}_n|) \\ &= \iiint_{V_R} \mathbf{j}^e(\mathbf{R})\mathbf{F}_n(\mathbf{R})d^3\mathbf{R} \qquad \text{(D.2)} \end{aligned}$$

Here $W = \sqrt{\mu_a/\varepsilon_a}$, $d\mathbf{E} = \text{grad divd}\mathbf{A}^e/i\omega\varepsilon_a - i\omega\mu_a d\mathbf{A}^e$, $d\mathbf{A}^e(\mathbf{R}_n,\mathbf{R}) = [\mathbf{j}^e(\mathbf{R})/4\pi]\dot{G}(|\mathbf{R}-\mathbf{R}_n|)d^3\mathbf{R}$, $d^3\mathbf{R} = dV$ is the volume element; $\dot{G}(|\mathbf{R}-\mathbf{R}_n|) = \exp(-iK|\mathbf{R}-\mathbf{R}_n|)/|\mathbf{R}-\mathbf{R}_n|$ is the free space Green's function,

$$\begin{aligned} \mathbf{F}_n(\mathbf{R}) &= z_n(z_{nn}+z_n)^{-1}\sqrt{2\pi G_n W \Re e z_{nn}}\mathbf{f}_n(\text{grad}|\mathbf{R}-\mathbf{R}_n|)\dot{G}(|\mathbf{R}-\mathbf{R}_n|) \\ &= F_n f_n(\text{grad}|\mathbf{R}-\mathbf{R}_n|)\dot{G}(|\mathbf{R}-\mathbf{R}_n|) \qquad \text{(D.3)} \end{aligned}$$

is the "Green's function for an array antenna element" [198].

Equation (D.2) mathematically formulates this problem. It is required to regenerate an unknown current distribution $\mathbf{j}^e(\mathbf{R})$ using measured voltages $\langle\dot{u}] = [\dot{U}_1, ..., \dot{U}_N]$ and the known Green's function (D.3).

D.1 The Reproduction Algorithm

Let us assume, as in chapter 4, that the array antenna elements are housed densely. Then (D.2) may be considered as an integral equation with respect to $\mathbf{j}^e(\mathbf{R})$ with $\mathbf{F}_n(\mathbf{R}_n)$ as its kernel. In many technical applications the solution to this equation is sought similarly to (4.3) as an integral of the measured voltage and the optimal element excitation amplitude phase distribution superposition, thus providing the array antenna focus on the object points with respect to the given criteria. Such type of a solution for the problem in consideration is expressed as [199]

$$\tilde{\mathbf{j}}^e(\mathbf{P}) = \sum_{n=1}^{N} \dot{U}_n \dot{J}_n(\mathbf{P}) = \langle \dot{\mathbf{u}}][\dot{J}(\mathbf{P})\rangle \tag{D.4}$$

where $[...\rangle = \langle ...]_T$ is a column matrix; T is the transpose symbol; $\dot{J}(\mathbf{P})$ is the optimum amplitude phase distribution (see below); **P** are the coordinates of the regenerated representation points.

Actually, taking into account (D.2), relationship (D.1) is transformed into

$$\tilde{\mathbf{j}}^e(\mathbf{P}) = \iiint_{V_R} \mathbf{j}^e(\mathbf{R})\langle \mathbf{F}(\mathbf{R})][\dot{J}(\mathbf{P})\rangle d^3\mathbf{R} = \iiint_{V_R} \mathbf{j}^e(\mathbf{R})\mathbf{q}(\mathbf{P},\mathbf{R})d^3\mathbf{R}$$

where $\mathbf{q}(\mathbf{P},\mathbf{R}) = \langle \mathbf{F}(\mathbf{R})][\dot{\mathcal{E}}(\mathbf{P})\rangle$ is the electrooptical array antenna imaging kernel, which, in the far array antenna region, concurs with its directivity diagram, and in this case, describes the field formed by the array antenna in the proximity of its focal point **P**.

The solution (D.4) is structurally similar to the known solutions for the far electrooptical array antenna region (2.7) or (4.3) and describes the weighted summation process of the reception results (D.2) with weighting factor $= [\dot{J}\rangle$. The amplitude phase distribution appearance is to a large extent defined by the choice of process optimization criteria for the array antenna focusing on its near field. As a rule [see (6.21)], the amplitude phase distribution is chosen to be equal to the complex conjugates of the Green's functions (D.3) $\dot{J}_n(\mathbf{R}) = \overset{*}{F}_n(P)$, ($*$ is the complex conjugation sign). A more substantiated criterion is suggested in [197]. There, such an amplitude phase distribution with which in the vicinity of the given point, P, in region (D.1), the square of the absolute value of one of the cartesian components of vector (D.4) reaches the maximum with the array antenna emitting constant power is taken to be the optimum one. Such a characteristic is named as the concentration parameter and coincides with the concept of partial array antenna directivity efficiency in the case of focusing on the far field. In accordance with [197, 199]

$$[\dot{J}(\mathbf{P})\rangle = [r']^{-1}[\overset{*}{F}_{X,Y,Z}(\mathbf{P})\rangle \tag{D.5}$$

where $[r']^{-1}$ is the inverse square matrix of the normalized array antenna elements reciprocal impedances real components; $\overset{*}{F}_{X,Y,Z}$ is one of the cartesian components of the complex conjugated Green's function (D.3). The cartesian component selection in relation (D.5) agrees with the specific current polarization in the emitting object. With noninteracting array antenna elements $[r'] = [E]$ is a unit matrix. In case of interacting elements the matrix $[r']$ diagonalization is achieved through the coordination of the array antenna elements with free spaces. Let us further suppose that the latter was carried out. Let us present algorithm (D.4) in integral form:

$$\hat{L}\{...\} = \langle\ldots][\dot{J}(\mathbf{P})\rangle = \iiint_{AA}\langle...]\dot{J}(\mathbf{P},\mathbf{R})[\delta(\mathbf{R}-\mathbf{R}_n)\rangle d^3\mathbf{R}$$

$$= \iiint_{AA}[\sum_{n=1}^{N} ...\delta(\mathbf{R}-\mathbf{R}_n)]\dot{J}(\mathbf{P},\mathbf{R})d^3\mathbf{R} \qquad \text{(D.6)}$$

where $\dot{J}(\mathbf{P},\mathbf{R}) = \dot{J}_n(\mathbf{P})$ with $\mathbf{R} = \mathbf{R}_n$, $\delta(\mathbf{R})$ is a δ-function. The variable $\mathbf{P} \in V_R$ is supposed to be current, i.e., the object regeneration has to be carried out for all of its points simultaneously. Let us study the possibilities of realizing algorithm (D.6) by Fourier optics for electrooptical array antennas of various geometries.

D.2 Planar Electrooptical Arrays

Let the array antenna consist of identical receiving elements, located in the XOY plane (Fig. 1.6). Then algorithm (D.6), keeping in mind the remarks made above, becomes:

$$\hat{L}\{\ldots\} = \iint\{\ldots\} \overset{*}{F}_X (\mathbf{P}-\mathbf{R}_\perp)d^2\mathbf{R}_\perp \qquad \text{(D.7)}$$

Here $\overset{*}{F}_X (\mathbf{P}-\mathbf{R}_\perp) = \overset{*}{F}_{X_n} (\mathbf{P})$ with $\mathbf{R}_\perp = \mathbf{R}_n$, $\mathbf{R}_\perp = \mathbf{x}_\circ X + \mathbf{y}_\circ Y$, $\mathbf{P} = \mathbf{P}_\perp + \mathbf{z}_\circ Z$, where $\mathbf{P}_\perp$ is a vector coplanar with XOY.

Weighting of the voltages being received by the δ-function is, in practice, realized by using sufficiently narrow pupils of the space and time light modulator channels, which carries out the signal feeding into the coherent optical processor (Section 5.2.2).

Considering the distance Z as a parameter, let us present algorithm (D.7) as:

$$\hat{L}\{\ldots\} = \iint_{AA}\{\ldots\} \overset{*}{F}_X (\mathbf{P}-\mathbf{R}_\perp, Z)d^2\mathbf{R}_\perp = \{\ldots\} \otimes\otimes \overset{*}{F}_X (\mathbf{P}_\perp, Z)$$

A similar operation can be realized by coherent optical methods because it can be written as

$$\hat{L}\{\ldots\} = \{\ldots\} \otimes\otimes \overset{*}{F_X} (\mathbf{P}_\perp, Z)\hat{\mathcal{F}}\{T(\mathbf{k}_\perp, Z)\hat{\mathcal{F}}^{-1}\{\ldots\}\} \tag{D.8}$$

Here $\dot{T}(\mathbf{k}_\perp, Z) = \hat{\mathcal{F}}^{-1}\{\overset{*}{F_X} (\mathbf{P}_\perp, Z)\}$, $\mathbf{R}_\perp = \hat{m}\mathbf{r}_\perp$ where $\mathbf{K}_\perp = \hat{m}^{-1}\mathbf{k}_\perp$ is the law of the array antenna elements addressing into the space and time light modulator channels [see (1.1), (1.2) and (2.2)].

Algorithm (D.8) is realized with a coherent optical processor similar to the one shown in Fig. 1.4, where instead of the multichannel acoustooptical light modulator a two-dimensional space and time light modulator (Fig. 1.6)is used, and where a fixed mask $\dot{T}(\mathbf{k}_\perp, Z)$, recorded in scale (2.13) was installed in the plane of the mask $\dot{T}$.

The light intensity distribution, corresponding to the structure of the emitting objects located in the $Z =$ const section, is reproduced at the processor output in the X, Y coordinate system. The resolution is defined by the electrooptical array antenna imaging kernel width—$\mathbf{q}(\mathbf{P}, \mathbf{R})$. A successive observation along the Z coordinate is possible by changing masks $\dot{T}(\mathbf{k}_\perp, z)$, coordinated with the current distance Z, where the $Z \longrightarrow \infty$ masks are equivalent to a spherical lens. The mask is the key element in the processor. Taking (D.3) into account, its transmission function can be presented as

$$\begin{aligned}\dot{T}(\mathbf{k}_\perp, Z) &= F_\circ\hat{\mathcal{F}}^{-1}\{\overset{*}{f_X}\} \otimes\otimes \hat{\mathcal{F}}^{-1}\{\overset{*}{G} (|\mathbf{P}_\perp + \mathbf{z}_\circ Z|)\} \\ &= F_\circ J_\circ(\hat{m}^{-1}\mathbf{k}_\perp) \otimes\otimes \hat{\mathcal{F}}^{-1}\{\overset{*}{G_z} (|\mathbf{P}_\perp|)\}\end{aligned} \tag{D.9}$$

where $F_\circ \equiv F_n$, $J = \hat{F}^{-1}\{\overset{*}{f_X}\}$ is the array antenna element amplitude phase distribution, and

$$\overset{*}{G_Z} (|\mathbf{P}_\perp|) = \exp\left(iK\sqrt{|\mathbf{P}_\perp|^2 + Z^2}\right) / \sqrt{|\mathbf{P}_\perp|^2 + Z^2}$$

is the Green's function for coherent sources in free space.

In the case of isotropic array antenna elements ($f_X \equiv$ const) $J_\circ \longrightarrow \delta$ and the mask (D.9) becomes

$$\dot{T}(\mathbf{k}_\perp, Z) = F_\circ\hat{\mathcal{F}}^{-1}\{\overset{*}{G_z} (|\mathbf{P}_\perp|)\}$$

D.3 Linear Electrooptical Arrays

Let the array antenna elements coincide with the Z-axis (Fig. 1.3). Then algorithm (D.6), keeping in mind the remarks made above, becomes

$$\hat{L}\{\ldots\} = \iint_{AA} \{\ldots\} \overset{*}{F}_X (\mathbf{P} - \mathbf{z}_o Z) \mathrm{d}Z \tag{D.10}$$

With isotropic elements in the XOY plane (obviously, this limitation is not a principal one), function (D.3) depends only on the Z coordinates and $\rho = \sqrt{X^2 + Y^2}$. Algorithm (D.10) is transformed into $\hat{L}\{...\} = \{\ldots\} \otimes \overset{*}{F}_X (Z, \rho)$ and is realized using a cylindrical electrooptical array antenna processor shown in Fig. 4.4b. The light intensity distribution in the processor output plane corresponds to the emitting object's Z coordinate and to the distance ρ. Mask $\dot{T}$ depends or ρ and the three-dimensional frequency ω_z. Its transmission function is

$$\dot{T}(\omega_z, \rho) = \hat{\mathcal{F}}^{-1}\{\overset{*}{F}_X (Z, \rho)\} = F_o \hat{\mathcal{F}}^{-1} \left\{\exp\left(iK\sqrt{\rho^2 + Z^2}\right) \Big/ \sqrt{\rho^2 + Z^2}\right\} \tag{D.11}$$

D.4 Linear Electrooptical Arrays with an Optical Spectroanalyzer

Using the space and time light modulator to conduct the space sweep of the time signals (see Chapter 3), it appears possible to make spectral analyses of wide band—in the space and time sense—signals along with isolating the two-dimensional (ρ, Z) coordinate information. Therewith, a well-known method of representing three-dimensional information as "fine" and a "rough" structures of the light field diffraction picture at the processor output (Section 3.3) can be used to condense the three-dimensional information onto a plane. This electrooptical array antenna's coherent optical processor can be realized according to the scheme of a circular electrooptical antenna array's (shown in Fig. 9.21) coherent optical processor. A complex structure mask which consists of a series of (D.11)-type masks (along the ω_z coordinate) is located in the spectrum plane π of the processor. The masks were recorded for discrete frequencies $\Omega_1, \ldots \Omega_N$ of the set frequency range $\Delta\Omega = 2\pi\Delta F$. Here, the scale of (D.11)-type masks must be chosen to be small enough compared with the optical spectroanalyzer's resolution. In this case, the Z coordinate and the frequency Ω are reproduced in the processor output plane as the rough structure along with the distance ρ as the fine structure.

D.5 Cylindrical Electrooptical Arrays

A wide angle, panoramic array antenna focusing into the Fresnel region without lowering the concentration parameter is possible with the receiving elements positioned on the cylindrical surface [197].

Let the cylinder axis coincide with the Z-axis of the cylindrical coordinate system (Fig. 4.4a), then $\mathbf{R} = \vec{\rho}_\circ R_\circ + \vec{\varphi}_\circ \alpha' + \vec{z}_\circ Z$, where $R_\circ$ is the cylinder radius. Substituting in the focal points in form $R = \vec{\rho}_\circ \rho + \vec{\varphi}_\circ \alpha' + \vec{z}_\circ Z'$ (ρ, α', Z' are the point $\mathbf{P}$ cylindrical coordinates), let us present algorithm (D.6) in the following way:

$$\begin{aligned}\hat{L}\{\ldots\} &= \int_\circ^{2\pi}\int_{-\Delta Z/2}^{\Delta Z/2}[\ldots]\,\overset{*}{F_X}\,[\vec{\rho}_\circ(\rho - R_\circ) + \vec{\varphi}_\circ(\alpha' - \alpha) + \vec{Z}_\circ(Z' - Z)]\mathrm{d}\alpha\,\mathrm{d}Z \\ &= [\ldots]\otimes\otimes\,\overset{*}{F_X}\,(\alpha', Z', \rho - R_\circ) \qquad \text{(D.12)}\end{aligned}$$

where

$$\begin{aligned}&\overset{*}{F_X}\,(\alpha', Z', \rho - R_\circ) \\ &= \overset{*}{f}\,(\alpha', \arctan Z'/\rho)\exp\left(iK\sqrt{(\rho - R_\circ)^2 + Z'^2}\right)\Big/\sqrt{(\rho - R_\circ)^2 + Z'^2}\end{aligned}$$

(D.13)

Algorithm (D.12) can be realized with the processor shown in Fig. 1.4b. The light distribution corresponding to the emitting object structure is formed at the processor output. The distribution also contains the information about the object's azimuth, α, and the Z coordinate of the object's points with fixed distance $(\rho - R_\circ)$.

The focus adjustment with respect to the object distance is achieved by exchanging masks

$$\dot{T}(\mathbf{k}_\perp, \rho - R_\circ) = F_\circ \mathcal{F}^{-1}\{\overset{*}{F_X}\,(\alpha', Z', \rho - R_\circ)\} \qquad \text{(D.14)}$$

Let us note that the considered masks' realization may be accomplished by using digital holographic methods (see A. 1).

Bibliography

1. Born, M., and E. Wolf. *Osnovy Optiki* [Principles of Optics]. Translated from English. G.P. Motulovich, ed. Moscow: Nauka, 1970.

2. Papoulis, A. *Teoria sistem i preobrazovannii v optike* [System and Processing Theory in Optics]. Edited by V. I. Alekseev. Moscow: Mir 1971.

3. Goodman, Joseph. *Vvedenie v fur'e-optiku* [Introduction to Fourier Optics]. Translated from English by G. I. Kosourov. Moscow: Mir. 1970.

4. Kondratenkov, G. S. *Obrabotka informatsii kogerentnymi optichskimi sistemami*. Moscow. Sov. radio. 1972.

5. Soroko, L. M. *Osnovy golografii i kogerentnoi optiki*. Moscow: Nauka. 1971.

6. Collier, R., K Burkhardt, L. Lin. *Opticheskaia golografia* [Optical Holography]. Translated from English by Iu. I. Ostrovskii. Moscow: Mir. 1973.

7. Safronov, G. S., and A. P. Safronova. *Vvedenie v radiogolografiu*. Moscow: Sov. radio. 1973.

8. Preston, K. *Kogerentnye opticheskie vychislitel'nye mashiny* [Coherent Optical Computers]. Moscow: Mir 1974.

9. Litvinenko. O. H. *Osnovy radiooptiki*. Kiev: Tekhnika. 1974.

10. Zverev. V. A. *Radiooptika*. Moscow: Sov. radio. 1975.

11. Ablekov, V. K., P. I. Zubkov, and A. V. Frolov. *Opticheskaia i optoelekronnaia obrabotka informatsii*. Moscow: Mashinostroenie. 1976.

12. Akaev., A. A., and S. A. Maiorov. *Kogerentnye opticheskie vycheslitel'nye mashiny*. Leningrad: Mashinostroenie. 1977.

13. Prokhorov, V. G., ed. *Akusticheskaia golografia* [Acoustic Holography]. Leningrad: Sudostroenie. 1975.

14. Vasilenko, G. I. *Golograficheskoe raspoznavanie obrazov.* Moscow: Sov. radio. 1977.

15. Potapov, O. A. *Opticheskaia obrabotka geofizicheskoi i geoloicheskoi informatsii.* Moscow: Nedra. 1977.

16. Gurevich, S. B., V. B. Constaninov, V. K. Sokolov, D. F. Chernykh. *Peredacha i obrabotka informatsii golograficheskimi metodami.* S. B. Gurevich, ed. Moscow: Sov. radio. 1978. pp. 304.

17. Kulakov, S. V. *Akustoopticheskie ustroistva spektral'nogo i korreliatsionogo analiza signalov.* Leningrad: Nauka. 1978.

18. Bakhrakh, L. D., and A. P Kurochkin. *Golografia i mikrovolnovoi tekhnike.* Moscow: Sov. radio. 1978.

19. Vasilenko., G. I. *Teoria vosstanovlenia signalov.* Moscow: Sov. radio. 1979.

20. Zvereva, V. A., and N. S. Stepanova, eds. *Eksperimental'naia radiooptika.* Moscow: Nauka, 1979.

21. Yu, F. T. S. *Vvedenie v teoriu difraktsii, obrabotku informatsii i golografiu* [Introduction to the Theory of Diffraction, Information Processing, and Holography]. Moscow: Sov. radio. 1979.

22. Gurevich, S. B., ed. *Opticheskaia obrabotka informatsii.* [Optical Processing of Information. D. Casasent, ed.] Moscow: Mir 1980.

23. Naryshkin, A. K. *Tsifrovye i opticheskie metody obrabotki radiolokatsionnykh signalov.* Moscow: Vysshaia shkola. 1982.

24. Lukoshkin, A. P., S. S. Karinskii, A. A. Shatalov et al. *Obrabotka signalov v mnogokanal'nykh RLS.* A. P Lukoshkin, ed. Moscow: Radio i sviaz'. 1983.

25. Ustinov, N. D., N. Mateev, and V. V Protopopov. *Metody obrabotki optichcheskikh polei v lazernoi lokatsii.* Moscow: Nauka. 1983.

26. Kremer, I. Ia., A. I. Kremer, V. M. Petrov, et al. *Prostranstvenno-vremennaia obrabotka signalov.* I. Ia. Kremer, ed., Moscow: Radio i sviaz'. 1983.

27. Goodman, J., ed. *Primenenie golografii.* [Applications of Holography. L. D. Bakhrakh, ed.] Moscow: Mir. 1973.

28. Gurevich, S. B., ed. *Opticheskii metodi obrabotki informatsii: Sb. statei.* Leningrad: Nauka 1974.

29. Bakhrakh, L. D. and E. Fridman, eds. *Sovremennoe sostoianie i perspektivy razvitiia golografii: Sb. statei.* Leningrad: Nauka. 1974.

30. Gurevich, S. B., ed. *Golografia i obrabotka informatsii: Sb. statei.* Leningrad: Nauka, 1976.

31. Korbukov, G. E., S. B. Kukakov, eds. *Radio—i akusticheskaia golografiia: Sb. statei.* Leningrad: Nauka. 1976.

32. "Opticheskaia vycheslitel'naia tekhnika." *TIIER.* [*Proc. IEEE*]. 1977. 65 (1): 211.

33. Gurevich, S. B., ed. *Opticheskaia obrabotka informatsii: Sb. statei.* Leningrad: Nauka. 1979.

34. Pilipovich, V. A., ed. *Opticheskie metody obrabotki informatsii: Sb. statei.* Minsk: Nauka i tekhniki. 1978.

35. Korbukov, G. E., and S. V. Kukalov, eds. *Akustoopticheskie metody informatsii: Sb. statei.* Leningrad: Nauka. 1978.

36. Bakhrakh, L. D., and A. P. Kurchkin eds., *Radiogolografiia i opticheskaia obrabotka informatsii mikrovolnovoi tekhnika: Sb. statei.* Leningrad: Nauka. 1980.

37. Gurevich, S. B., and G. A. Gavrilov. *Optiko-Elektronnye metody obrabotka izobrazhenii: Sb. statei.* Leningrad: Nauka. 1983.

38. Bakhrakh, L. D., and A. P. Kurochkin, eds. *Metody i ustroistva radio-i akusticheskoi golografii: Sb. statei.* Leningrad: Nauka. 1983.

39. Svet, V. D., V. I. Teliatinkov, and V. M. Komarov. "Golograficheskaia metody v gidroakustike." *Zarubezhnaia radioelektronika.* 1977. No. 9: 48–76.

40. Grinev, A. Iu., and E. N. Voronin. "Antennye reshetki s obrabotkoi signala metodami kogerentnoi optiki–radioopticheskie antennye reshetki." *Zarubezhnaia radioelektronika.* 1977. No 9: 69–86.

41. Tarasov, L. V., and V. A. Ezhov. "Kogerento-opticheskaia obrabotka radiosignalov." *Zarubezhnaia radioelektronika.* 1980. No. 2. pp. 3-36.

42. Lazarev. S. V., Onishchenko, T.A. "Opticheskie metody obrabotki radiolokatsionnykh signalov." *Radioelectronik za rubezhom.* 1980. No. 16: 1–34.

43. Gribnev, A. Iu. "Radio–opticheskie antennye reshetki." *Izv. vuzov. Ser. Radio-elektronika.* 1981. No. 3: 51–70.

44. "Akustoopticheskaia obrabotka signalov." *TIIER. [Proc. IEEE]*. 1981. No. 1: 163.

45. Beloshitskii, A. P., V. M. Kamarov, B. P. Krekoten', and B. T. Sopozhnikov. "Akustoopticheskie analizatory spektra radiosignalov." *Zarubezhnaia radioelektronika.* 1981. No. 3: 51–70.

46. Ezhov, B. V., and L. V. Tarasov. "Akustoopticheskaia obrabotka radiosignalov." *Zarubezhnaia radioelectronika.* 1982. No. 7: 3–35.

47. Reutov, A. L., V. A. Mikhailov, G. S. Kondratenkov, and B. V. Boiko. *Radiolokatsiionnye stantsii bokogo obzora.* A. L Reutov, ed. Moscow: Sov. radio. 1979.

48. Kondratenkov, G. S., ed. *RLS obzora zemli.* Moscow: Radio i sviaz'. 1983.

49. Johnson, R. *Scanning Antenna Systems (UHF).* Edited by G. T. Markov, and A. F. Chaplin. Moscow: Sov. radio. Vol. 1: 1966. Vol. 2: 1968. Vol. 3: 1972.

50. Gostiukhin, V. L., K. I. Grinev, and V. N. Trusov. *Voprosy proektirovania aktivnykh FAR s ispol'zovaniem EVM.* Moscow: Radio i svaiz'. 1983.

51. Oliver, A. Olinar, and G. H. Knittel, eds. *Phased Array Antennas.* Dedham, Mass.: Artech House. 1972.

52. Markov, G. T., and D. M. Sazonov. *Antennas.* Moscow: Energiia. 1975.

53. Bakhrakh, L. D., and D. I. Voskeresenkii, eds. *Antenny (sovremmenoe sostoianie i problemy).* Moscow: Radio i sviaz'. 1979.

54. Voskresenkii, D. I., ed. *Antenny i ustroistva SVCh (raschet i proektriovanie FAR).* Moscow: Radio i sviaz' 1981.

55. Mailloux, R. J. "Teoria: Tekkhnika fazirovannykh antennykh reshetok." *TIIER.* [*Proc. IEEE*]. 1982. 70(3): 5–62.

56. Skolnik, M., ed. *Spravochnik po radiolokatsii.* Moscow: Sov. radio. 1976–1979.

57. Shirman, Ia. D., and V. N. Manzhos, V. H. *Teoriia i tekhnika obrabotki radiolokatsionnoi informatsii na fone pomekh.* Moscow: Radio i sviaz'. 1981.

58. Rabiner, L. R., and B. Gold. *Teoriia i primenie tsifroviia obrabotka signalov.* Moscow: Mir. 1978.

59. Knight, W. C., R. G. Pridham, and S. M. Kay. "Tsifroviia obrabotka signalov v gidrolokatsionnykh sistemakh." *TIIER.* [*Proc. IEEE*]. 1981. 69(11): 84–155.

60. Preston, K. "Sravenie analogovykh i tsifrovykh metodov raspoznavaniia obrazov." TIIER. [*Proc. IEEE*]. 1972. 60(10): 141–160.

61. Cutrona, L. J., E. N. Leith, E. N. Palermo, and L. J. Parcello. "Optical Data Processing and Filtering Systems." *IRE Trans.* 1960. IT-6(3): 386–400.

62. Grinev, A. Iu. "Antenny s obrabotki signala." *Antenny (Sovremenoe sostoianie i problemi).* L. D. Bakhrakh, D. I. Voskresenskii, eds. Moscow: Sov. radio. 1979: 102–139.

63. Lambert, L., M. Arm., and A. Aimette. "Electro-optical signal processor for phased array antennas." *Optical and electro-optical information processing.* J. T. Tippet et al., eds. Cambridge: Inst. of Tech. Press. 1965: 715–748.

64. McLean, D. J., L. B. Lambert, M. Arm., and H. Stark. "An Electro-optical Processor for Radio-geliograph." *Proc. IREE.* Australia. 1967. No 9: 375–380.

65. Bakhrakh, L. D., A. P. Kurochkin, and S. G. Rudneva. "Isopol'zovanie ul'trazvukovogo modul'ator sveta dla parallel'noi obrabotki signalov A.R." *Voprosy elektroniki. Ser. Obshchetekhn.* 1972. No 1: 36–48.

66. Aksenov, E. T., V.A. Grigor, N.A. Esepkin, et al. "Mnogokanal'nyi ul'trazvukovoi modul'ator sveta rabotaiushchii v rezhime difraktsii Bregga, dlia sistem opticheskoi obrabotka." *Voprosy radioelektroniki. Ser. Obshchetekhn.* 1973. No 5: 58.

67. Bannov, V. Ia., V. A. Gusev, R. I. Kiper. "Optiko-mikroelektronika ustroistvo obrabotki signalov mnogoluchevoi AR." *Izv LEIS.* 1981. No 294: 30–36.

68. Esenkina, N. A., Iu. V. Petrun'kin, E. T. Aksenov, et al. "Mnogokanal'nye akustoopticheskie ustroistva." *ZhTF.* 1975. 26(11): 2353–2360.

69. Lee, J. W. "Interferometric Acoustooptic Signal Processor for Signal Processor for Simultaneous Direction Finding and Spectrum Analysis." *Applied Optics.* 1983. 22(6): 867–873.

70. Krupitskii, E. N., V. N. Iakovlev. "Akustoopticheskie protsessory radiosignalov." *Akustoopticheskie metody obrabotki informatsii.* G. E. Korbukov, S.V. Kukalov, eds. Leningrad: Nauka. 1978. 30–46.

71. Bondarenko, V. S., N. A. Bukharin, V. A. Grigorov, N. A. Esepkina, et al. "Mnogokanal'nye zhidkotnye ul'trazvukovye modulator sveta." *Trudy LPI.* 1975. No 344: 38–44.

72. Aksenov, E. T., N. A. Esepkina, V. A. Markov, et al. "Tverdotel'nye ul'trazvukovye modulatory sveta co zvukprodami iz tazhelykh flintov." *Trudy LPI.* 1975. No 344: 44–48.

73. Bukharin, N. A., V. A. Grigorev, N. A. Esepkina, et al. "Ispol'zovanie mnogokanalnykh ul'trazvukovykh modul'ator v sistemakh opticheskoi soglasovannoi filtratsii." *Avtometria.* 1976. No 6: 18–24.

74. Esepkina, N. A., Iu. V. Petrun'kin, N. A. Bukharin, et al. *Akustoopticheskie analizatory spektra dla radioastronomii.* Izv. Vuzov: Sov. radio-fizika. 1979. 19(11): 1732–1739.

75. Espkina, N. A., Iu. V. Petrun'kin, E. T. Aksenov, et al. "Mnogokanal'nye akustoopticheskie modulatory." *Golografia i obrabotka informatsii.* Leningrad: Nauka. 1976: 96–105.

76. Gusev, O. B., S. V. Kukalov, A. V. Mel'nikov, et al. "Mnogokanal'nye akustoopticheskie modul'atory dla ustroistva vvoda i opticheskoi obrabotki informatsii v real'nom vremeni." *ZhTF.* 1978. 8(1): 163–176.

77. Bondarenko, V. S., B. V. Zorenko, I. N. Kuligin et al. "Mnogokanal'nye akustoopticheskie modul'atory sveta dlia ustroistva obrabotki informatsii." *Voprosy radioelektoniki. Ser Obshchetekhn.* 1978. No. 8: 82–93.

78. Rhodes, W. T. "Akustoopticheskaia obrabotka signalov." *TIIER*. [*Proc. IEEE*]. 1981. 69(1): 74–91.

79. "Golograficheskii metod kompensatsii raskhodimosti v mnogokanal'nykh akusoopticheskikh ustroisvakh." Vodovatov, I. A., N. A. Esepkina, V. Iu. Petrun'kin, et al. Letters to *ZhTF*. 1981. 7(6): 369–373.

80. Lewis, P. "An Optical Signal Processing Technique for Directional System Using Circular Arrays." *Optica Acta*. 1970. 17(1): 19–36.

81. Vodovatov, I. A., M. G. Vysotskii, N. A. Esepkina, and S. V. Rogov. "Eksperimental'nye issledovanie opticheskoi obrabotki signalov v kol'tsevykh reshetkakh." *Trudy LPI*. 1979. No. 336: 62–65.

82. Grinev, A. Iu., L. P. Iaroslavskii, N. S. Merzl'akov, et al. "Sintez diagrammoobrazuiushchego transpraranta kol'tsevykh AR metodami tsifrovoi golografii." *Radio tekhnika i elektronika*. 1984. 29(7): 1266–1272.

83. Grinev, A. Iu., and E. N. Vororin. "Kogerentno-opticheskaia obrabotka signala krougovykh antennykh reshetok s ispol'zovaniem ul'trazvukovykh modul'atorov sveta." *Proektirovanie antenn i ustroistv SVCh s primeneniem EVM. Trudy MAI*. 1980: 33–39.

84. Casasent, D., and J. Stephenson. "Electrooptical Processing of Phased Array Antenna Data." *Applied Optics*. 1972. 11(5): 1269–1271.

85. Casasent, D., and F. Casasayas. "Electrooptical Processing of Phased Array Antenna Data." *IEEE Trans*. 1975. AES-11(1): 65–75.

86. Casasent, D. "Kogerentnye opticheskie preobrazovateli." *Prostranstvennye modul'altory sveta*. S. B. Gurevich, ed. Leningrad: Nauka. 1977: 18–41.

87. Dun, A. Z., A. E. Tolmachev, A. N. Krivoruchko, et al. "Elektroopticheskii prostranstvennyi modul'ator sveta s elektronno-luchevoi adresatsii." *Prostranstvennye modul'ator sveta*. S.B. Gurevich, ed. Leningrad: Nauka. 1977: 119–124.

88. Casasent, D., and E. Klimas. "Multichannel Optical Correlator for Radar signal Processing." *Applied Optics*. 1978. 17(13): 2058–2063.

89. Mustel', E. R., and V. I. Parygin. *Metody modul'atsii i skanirovania sveta*. Moscow: Nauka. 1970: 275.

90. Currie, G. D., C. Leonard, P. Either, and G. Orbits. "Recent Progress in Thermoplastic Light Modulators." *Devices and Systems for Optical Processing. Proceedings of the Society of Photo-optical Instrumentations Engineers (SPIE).* 1980. 218: 48–53.

91. Rhodes, J. Elmer, Jr. "Requirements for Optical Processing Information from Arrays." *IEEE/AP-S Intern. Microwave Symp.* New York:IEEE. 1974: 46–49.

92. Meyer, R. A. "Optical Beam Steering Using a Multichannel Lithium Tantalate Crystal." *Applied Optics.* 1972. 11(3): 613–616.

93. Grinev, A. Iu., V. M. Pankratov, V. S. Temchenko, et al. "Mnogokanal'nyi elektroopticheskii modul'ator sveta na osnove tantalata litia dla parallel'noi obrabotki signalov antennykh reshetok." *Kvantovaia electronika.* 1981. 8(1): 209–211.

94. Grinev, A. Iu., V. I. Osinskii, E. N. Voronin, et al. "Kogerentno–opticheskii prosessor dlia prostranstvenno-vremennoi obrabotki antennykh reshetok." *Izvestiz vuzov. Radioelektronika.* 1981. 24(2): 27–33.

95. Pankratov, V. M., N. N. Fomichev, T. P. Demina, and T. B. Egina. "Shestnadtsatikanal'nyi elektroopticheskii modul'ator lazernogo izlucheniа na osnove tantalata litiia." *Elektronnaia tekhnika Ser. Kvantovaia elektronika.* 1975. No. 1: 209–211.

96. Volkov, V. V., N. C. Karaseva, L. L. Lukasevich, et al. "Lineinyi transparant na osnove niobata litiia." *Avtometriia.* 1978. No. 1: 108–111.

97. Meyer, R. A., and D. G. Grant. "Two-Dimensional Optical Phased-Array Beam Steering." Proc. Electrooptical Syst. Design Conference. 1972: 107–109.

98. Beste, D. C., and E. N. Leith. "An Optical Techique for Simultaneous Beamforming and Cross-Correlation." *IEEE Trans.* 1966. AES-2(4): 376–381.

99. Williams, R. E., and K. Bieren. "Combined Beam-Forming and Cross-Correlation of Broad Band Signal from a Multidimensional Array Using Coherent Optics." *Applied Optics.* 1971. 10(6): 1386–1392.

100. Leith, E. N. "Optical Processing Techniques for Simultaneous Pulse Compression and Beam-sharpening." *IEEE Trans.* 1968. AES-4(6): 879–885.

101. Mitrofanova, L. I., A. S. Ostrovskii, I. M. Pocherniaev, and E. K. Shmarev. "Metody fotoplasticheskoi zapisi v zadachakh fil'tratsii izobrazhenii." *Avtometriia.* 1976. No. 3: 16–21.

102. Casasent, D. "Opticheskaia obrabotka signalov" ["Optical Signal Processing"]. *Opticheskaia obrabotka informatsii* [Optical Information Processing]. Moscow: Mir. 1980: 289–307.

103. Libreton, G., and E. Bazelaire. "Holographic Processing of Wideband Antenna Data." *Optical Eng.* 1980. 19(5): 739–747.

104. Penn, W. A., and J. Chovan. "Primenenie golograficheskikh metodov v gidrolokatsii." ["Applications of Holography in Sonar"]. *Akusticheskaia golografia* [Acoustic Holography]. Translated from English by V. G. Prokhorova. Leningrad: Sudostroenie. 1975: 183–219.

105. Gavrilov, G. A., V. I. Marokhonov, A. V. Khomenko, and M. S. Cheber'ak. "Primenenie transparanta tipa PROM v sistemakh golograficheskikh zapisi v real'nom vremeni." *ZhTF.* 1981. 51(1): 97–101.

106. Gurevich, S. B., ed. *Prostranstvennye modul'atory sveta.* Leningrad: Nauka. 1977.

107. Petrov, M. P., S. I. Stepanov, and A. V. Khomenko. *Fotochuvstvitel'nye elektroopticheskie sredy v golografii i opticheskoi obrabotke informatsii.* Leningrad: Nauka. 1983.

108. Kompanets, I. N. "Upravliaemye transparanty." *Zarubezhnaia radioelektronika.* 1977. No. 4: 46-76.

109. Balakshii, V. I., V. N. Parygin, and L. E. Chirkov. *Fizicheskie osnovy akustooptiki.* Moscow: Radio i sviaz'. 1985.

110. Magdich, L. N., and V. Ia. Molchanov. *Akustoopticheskie ustroistva i ikh premenenie.* Moscow: Sov. radio. 1978.

111. Young, E. H., and Shi-kay Yao. "Raschet aiustoopticheskikh ustroistv." *TIIER.* [*Proc. IEEE*]. 1981. 69(1): 62–74.

112. Demidov, A. Ia., L. Ia. Serebrennikov, S. M. Shandarov. "Shirokopolosnye akustoopticheskie iacheiki na osnove kristallov $LiNbO_3$, Si, $PbMoO_4$." *Akustoopticheskie metody obrabotki informatsii.* G. E. Korbukov, and S. B. Kukalov, eds. Leningrad: Nauka. 1978.

113. Anisimova, I. D., I. M. Bikulin, F. A. Zaitov, and Sh. D. Kurmashev. *Poluprovodnikovye fotopriemniki.* V. I. Stafeeva, ed. Moscow: Radio i sviaz'. 1984.

114. Seken, K, and M. Thompsett. *Pribory s perenosom zariada.* Moscow: Mir. 1978.

115. Kotov, Iu. A., A. V. Mikhailov, and A. P. Novitskii. "Ustroistvo sopriazhenia opticheskoi i tsifrovoi sistem obrabotki informatsii s ispol'zovaniem lineinykh i matrichnykh priborov s perenosom zariada." *Trudy LPI.* 1979. No. 366: 41–44.

116. Borsuk, G. M. "Fotodetektory dlia akustoopticheskikh sistem obrabotki signalov." *TIIER.* [*Proc. IEEE*]. 1981. 69(1): 117–137.

117. Esepkina, N. A., H. A. Bukharin, Iu. A. Kotov, et al. "Gibridnaia optiko-tsifrovaia sistema obrabotki signalov pul'sarov." *Radiogolografiia i opticheskaia obrabotka informatsii v mikrovolnovoi tekhnike.* L. D. Bakhrakh, and A. P. Kurochkin, eds. Leningrad: Nauka. 1980.

118. Turpin, T. M. "Spektral'nye analiz signalov opticheskimi metodami." TIIER. [*Proc. IEEE*]. 1981. 69(1): 92–108.

119. Osinskii, V. I., L. L. Brublevskii, V. M. Blynskii, et al. "Bystrodeistvuiushchie monolitnye matritsy fotopriemnikov s vnutrennei kommutatsiei." *Opticheskie metody obrabotki informatsii.* B. L. Pilipovich, ed. Minsk: Nauka i tekhnika. 1978: 85–106.

120. Kruglikov, S. B., and S. N. Naimark. "Intergralnye MDP fotodiodnye ustroistva i ikh primenenie–Obzory po elektronnoi tekhnike." *Ser. Mikroelektronika.* 1980. No. 2: 3–63.

121. Borsuk, G. M. "Fotodetektory dlia akustoopticheskikh sistem obrabotki signalov." *TIIER.* [*Proc. IEEE*]. 1981. 69(1): 117–137.

122. Grinev, A. Iu. and E. N. Voronin. "Preobrazovanie prostranstvenno-vremennogo spektra antennymi reshetkami s obrabotkoi signala metodami kogerentnoi optiki." *Izv. vuzov. Ser. Radioelektronika.* 1978. 21(2): 74–83.

123. Grinev, A. Iu. E. N. Voronin, and A. P. Kurochkin. "Ploskie radioopticheskie antennye reshetki." *Radiogolografiia i opticheskaia obrabotka informatsii v mikrovolnovoi tekhnike.* L. D. Bakhrakh, and A. P. Kurochkin, eds. Leningrad: Nauka. 1980.

124. Voronin, E. N., A. Iu. Grinev, and V. S. Temchenko "Kogerentno-opticheskii protsessor radiosignalov antennykh reshetok." *Avtometria.* 1981. No. 3: 32.

125. Voronin, E. N., and A. Iu. Grinev. "Osobennosti kogerentno-opticheskogo formirovaniia prostranstvennogo i vremennogo spektrov s pomoshch'iu lineinykh antennykh reshetok." *Fazirovannye antennye reshetki i ikh elementy. Trudy MAI.* 1980: 63–69.

126. Rytov, S. M., Iu. A. Kravtsov, and V. I. Tatarskii. *Vvedenie v statich-eskuiu radiofiziku (sluchainye polia).* Moscow: Nauka. 1978.

127. Akhmanov, S. A., Iu. E. D'iakov, and A. S. Chirkin. *Vvedenie v staticheskuiu radiofiziku i optiku.* Moscow: Nauka. 1978.

128. Grinev, A. Iu., V. S. Temchenko, and E. N. Voronin. "Formirovanie chastotno-uglovogo spektra signalov lineinykh antennykh reshetok ko-gerentno- opticheskimi protsessorami na prostranstvennykh modula-torakh sveta s mnogokanal'noi opticheskoi adresatsiei." *Izv vuzov. Ser. Radioelektronika.* 1983. 26(2): 17–23.

129. Nakhmansok, G. S., and V. M. Ianyshev. "Akustoopticheskaia obrab-otka shirokopolosnykh signalov dvumernykh fazirovannykh antennykh reshetok." *Izv. vuzov. Ser. Radioelektronika.* 1979. 22(2): 76–79.

130. Grinev, A. Iu., V. F. Trukhin, V. S. Temchenko, and O. F. Iankovskii. "Kogerentno–opticheskii protsessor signalov na PVMS s mnogokanal'-noi opticheskoi adresatsiei." *Kvantovaia elektronika.* 1983. 10(5): 1036–1038.

131. Grinev, A. Iu., E. N. Voronin, and V. S. Temchenko. "Kogerentno-opticheskii protsessor dvumernykh AR so slozhnym formatom zapisi signalov." *Izv. vuzov. Ser. Radioelektronika.* 1983. 26(2): 86–88.

132. Shmarev, E. K., G. Kh. Zelenskii, and S. V. Levyi. "Opticheskoe vy-chislenie funktsii neopredelennosti signalov." *Zarubezhnaia radioelek-tronika.* 1981. No. 10: 29–46.

133. Voskresenskii, D. I., L. I. Ponomarev, and V. S. Filippov. *Vypuklye skaniruiushchie antenny.* Moscow: Sov. radio. 1978.

134. Khurgin, Ia. N., and V. P. Iakovlev. *Finitnye funkstii v fizike i tekhnike.* Moscow: Nauka. 1971.

135. Zelkin, E. G., and V. G. Sokolov. *Metody sinteza antenn.* Moscow: Sov. radio. 1980.

136. Grinev, A. Iu., and E. N. Voronin. "Formirovanie priemnykh luchei neploskikh antenn radioopticheskimi metodami." *Izv. vuzov. Ser. Radioelektronika.* 1979. 22(2): 25–33.

137. Grinev, A. Iu., and E. N. Voronin. "Neploskie antennye reshetki sformirovaniem priemnykh luchei metodami kogerentnoi optiki." *Radiografiia i opticheskaia obrabotka informatsii v mikrovalnoboi tekhnike.* L. D. Bakhrakh, and A. P. Kurochkin, eds. Leningrad: Nauka. 1980: 118–135.

138. Gonorovskii, I. S. *Radiotekhnicheskie tsepi i signaly.* Moscow: Sov. radio. 1971.

139. Grinev, A. Iu., and E. N. Voronin. "Formirovanie priemnykh luchei tsilindricheskikh antennykh reshetok radioopticheskimi metodami." *Izv. vuzov. Radioelectronika.* 1979. 22(5): 29–34.

140. Iaroslavskii, L. P., and N. S. Merzliakov. *Tsifrovaia golografia.* Moscow: Nauka. 1982.

141. Grishin, M. P., Sh. M. Kurbanov, and V. P. Markelov. *Avtomaticheskii vvod i obrabotka izobrazhenii na EVM.* Moscow: Energia. 1976.

142. Fedorov, B. F., and R. I. El'man *Tsifrovania golografiia.* Moscow: Nauka. 1976.

143. Gurevich, S. B., ed. *Opticheskaia golografia.* [Optical Holography. G. Caulfield, ed.] Moscow: Mir. 1982. (Vol. 2).

144. Bateman, G., and A. Erdin. *Vysshie transtsendentnye funktsii.* N. Ia. Vilenkin, ed. Moscow: Nauka. 1966. (Vol. 2).

145. Jahnke, E., F. Emde, and F. Losh. *Spetsial'nye funktsii.* L. I. Sedova, ed. Moscow: Nauka. 1968: 344.

146. Grinev, A. Iu., and V. S. Temchenko. "Formirovanie prostranstvennykh kharakteristikh naprovlennosti neploskikh osesimmetrichnykh antennykh reshetok kogerentno-opticheskimi metodami s pomoshch'iu ob"emnykh gologramm." *Izv. vuzov Ser. Radioelektronika.* 1982. 25(2): 29–34.

147. Krasnov, A. E. "Funktsional'nye svoistva ob"emnykh gologramm i ikh primenenie." *Avtomatika i telemekhanika.* 1978. No. 11: 183–187.

148. Krasnov, A. E. "Prostranstvenno-neinvariantnye fil'try opticheskikh signalov na osnove ob'emykh gologramm." *Kvantovaia elektronika.* 1980. 7(4): 818.

149. Kogelnic, N. "Coupled Wave Theory of Thick Hologram Gratings." *Bell Sys. Tech. Journ.* 1969. 48(9): 2909–2919.

150. Denisiuk, Iu. N., ed. *Opticheskaia golografia.* Leningrad: Nauka. 1979.

151. Lewis, M. D., J. F. Walkup, and M. O. Hagler. "Representations of Space Varient Optical Systems Using Volume Holograms." *Applied Optics.* 1975. 14(10): 2438–2448.

152. Bakhrakh, L. D., and V. A. Makeev. "Korrektsiia fazovykh neodnorodnostei v raskryve antenn s ispol'zovaniem golograficheskikh soglasovannykh fil'trov." *Radiotekhnika i elektronika.* 1973. 18(4): 741.

153. Voskresenii, D. I., A. Iu. Grinev, E. N. Voronin. "Ostronapravlennyi parallel'nyi priem radioizlucheniia konformnymi antennymi reshetkami s kogerentno-opticheskoi obrabotkoi." *Izv. vuzov. Ser. Radiofizika.* 1980. 23(2): 197–201.

154. Voskresenskii, D. I., E. N. Voronin, A. Iu. Grinev, et al. *Ustroistvo dlia obrabotki signalov priemnoi antennoi reshetki.* Autorskoe svidetel'stvo [Patent] 558822 (USSR).

155. Grinev, A. Iu. "Prostranstvenno–zavisimaia kogerento–opticheskaia obrabotka radiosignalov neploskikh antennykh reshetok." *Fazirovannye antennye reshetki. Trudy Mosk. aviats. in-ta.* 1981: 35–47.

156. Voronin, E. N., and A. Iu. Grinev. "Osobennosti preobrazovaniia prostranstvenno–vremmenogo spektra ploskimi antennymi reshetkami s kogerentno–opticheskoi obrabotkoi signala." *Izv. vuzov. Ser. Radioelektronika.* 1979. 22(11): 3–8.

157. Il'inskii, A. S., A. Iu. Grinev, and Iu. B. Kotov. "Chislennye metody resheniia zadach izlucheniia antennykh reshetok." *Vycheslitelnye metody i programmirovanie.* Moscow: Izd. MGU. 1980. No. 32: 104–130.

158. Fel'dbush, V. N., Iu. V. Chugui. "Kogerentno–opticheskie sistemy obrabotki signalov no osnove primeneniia siluetinykh fil'trov." *Avtometria.* 1976. No. 6: 53–67.

159. Brown, B. R., and A. W. Lohmann. "Complex Spatial Filtering With Binary Masks." *Applied Optics.* 1966. 5(5): 967–968.

160. Abramovich, Iu. N., and B. G. Danilov. "O minimizatsii diagrammy napravlennosti v zadannykh napravleniiakh pri kommutatsionnom upravlenii." *Radiotekhnika i elektronika.* 1978. 23(2): 257.

161. Abramovich, Iu. N., and M. B. Sverdlik. "Nekotroye zadachi sinteza diatramm napravlennosti antennykh reshetok po srednekvadraticheskomu kriteriu." *Radiotekhnika i elektronika.* 1973. 18(7): 1341–1346.

162. Sudakov, O. A., "Korrektsiia diagrammy napravlennoski antenn metodom fil'tratsii." *Radiotekhnika i elektronika.* 1980. 25(4): 696–702

163. Burakov, V. A., L. A. Zorin, and M. V. Ratynskii. "Adaptivnaia obrabotka signalov. AR/V." *Zarubezhnaia radioelektronika:* 1976. No. 8: 35–80.

164. Grinev, A. Iu., E. N. Voronin, and V. S. Temchenko. "Ploskie radioopticheskie antennye reshetki s rezhektsiei meshaiushchikh signalov po napravleniu prikhoda." *Izv. vuzov. Ser. Radiofizika.* 1980. 23(7): 849–863.

165. Temchenko, V. S. "Formirovanie upravliaemykh provalov v diagrammakh napravlennosti ploskikh radioopticheskikh reshetok s krugloi aperturoi." *Fazirovannye antennye reshetki. Trudy MAI.* 1982: 47–52.

166. Grinev, A. Iu., I. N. Kompanets, V. S. Temchenko, A. A. Vasil'ev, and V. A. Ezhov. "Eksperimental'nye rezul'taty po rezhektsii meshaiushchikh prostranstvennykh signalov v radioopticheskikh antennykh reshetakh." *Izv. vuzov. Ser. Radiofizika.* 1981. 24(11): 1392–1397.

167. Grinev, A. Iu., I. N. Kompanets, B. S. Temchenko, A. A. Vasil'ev, and V. A. Ezhov. "Antennye reshetki s fil'tratsiei meshaiushchikh prostranstvennykh radiosignalov kogerentno–opticheskimi metodami s ispol'zovaniem upravliaemykh zhidkokristallicheskikh transparantov." *Radiotekhnika i elektronika.* 1983. 28(4): 785–791.

168. Grinev, A. Iu., and V. S. Temchenko. "Vliianie pogreshnostei rezhektornogo fil'tra na glubinu formirovaniia provalov v diagrammakh napravlennosti antennykh reshetok s kogerentno–opticheskii obrabotkoi." *Izv. vuzov. Ser. Radioelektronika.* 1982. 25(5): 14–21.

169. Lavrentov, M. M., V. G. Romanov, and S. P. Shishatskii. *Nekorrektnye zadachi matematicheskoi fiziki.* Moscow: Nauka. 1980.

170. Kolmogorov, A. N., and S.V. Fomin. *Elementy teorii funktsii i funktsional'nogo analiza.* Moscow: Nauka. 1976.

171. Korn. G, T. Korn. *Spravochnik po matematike.* Moscow: Nauka. 1974.

172. Razmakhin, M. K., and V. P. Iakovlev, eds. *Funktsii s dvoinoi ortogonal'nostiu v radioelektronike i optike* [Functions with Two-dimensional Orthognality in Radioelektronics and Physics]. Moscow: Sov. radio. 1971.

173. Khurgin, Ia. I., and V. P. Iakovlev "Progess v SSSR po teorii finitnykh funktsii i ee primeneniiam v fizike i tekhnike." *TIIER.* [*Proc. IEEE*]. 1977. 65(7).

174. Gurevich, M. S. "Signaly konechnoi prodolzhitel'nosti soderzheshchie maksimal'nuiu doliu energii v zadannoi polose chastot." *Radiotekhnika i elektonika.* 1956. 1(3): 313–319.

175. Venevtsev, M. K., and B. A. Khadzhi, "K Voprosu o sinteze diagramm napravlennosti antennykh reshetok s naimen'shim srednekvadratichnym otkloneniem ot nulia v sektore bokovykh lepestkov." *Radiotekhnika i elektronika.* 1970. 15(11): 2372–2374.

176. Basov, N. G., ed. *Upravliaemye transparanty i reversivnaia zapis' opticheskikh signalov.* Moscow: Fiz. in-t AN SSSR. 1981. No. 126: 174.

177. Rucinov, V. V. *Tekhnicheskaia optika.* Leningrad: Mashinostroenie. 1979.

178. Grinev, A. Iu., and E. N. Voronin. "Vlianie kogeretno-opticheskogo protsessora na diagrammoobrazuiushchie svoistva radioopticheskikh antennykh reshetok." *Izv. vuzov. Ser. Radioelektronika* 1980. 23(2): 16–24.

179. Grinev, A. Iu., and E. N. Voronin. "Vlianie prostranstvenn-vremennogo moduliatora sveta i shumov na kharakhteristiki radioopticheskikh antennykh reshetok." *Izv. vuzov. Ser. Radioelektronika*: 1980. 23(12): 3–10.

180. Shifrin, Ia. S. *Voprosy statisticheskoi teorii antenn.* Moscow: Sov. radio. 1970.

181. Schulman, A. R. *Optical Data Processing.* New York: John Wiley and Sons., Inc. 1970.

182. Perina, Ia. *Kogerentnost' sveta.* Moscow: Mir. 1971.

183. Prokhorov, A. M., ed. *Spravochnik po lazeram* [Handbook on Lasers]. Moscow: Sov. radio. 1978. (Vol. 1 and 2).

184. Zolotarev, A. I., V. A. Zubov, A. D. Kovalevskii, et al. "Inzhektsionnye poluprovodnikovye lazery v sistemakh korreliatsionoi obrabotki informatsii." *Kvontovaia elektronika.* 1979. 6(11): 2460–2463.

185. Hecht, D. L. "Multifrequency acoustooptic diffraction." *IEEE Trans.* 1977. SU-24(1): 7–18.

186. Kukalov, S. V., V. V. Molotok, and B. P. Razzhivin. "Nelineinye iskazheniia v akustoopticheskom analizotore spektra." *Izv. vuzov. Ser. Radioelektronika.* 1980. 23(11): 38–42.

187. Ainbinder, I. M. *Shumy radiopriemnikov.* Moscow: Sviaz'. 1974.

188. Ross, M. *Lazernye priemniki.* Moscow: Mir. 1969.

189. Tseitlin, H. M. *Antennaia tekhnika i radioastronomiia.* Moscow: Sov. radio. 1976.

190. Frizer, Kh. *Fotograficheskaia registratsiia informatsii.* Moscow: Mir. 1978.

191. Ginzburg, V. M., and B. M. Stepanov, eds. *Opticheskaia golografiia (prakticheskie primeneniia).* Moscow: Sov. radio. 1978.

192. Upatnieks, J. "Portable Real-time Coherent Optical Correlator." *Applied Optics.* 1983. 22(18): 2798–2803.

193. "Real Time Signal Processing." *Proceedings of the Society of Photo-optical Instrumention Engineers (SPIE).* 1982. 341: 1–351.

194. "Opticheskaia vychislitel'naia tekhnika." *TIIER.* [*Proc. IEEE*]. 1984. 72(7): 269.

195. Grinev, A. Iu., E. N. Voronin, and O. F. Iankovskii. "Radioopticheskie antennye reshetki, sfokusirovannye v zony Frenelia." *Izv. vuzov. Ser. Radioelektronika.* 1985. 28(2): 55–61.

196. Pon'kin, V. A., and N. A. Potapov. "Nekotorye voprosy realizatsii algoritmov prostranstvenno-vremennoi obrabotki signalov." *Prostransvenno-vremmonaia obrabotka signalov.* I. Ia. Kremer, ed. Moscow: Radio i sviaz'. 1984: 195–212.

197. Sazonov, D. M., and K. A. Tishchneko. "Optimizatsiia parametrov sfokusirovannykh antennykh reshetok proizvol'noi geometii." *Izv. vuzov. Ser. Radioelektonika.* 1978. 21(21): 84–90.

198. Voronin, E. N. "Elektrodinamicheskaia obshchnost' radiogolograficheskikh zadach." *Radiotekhnika i elektronika.* 1984. 29(10): 1906–1916.

199. Voronin, E. N. "Optimal'nye v srednem resheniia zadach selektivnoi golografii." *Izv. vuzov. Ser. Radioelektronika.* 1986. 29(2): 16–29.

Index